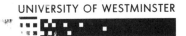

SCREENWRITER'S
HANDBOOK 2009

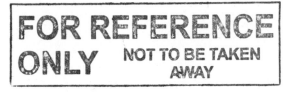

arry Turner has worked on both sides of publishing, as a journalist and author, editor and arketing director. He started his career as a teacher before joining *The Observer* and then oving on to radio and television. His first book, a study of British politics in the early ventieth century, was published in 1970. He has written over twenty books including *A lace in the Country*, which inspired a television series, and a best-selling biography of the ctor, Richard Burton. For many years he wrote on travel for *The Times* and now reviews d serialises books for the paper. Among his recent books are *Countdown to Victory* on e last months of World War II and *Suez 1956: The First Oil War*. As founding editor of *he Writer's Handbook* he has taken this annual reference title through to its twenty-first lition. For ten years, Barry has edited the annual *Statesman's Yearbook*. He is a founder d current chairman of the National Academy of Writing at Birmingham City University here he is a visiting professor.

THE SCREENWRITER'S HANDBOOK 2009

2nd EDITION

EDITOR

BARRY TURNER

MACMILLAN

© Macmillan Publishers Ltd 2008
All rights reserved. No reproduction, copy or transmission of this
publication may be made without written permission.

No paragraph of this publication may be reproduced, copied or transmitted
save with written permission or in accordance with the provisions of the
Copyright, Designs and Patents Act 1988, or under the terms of any licence
permitting limited copying issued by the Copyright Licensing Agency, 90
Tottenham Court Road, London W1T 4LP.

Any person who does any unauthorized act in relation to this publication
may be liable to criminal prosecution and civil claims for damages.

The editor has asserted his right to be identified as the editor of this work
in accordance with the Copyright, Designs and Patents Act 1988.

First published 2008 by
MACMILLAN PUBLISHERS LTD
Houndmills, Basingstoke, Hampshire RG21 6XS and
175 Fifth Avenue, New York, N.Y. 10010
Companies and representatives throughout the world

ISBN-13: 978-0-230-57326-0
ISBN-10: 0-230-57326-6

Inclusion in *The Screenwriter's Handbook* is entirely at the editor's discretion. Macmillan Publishers Ltd makes no
recommendation whatsoever by inclusion or omission of any agency, publisher or organization.

While every effort has been made to ensure all of the information contained in this publication is correct
and accurate, the publisher cannot accept any responsibility for any omission or errors that may occur or for any
consequences arising therefrom.

This book is printed on paper suitable for recycling and made from fully
managed and sustained forest sources. Logging, pulping and manufacturing
processes are expected to conform to the environmental regulations of the
country of origin.

A catalogue record for this book is available from the British Library.

A catalog record for this book is available from the Library of Congress.

10 9 8 7 6 5 4 3 2 1
17 16 15 14 13 12 11 10 09 08

Credits

Editor	*Barry Turner*
Assistant Editor	*Daniel Smith*
Consultant	*Pete Daly*
Reference Editor	*Hazel Woodbridge*
Web Editor	*Lisa Hayden*
Marketing	*Ciara O'Connor*
Production	*Phillipa Davidson-Blake*
	Tim Fox
	Wayne Hayward
	Andrew Joseph
	Gail Robins

Printed and bound in Great Britain by
Creative Print and Design, Wales

Contents

vii Introduction
Barry Turner

ix Foreword
Jake Eberts

Articles

1 An Ever Changing Scene
Barry Turner

4 Film, Television...It's all the Same
Isn't it? The Differences in Writing
for the Two Media
Nick Moorcroft

7 Just a Mamet
Barry Turner

9 From Stage to Screen: Why
Theatrical Drama Rarely Finds a
Home on the Big Screen
James Christopher

12 The Art of Adaptation
Ronald Harwood

16 Learn from the Classics:
Assessing Why Five Classic Scenes
Work
Pete Daly

19 Helping the Baby Grow Up: The
Art of Good Development
Jon Croker

22 'Play it Again, Sam':
Screenwriting in a Virtual World
Andrew S. Walsh

31 Frequently Asked Questions
About How to Develop a Script
Neil Peplow

Directory Listings

Production Companies

37 USA

76 UK

111 Canada

124 Ireland

134 Australia

142 New Zealand

Representation

151 USA

158 UK

167 Canada

168 Ireland

169 Australia

169 New Zealand

Courses

173 USA

179 UK

189 Canada

190 Ireland

191 Australia

193 New Zealand

Societies and Organizations

197 USA

200 UK

207 Canada

208 Ireland

210 Australia

210 New Zealand

Festivals

215 USA

229 UK

241 Canada

245 Ireland

246 Australia

248 New Zealand

Awards and Prizes

253 USA

263 UK

266 Canada

268 Ireland

269 Australia

270 New Zealand

271 **Recommended Reading**

277 **Index**

Introduction

Barry Turner

Feedback from readers of the first edition of *The Screenwriter's Handbook* suggests that interest in this branch of the creative arts remains strong and may even be gaining strength. This despite the tendency of the circuit bosses to favour Hollywood blockbusters and for producers to look two or three or even thirty times at a pitch that cries out quality but may, in consequence, be destined for DVD via the art houses. The consolation is in knowing that, in the end, quality will be vindicated. Public interest goes way beyond the latest special effects, while the opportunities for viewing movies of choice will inevitably increase to meet demand. To help the process along, this latest edition of *The Screenwriter's Handbook* has expanded lists of people, companies and organizations central to the film industry, with contributions by leading filmmakers to show how newcomers can break into the market.

International telephone codes have been omitted for brevity's sake and because the book is itself international and thus unable to keep track of regional variations of codes.

Suggestions for further developments are welcome. Please do write or email: screenwriters@macmillan.com.

Foreword

Jake Eberts

We're still young at heart. With Sidney Lumet making a feature (and a very good one at that) in his mid-eighties, and Martin Scorsese still churning out critically acclaimed movies in his sixties (the latest about a rock group whose members are about the same age), you might start believing that the film business is 'no country for young men', to paraphrase the Cohens. Some say that the cinemas are now full of remakes and rehashes, citing franchises like the Scary Movie films which curiously try to lampoon a lampoon. With the studios all part of bigger corporate conglomerates, and controlling so much of the output, we often hear there is no room for the up-and-coming auteur or a new creative brat pack. But nothing could be further from the truth.

Film is our youngest and most vibrant art form. Its energy and popularity comes from the constant influx of young filmmakers with their new ideas and overpowering enthusiasm. The beauty of our industry is that it is always looking for something new and different. The reason why the business has managed to see off predicted assaults from television, home entertainment, computer games et al. is that we can still surprise and delight.

Every year the Sundance Festival makes us keenly aware that film is still producing new ideas and new faces. And look at the wit and originality of *Juno*, for which debutant Diablo Cody won a screenplay Oscar. Or even the imagination and amazing ability to create warmth in a wintry Minnesota as displayed by Nancy Oliver's *Lars and the Real Girl*, another big screen first-time writer who received a nomination for her screenplay. While the great talent that is Lumet was the power behind the camera of *Before the Devil Knows You're Dead*, don't forget the labyrinthine script came from another new screenwriter (Kelly Masterson, who used to be a Franciscan brother).

It all just goes to prove the cliché that the only barriers to success in this industry are in one's mind. Richard Attenborough had been trying to make the film *Gandhi* for over fifteen years, repeatedly being turned down by all the traditional sources of finance, in spite of his excellent track record as a director, actor and producer. But he never gave up, because he had a vision and a dream.

The same could be said for Kevin Costner. Although a rising star at the time, nobody had the confidence to let him direct; especially to direct a three hour epic, in which the Native Americans were the good guys. Trying to put *Dances With Wolves* together was a long hard struggle, but it also turned into the most rewarding experience for all concerned.

Film is where dreams come true ('you will believe a man can fly'!). And if you have the talent and the imagination, there is always room for another break-through. With the advent of High Definition and the Internet, the industry has become even more of a meritocracy. Without a shadow of a doubt, you still need determination and ability. Millions were taken on magical journeys around the world last week. And this week. And undoubtedly, next week. There is always a newer and better experience just over the horizon. As long as your vision can provide this then there is room for you in the film business!

Jake Eberts is the Executive Producer of Dances With Wolves, Chicken Run *and* A River Runs Through It.

An Ever Changing Screen

Barry Turner

There is a Godard film noir from the early sixties (*Bande à part*) which ends with the director promising in voice-over that his next feature would be in Cinemascope and Stereophonic Sound. Nobody believed him, of course. Jean-Luc Godard was one of the original auteurs, a brotherhood of young French filmmakers led by François Truffaut, who believed that advanced technology was incidental to the story on screen, a story moreover that was created solely by the director, who used the camera as a writer used a pen. So not only were gimmicks like Cinemascope and Stereophonic Sound superfluous to movie creativity, so too was the screenwriter who was, at best, an unnecessary appendage and at worst a distraction from the job in hand. It was lucky the actors got a look in.

Looking back

The debate over the artistic authorship of a film is as old as the movie industry itself. In the early days when the only sound accompanying a film was the mood music from the cinema pianist, the idea of engaging a screenwriter seemed faintly ludicrous. As Marc Norman tells us in his history of American screenwriting (*What Happens Next*, Aurum), when D. W. Griffith made *Birth of a Nation* in 1915 there was not a screenplay in sight:

> The scenes were improvised according to Griffith's usual casual manner – turn the actors loose on the set, let them block themselves, come up with their own lines, try it one way, try another, make suggestions, take suggestions from anyone, and finally tell them exactly what he wanted them to do.

Rumour had it that the novel on which the movie was based had been broken down into a list of major scenes; but if this was so Griffith never referred to it.

Screenplays and sound

Birth of a Nation was a mammoth, if controversial, success, but its cost and the prolific extravagance, that were part and parcel of Griffith's directorial technique, frightened producers. While it is said that, as late as 1937, Raoul Walsh shot a feature from an outline scribbled on the back of a shirt cardboard, post-Griffith few movies were made without an approved screenplay. From now on, Marc Norman tells us, 'somebody would dream up movies and write them down before they were filmed'.

The rise of the screenwriter in the movie pecking order was accelerated by the advent of sound. But the writer's status in what was now seen as a collaborative enterprise did not remain fixed for very long. Depending on their artistic reputation and financial muscle directors made claims for precedence that put the screenwriter at a disadvantage.

Directors and auteurs

In popularising the notion that the director is the true author of work that is essentially visual, the auteurs were simply raising the stakes in a long-running debate. Their argument was strengthened by the invention of the hand-held camera which gave Truffaut, Godard and the rest much greater freedom to impose their personalities on movies that relied heavily on location footage.

Auteur theory had an immediate appeal in American film schools where directors were only too happy to lecture on their indispensability in the creative process. For Roman Polanski, 'the director is always a superstar ... He makes the film, he creates it'. Inevitably, there was a backlash. Carl Foreman famously declaimed that 'a director without a screenplay was like a Don Juan without a penis'.

Even so, auteur theory made the 1970s the decade of the screenwriter director and, it has to be said, of some outstanding movies by auteurs such as Francis Ford Coppola, George Lucas and Steven Spielberg. But if some multitalented directors were writing their own scripts, the fact remained that they had to have scripts; they could not simply let the camera do the writing for them as the French New Wave would have wished.

Changing roles

So it was that while the Hollywood auteurs were collecting their Oscars, a new generation of screenwriters was emerging into the limelight, screenwriters moreover who often harboured ambitions to direct but were content to let their writing take precedence. Meanwhile, a string of expensive flops, *Heaven's Gate* being the prime example, brought the curtain down on the director as stand-alone creative genius. Henceforth when director and writer were the same person it was usually because the studio was trying to save money, two for the price of one.

Home entertainment

Which brings us to the latest state of play. As for every other medium, the movie industry is having to adapt to the age of the Internet. Cinema attendances are way down because core audiences – those under the age of thirty – are increasingly turning to home entertainment, either online downloading or the big screen TVs, for the latest DVD releases. None of this is necessarily bad news, except perhaps for cinema owners who will have to rely on a dwindling supply of sure-fire Hollywood blockbusters.

Challenges to screenwriting

For screenwriters, there are two challenges. The first is to secure fair payment for the use of their work in the new media. It is a chilling thought that until the Writers' Guild of America voted for an outright strike lasting over two months, their members were entitled to just four cents for every $20 DVD sold. For once, British writers are ahead, having, some time ago, through the good offices of their own Writers' Guild, secured the right to decent royalties on subsidiary sales. But the industry is fluid, and as new opportunities for money making arise producers may well revert to their bad old ways of grabbing whatever is there for themselves.

The second challenge is to become active participants in the Internet revolution. It is a frequent complaint of mainstream movie making that it rarely focuses on serious issues or engages in ideas that are anything but mundane. This could change with the opportunity of harnessing the computer generation who are forever looking for something new and compelling to download.

Calling the shots

Outside Hollywood, where it is rare for a movie to be budgeted at less than $100 million, filmmaking can be reduced to a modest cost by cutting out the star names, special effects and mega-promotions. Now, more than ever before, the screenwriter, wielding the hand-held camera, can give visual substance to a movie proposal that otherwise might never shift from the printed page. This is nothing less than auteur theory updated, but this time with the screenwriter calling the shots. Someone should ask Jean-Luc Godard if he approves.

Film, Television … It's All the Same, Isn't it? The Differences in Writing for the Two Media

Nick Moorcroft

Unlike a lot of writers in the UK I was lucky enough to start my career penning feature screenplays. My writing partner and I hit the jackpot early – we managed to sell our very first script to Ealing Film Studios. But the established 'wisdom' is that writers learn their craft in TV and then aspire to 'graduate' into film. It is almost as if film is the higher art form – and television some motley apprenticeship. Nothing could be further from the truth. A good TV scribe will not necessarily write good movies. Or vice versa. It all really comes down to how the script is executed and how much you understand about the art form you are writing for.

Different ways of working

There is much to be said for starting off in TV. You learn a very rigid structure and have to work within budgetary constraints, time constraints and also commercial constraints in that you have to write towards a break (unless you're working for the BBC). It is a great learning curve. But in a way it can also impede you when you come to write a film, which is a lot more fluid.

There are few hard-and-fast rules when comparing writing for the two media; and size really is not important! TV is supposed to be the poorer relation in terms of budget, but that should never inform how you write in terms of ambition. In either film or television you should not be limited just because the producer cannot afford to realise that vision. Making the dream fit the budget can always come later in the development process.

Writing styles

Rather than having to think small, in TV your writing has to be that bit more succinct – and that's an incredible discipline. I always think that television writing requires banner headlines. It's almost like you're writing shorthand at times: you have to communicate important details quickly and constantly update the viewer. This is very different to writing for film where you might take the whole movie to explain and describe a character and watch his or her arc.

The beats (for drama on the commercial channels) are also very different, channelled as they are towards writing to the breaks. The producers, and the advertisers, need to know that the drama is interesting enough to lead into an advert. The plot point has to ask a question at the end of each part, with the answer at the start of the next. In film screenwriting you're working within your own self-imposed sequences and structures. Everybody talks about the three-act structure and the inciting incident. But the truth is that there are no rules – these are just guidelines (to paraphrase Jack Sparrow!). You can take them or leave them. With TV you have no choice. Your script constantly has to build and then build again – usually to four incredibly big moments that will keep the viewer enthralled (including the last one that hopefully will bring them back for episode two). The breaks affect the dramatic tension, the sequencing within those mini-acts, and the time in which you can set up a character.

What the medium requires

Television is also a faster medium in terms of turnaround and decisions. The process can be incredibly frustrating for an aspiring film screenwriter because of the few films that are made, and the length of time it takes from pitch to (hopefully) release. This is very different to the soap writers, who write one week and see their work on screen the next. But as a rule of thumb, when you do get commissioned, film gives you more room and time to explore your story.

However, that's where the easy generalisations basically finish. Overall it is much more about your grasp of the medium which you are attacking. There are just too many different types of television, and for that matter of film, to pinpoint succinctly all the differences.

Making the transition

The question is not whether you can cross over from one medium to another, but whether you can learn to adapt to any of the many different situations. The transition from television to film can sometimes be quite disastrous. Just because someone writes fabulous TV drama does not mean that they can handle movies – or vice versa. There are great TV writers out there, and there are great movie writers out there. But for every Allan Ball, who can write a movie so filmic and ambitious as *American Beauty* and then can take that skill and adapt it to an amazing series like *Six Feet Under*, there is a Dennis Potter whose work simply came alive on the small screen but refused to work in the cinema.

In the end, it is all down to structure. I know of a very respected television writer who was commissioned to write a movie and ended up submitting 240 pages. His overall awareness of structure was so different because he had spent years working towards a different format. I hope I'm fully aware of my own limitations. It's about having an awareness of the differences not only between television and film, but of soap operas, drama series, sitcoms, horror movies, romcoms, etc.

Getting it right

There used to be a tradition of taking sitcom writers in the UK and, because they're funny, believing that they can write movies. History has taught us that this doesn't work. A sitcom is a completely different form in terms of pace and structure to a movie. Most of the writers were unable to tell the jokes differently.

They've learned how to deliver a gag one way, but to tell it within the context of 110 minutes is a whole different form of delivery.

Richard Curtis is a great example of a talented writer who can straddle the art forms. He started out writing sketch shows, great sketch shows. From that he went on to write sitcom classics like *Blackadder*, and then was able to move again into the world of romantic comedy – some say he has created his own subgenre within romcoms. But very few people are able to do that.

Mastering the tricks

Within television and film there are separate rules – tricks let's say. And then there are tricks within tricks – all needing very different skills. The key is identifying the individual structures of each medium, and hopefully mastering them. It's not easy and you can't just jump from one to another. It takes a long time to adapt to the differences in pace, density, timing, etc. It's about recognising the different forms, realising that you have to learn about them and understand how they work. A great novelist would not have the arrogance to think that he or she could automatically write a problem page or a trade magazine, so why should a television writer expect to pen an Oscar winner – or even want to?

Nick Moorcroft is an established UK screenwriter for television and film, whose credits include the recent box-office hit St Trinians.

Just a Mamet

Barry Turner delights in David Mamet's reflections on the movie business in Mamet's new book, Bambi vs. Godzilla

David Mamet is a maverick among filmmakers in that he tells the truth as he sees it and to hell with the consequences. If for no other reason, this makes him required reading for anyone hoping to make a living in movies, with aspiring screenwriters having most to gain. The starry-eyed will find much that Mamet has to say hard to take. For a start, he does not warm to Hollywood, dedicated as it is to 'fewer and fewer films, of diminishing worth and ever-inflated production costs'. These blockbusters, he argues, have little to do with genuine entertainment. Instead, they appeal to a part of the American psyche that takes seriously anything with large wads of money attached. 'We are reassured by their presence rather than their content or operations. As examples of waste (big and bad films) appeal to our need – not for entertainment but for security.'

In the Mamet lexicon, a producer equates with villainy – on a massive scale. The money comes in and rather more rapidly goes out as the producer deducts a fanciful list of personal expenses. 'I have seen producers bill the movie salaries for their mistresses, for their absent yes-man, for travel and lodging never used, for services never proffered, for inedible cast and crew meals charged off as gourmet fare, for imaginary bank fees interest, and so on, and so on.' Are there then any good producers? Yes, says Mamet, a few, 'but we must remember that even Diaghilev went into the ballet because he wanted to screw Nijinsky'.

And who suffers most from the producer's machinations? Well, of course, it is the screenwriter. Having put up the warning signal that most scripts are 'written by that tiny coterie of Mamelukes, harem detainees or house slaves who constitute the chosen among the Hollywood faithful', Mamet tells the newcomer who is ready to defy the odds that his purpose in life, as dictated by the producer, is to write a duplicate of the script for the hit of last year. Then, just at the point where the reader is primed to accept any occupation – grave digging, say, or pimping – as a welcome alternative to a career in movies, Mamet switches tack with a reminder that the bank busting ways of Hollywood are only part of the story. There is good screen entertainment to be seen; it is just that few movies worthy of the name originate on Sunset Strip.

Mamet has constructive advice for those who put pride in the job before the sea-view mansion. He clearly enjoys teaching (he is a doyen of the Yale School of Drama) using his own formidable writing skills to reduce standard classroom jargon to simple, time-tested propositions. 'Almost any film can be improved by throwing out the first ten minutes.' And don't we know it. Don't we know too, 'if you can't figure out what the scene's about, it's probably unnecessary'. Best of all, 'if a director or writer wants to find out if a scene works, he may remove the dialogue and see if he can still communicate the idea to the audience'. There's much more where that came from. Add blood, sweat and tears for a promising future as a screenwriter.

Bambi vs. Godzilla. On the Nature, Purpose and Practice of the Movie Business *by David Mamet (published by Simon & Schuster).*

From Stage to Screen: Why Theatrical Drama Rarely Finds a Home on the Big Screen

James Christopher

The stage and the silver screen have always had a rocky relationship. Superficially they appear to be two halves of the same coin: we watch actors on stage, or some nice person films them acting so we can see them in our local cinema. But stage adaptations rarely, if ever, work.

Perhaps it is all in the writing. Stage plays hinge very much on the script itself for them to succeed, with the other components coming much further down the list. Cinema is a much different way of engaging with what is on the page, which is why writers are neglected in Hollywood. More important for a film is the sensation – the images, the music. Theatre is more to do with the imagination.

Different worlds

It may be the way we view the two. In the theatre the audience watches live people performing in a physical space. The actors have to create a world out of this limited space, and the audience have to buy into that. Conversely film can actually take you to that place and show you physically what the trees look like, how the wind blows, etc. So they are actually two very different assaults on the senses. Cinema is very much a visual sensation; with theatre it's much more an imaginative engagement – you have to make a much bigger leap of imagination. Overall the audience must make a contract with what is happening on stage.

An engaging story

Without doubt the one thing that you desperately need for either is a cracking good story. The crux is how that story brings you to a certain level of anxiety. When you begin to feel anxious for the characters in the theatre you begin to really engage with the play. That's when the hook is in. It's perhaps even more magical in the theatre because you are in such an artificial environment; whereas in cinema you can sink yourself into that big screen. When the character is in the car, you're in the car. That's why actors love performing on stage. They feel that it is their performances alone that brings you on the journey.

It's quite interesting when watching a very good stage play to work out why it is a very good stage play. Generally it is because it works and lives inside your own head. It is what the play is actually painting – what the audience believe they're looking *at* rather than looking *into*. But be warned: it is then extremely difficult to take the pictures that you have imagined from a play and put them onto screen.

The great divide

I feel that now cinema is so graphic it provides something the theatre can never recreate. Hitchcock really broke the mould with *Psycho*. When that knife went in it broke the innocence of the gas lights. We didn't need to use our imaginations at all any more. A whole new genre of cinema exploded from *Psycho* – you were suddenly allowed to see people stabbing each other. That really took cinema to a place that theatre could never go, and created a huge divide between the two. Hitchcock upped the bar, and theatre has never managed to replicate the experience, no matter how much people like Sarah Kane tried. I've never seen a sex scene on stage that has worked. Cinema can film the taboos and make them seem real, which you just can't do in the theatre. That is what created the biggest rift between the two art forms.

Adapting a script

The worst adaptations are those that change the least in their crossover. A lot of the difference between scriptwriting and screenwriting is in the dialogue. Take Mike Nicols's *Closer* for instance. It doesn't work because the original dialogue was written in such a way to fit a physical space and real time. Stage language is usually closer to poetry than real life dialogue. Dialogue is so much more mannered on stage. The idea is for the actors to pronounce the words so the back row can hear them.

Glengarry Glen Ross is a strange example of getting over this problem and making it work on screen. Probably the two American writers whose work has transferred most successfully (critically, if not always commercially) are Sam Shephard and David Mamet. But that is because they initially brought the cinema to the stage, and then basically returned it. Cinema is so much an American art form that their theatre imitates film. It is the opposite in Europe. Shephard and Mamet brought cinematic sensibilities to their stage plays – their hard-boiled dialogue was learned from Howard Hawks and Jimmy Cagney, not Oscar Wilde.

Musical crossovers

Musicals are different again. They come out of a concept that is much more cinematic anyway. Musicals are built upon spectacle, almost pantomime. People like musicals because they are simple candyfloss concoctions. That's why they work so well when transferred to cinema. The spectacle can be even bigger and the visceral sense of entertainment made even greater. Sondheim is interesting because, conversely, his musicals fail to work on the big screen (barring Tim Burton's *Sweeny Todd*, which I have to admit is the only musical I've ever liked). But this is because Sondheim doesn't actually write musicals: he writes plays with music.

Early movies were very simple visuals that told a story – cowboys chasing trains etc. When sound came in the producers realised that many of their biggest stars couldn't actually talk convincingly! So it's no coincidence that the first 'talkie' was Al Jolson's *The Jazz Singer*. Cinema's evolution was trying to put songs to the visuals – melodrama with music, if you will.

Remember the differences

Stage scriptwriters have to remember that they are writing physically for a different space. It's a different time limit with greater spatial constraints. All in all it's actually pathetically limited, which is why writers are so important. They are the only people who can put colour into it. On-screen the writer is the last person you want putting colour into the proceedings because you've got Mr Stuntman, Mr Special Effects, Mr Music Composer, et al. So the writer gets shunted right down the pecking order – and quite rightly because you can actually just tell the story with visual images. A lot of theatre writers cannot grasp this very simple fact; they fail to understand why their brilliant pieces don't work on cinema.

They have to remember that stage drama relies entirely on the writer to paint the scenery and the suspense. In film the point of view is controlled exclusively by the camera lens, which is therefore controlled by the director. That is why the director is God in cinema and the writer merely one of the crew!

James Christopher is Chief Film Critic for The Times. *He began his career working under legendary theatre reviewer Jack Tinker at* The Daily Mail, *and has been theatre critic for* The Daily Express *and* Time Out.

The Art of Adaptation

Ronald Harwood

As a screenwriter with a string of awards, including an Oscar for the Roman Polanski movie, The Pianist, the creative talents of Ronald Harwood are much in demand. Yet he has never written an original screenplay.

There are two distinct strands to screenwriting. The first is pursued by those who write original material for the movies, stories they feel compelled to tell through the medium of cinema and no other. The second, and most common in the English-speaking world, is adaptation, transferring a book or a play or some other source material to the cinema screen. This is what I am most practised in. In a very long career, I have adapted for the cinema novels, plays, two of them my own, and works of non-fiction.

Subject matter

There are several ways in which a subject for a film may come to the screenwriter's notice. A producer or director or a studio may present the writer with the subject and ask if it is something that may be of interest. In my own case, I now know immediately if the book or play appeals to my creative and technical process.

It has been my lot to work over the years on several subjects concerning the Second World War and the Holocaust. I say 'my lot' because I have come to the conclusion that my preoccupation with this horrific period in history is something of a burden, an obsession, and I have tried to analyse why this should be.

Early years

I was born in South Africa, which, in my time, was a totalitarian state. I cannot pretend I was aware of that when I was growing up, but when I left South Africa at the age of seventeen and came to England, I became aware of it.

I was five years old when war was declared on Nazi Germany, and Cape Town was 6,000 miles from the centre of things. Yet I remember vividly the excitement of being a child during that time: the convoys carrying British and Commonwealth troops on their way to the Far East, the BBC news bulletins and, of course, Winston Churchill's speeches.

My father, an emigrant from Lithuania, was in his early forties then, and had a lame left hand. Yet one evening he came home in military uniform, having enlisted that afternoon in the South African army. He had apparently passed the medical. My mother said, 'What did the doctor say when you showed him your bad hand?' 'I didn't show it to him,' he replied. 'He didn't ask, so I didn't show it.'

Haunting images

A belief in a just cause and the growing awareness of a great battle being fought against a barbaric enemy of the Jews informed my daily life. But in 1945, when the Nazi atrocities were revealed, I was taken with other Jewish schoolchildren to see the newsreels of Belsen and Auschwitz. Those dreadful images – the skeletons passing for human beings, the bulldozers shifting mounds of corpses into mass graves – have haunted me ever since. And I remember, too, a photographic essay in *Life* magazine, showing the bodies of the Nazi war criminals after they'd been hanged at Nuremberg. I can't deny the sense of satisfaction I felt at the sight of them.

Now, there may have been other children in the group that were taken to the cinema that day who were similarly scarred. But, as far as I know, none of them became writers and that is the critical difference.

Inner voice

I became aware, really politically aware, in 1959-60 when there was an incident at a place called Sharpeville, when black people protesting against the Pass Laws were shot in the back by South African police. I wrote a book as a result of that, my first novel. And I think that's what politicised me in that sense, against that kind of injustice, which Nazism and Communism were great examples of in the twentieth century.

The war defined my childhood, the Holocaust my adolescence. That a synthesis has dominated much of my creative life ever since, which is why I have come to realise that I do not look for the themes I write about: they look for me.

It will be the same to some degree or other, for every writer. What he or she writes about will be dictated to them by forces outside of their conscious control. Of course, mistakes will be made. One's dear old conscience tries to say, 'no, don't do this one', but the conscious mind says, 'ah, but the director's a brilliant man', 'they tell me Nicole Kidman or Johnny Depp is going to be in the film', 'this will do my career some good' and, be sure of it, disaster will ensue. But if you obey without question your inner voice, no harm will befall you.

Risks of historical writing

Writing about the Holocaust is a sensitive and often risky business and one has to tread warily. As we all know, there is no such thing as undeniable historical facts. All history is subject to dispute. Revisionism has always been a fashionable way for an historian to make a reputation and a living. From the story of Creation in the Book of Genesis to the recent invasion of Iraq, history is disputed, interpreted, misinterpreted, falsified and ignored. Even the most conscientious and honest researchers are prone to error. Witnesses, especially eyewitnesses, are accused of faulty memories, of being self-serving, vengeful and often naïve. All of which tends not to illuminate the past but to obfuscate it, to make it more and more difficult to penetrate the events under investigation.

The nature of truth

There is a further problem for the analyst of historical truth, which I can best illustrate, immodestly, of course, by quoting from my play, *Taking Sides*. A young American intelligence officer, in trying to placate a distressed witness, says to her, 'We're just trying to find out the truth.' To which she replies, 'How can you find out the truth? There's no such thing. Who's truth? The victors? The vanquished? The victims? The dead? Whose truth?'

Now, I don't happen to agree with her, but nevertheless that is what she says. I don't agree with her because I think it is possible to reveal the truth yet not necessarily through accounts written by historians, however brilliant or worthy, or by searching the archives for hitherto undiscovered facts. The truth, I believe – and here is a genuine paradox – can best be revealed through fiction. Because it is the individual artist's truth that allows us to accept the validity of the past.

Inspiration

My personal understanding of history has been immeasurably enriched by novels and plays and films. I suspect it may be the same for a great many people. For me, the most vivid insight into the Russia of the early nineteenth century was derived from Tolstoy's *War and Peace*. My understanding of the declining years of the Austro-Hungarian Empire came from *The Radetzky March* by Joseph Roth. My consciousness was first made aware of the evil of apartheid in the country of my birth by Alan Paton's *Cry, The Beloved Country*. To imagine Vienna in 1945, it is difficult for me not to be bombarded with images of Carol Reid's and Graham Greene's film *The Third Man*. The list is endless and I haven't even touched on the momentous achievements of Victor Hugo, Charles Dickens or William Shakespeare.

Rules to apply

There are, I believe, at least two rules that can be applied to all dramatisations dealing with the Holocaust. The first one is to shun any temptation to manipulate. Manipulation must be ruthlessly avoided simply because manipulation obviously is in itself a distortion of the truth. And because the events speak loudly enough for themselves.

The second rule, and perhaps the more important, is that sentimentality must also be mercilessly shunned because the events are burdened with genuine suffering and therefore capable of provoking genuine feeling.

Final thoughts

What conclusions, if any, are to be drawn from my experience as a screenwriter? Certainly, I have learned that the screenwriter's relationship with the director is at the very heart of filmmaking. But the screenwriter must learn, and it is sometimes a painful lesson, that he is not an equal partner, indeed he is somewhat subservient. Once the film goes into preparation, the writer is distanced from the centre of things. Some writers, I am told, visit the set and watch the filming. I avoid it like the plague. I would rather check-in at an airport than watch a day's filming.

I have also learned, at times painfully, that the writer must be sure of the world he's writing about, and has to approach the screenplay with the same degree of commitment as he or she would any other work. I have learned that form is of secondary importance to content. What a film is about stands above all else. And the screenplay, besides supplying all the information that it needs to supply, must be enjoyable to read. And, finally, the director should shoot the film laid down in that document and no other. I talk, of course, of an ideal world.

Extracted from Ronald Harwood's Adaptations from Other Works into Films. *Edited by David Nicholas Wilkinson and Emlyn Price. Published by Guerilla Books; £12.99.*

Learn from the Classics: Assessing Why Five Classic Scenes Work

Pete Daly

My working week usually entails reading between seven and ten scripts. Many contain the same problems – they do not know who the protagonist is, or there is no stated intention for the drama, perhaps the premise fails to develop and add layers of complexity, or it could be even that the writer has a poor grasp of the English language (happens more than you would expect!).

However, the most common (and fatal) error is to believe that a screenplay is merely about the dialogue. Not even a radio script is *just* about the lines. And movies are about so much more. The joy of cinema is the assault on all the senses that a good film can achieve. You can always tell an immature or unpolished writer because their work has pages and pages of dialogue.

Consider what your most memorable film moments are – the classic scenes that first made you want to write for the big screen. I've listed five below (in no particular order), and have tried to express briefly why these work. These lists are always going to be subjective, so I make no apologies for the choices. Even if you don't think they are classic scenes, or classic movies, hopefully it will make you think about what your inspirations are. Whatever makes it onto your list, I'm betting now that the dialogue is only a small part of the reason why these scenes stick in your mind so.

The shower scene, *Psycho*

This is a moment that encapsulates the power that good cinema has. Even when compared to today's torture porn and slasher-fests nothing comes quite so close to scaring you witless as this scene. And it's in black and white! The concept, that you would be oblivious and extremely vulnerable because of the water cascading on your head, is perhaps the scariest thing (come on, you must have had a Hitchcock flashback at least once when in the shower!). The power also came from the swaying back and forth of the perspectives, the drilling music, the genius of using the shower curtains, and the blood pouring down the plughole. But the other masterstroke that many of us take for granted now is that the lady in the shower was the star. No one predicted that Janet Leigh was going to be killed off half an hour into the movie. It was the shock of such an audacious move, throwing the audience out of their sense of security, that made this scene a classic. And not a word spoken!

Quarter Pounders, *Pulp Fiction*

That's not to say that the spoken word does not have its part to play. Tarantino is a master of inserting the ordinary into extraordinary situations. Sometimes his dialogue can be overly mannered and stylised, but this exchange between Samuel L Jackson and John Travolta is pure genius. The beauty of the scene is how the lines work on different levels simultaneously. The two hit men are basically torturing Brett (Frank Whaley), drawing out the tension before his inevitable murder. But the scene is also extremely funny in both a surreal and a macabre sense. You do not expect these two heavies to be wittering on about something so banal as burgers (and how they are named), and so it puts us off centre – we really don't know what is coming next (heightening the tension). It is a classic scene because it leaves an impression. Everyone now knows what Quarter Pounders are called in France, and why!

The Richest Man in Bedford Falls, *It's a Wonderful Life*

It's a Wonderful Life has shaped modern Christmas as much as Coke's red-suited Santa. The final scene is a masterclass in manipulating the emotions. This is how to build to a climax. Despite being a long film with an unconventional structure (a very brave 20-minute prologue), the screenplay never releases the valve on the tear ducts until the very end. Put this against something like *Vera Drake*, where the protagonist starts to cry halfway through, and work out which one you are more emotionally connected to. By the time everyone arrives and George finally understands what a great life he's had, the relief unconsciously opens the audience's floodgates. Even without dialogue this scene is a classic, but the incredibly tight script manages to sum up the movie with his brother's one-line toast. Cleverly the film does not milk it too long or outstay its welcome. One more line from a child, referencing the angel, and the movie is finished – embedded in the hearts of the audience forever.

Hair gel, *There's Something About Mary*

As audacious as Hitchcock killing off his star, this scene also works because you cannot believe it is happening. It pushes the boundaries of mainstream comedy. From John Walters you would not be surprised – but you may not have laughed as much either. It would be expected. But to see a girl mistake sperm for hair gel in the multiplex? When you're with your family or partner? Half the comedy comes from the basic idea (a nice, if crude, little gag); the other half comes from the embarrassment of witnessing this in a shared environment. It shows why you should always consider how the audience will react. The Farrellys were called the 'gross-out kings'. But it is actually what they don't show that allowed them to pull this off (excuse the pun). The humour is controlled and takes the audience to the brink of taste, and no further. They are also masters of economy – never labouring the joke. In the wrong hands the scene could have been much too prurient and thus would have fallen flat.

D-Day, *Saving Private Ryan*

As a screenwriter you have to accept that film is a director's medium. The movie was officially written by Robert Rodat. But does anyone refer to it as Robert Rodat's *Saving Private Ryan*? Do they hell as like! The vision is widely acknowledged as Spielberg's alone, fully credited to him. And let's be honest, that's all we really remember from the film. The first epic thirty minutes on the beach changed the whole landscape of war films. There was no John Wayne single-handedly mowing down the opposition. This was war as documentary, showing scared young men getting brutally killed. It was also on such a scale that it filled the big screen and captivated the senses. Sorry Mr Rodat, but the rest of the film was not that good. Once Tom Hanks leaves the beach the movie pales into insignificance. It is the spectacle and innovation of the opening that makes the experience. I'd like to think that this was all part of the writer's vision, but don't think he'll ever get the credit for it even if it was!

So what can we learn from the classic scenes? Perhaps that the best (and most memorable) movie moments are when the audience's emotions are engaged as much as their intellect. It is also important to consider trying something new – taking a risk and surprising the viewer. Dialogue is important in a screenplay, but it is only a part of what makes up the spectacle.

With nearly twenty years' experience in the film business Pete Daly reads and develops scripts for a wide range of funding bodies, production companies, distributors and established writers.

Helping the Baby Grow up: the Art of Good Development

Jon Croker

Just what is development? You *conceive* an idea; you *write* a script of that idea; you *make* a film of that script. All good, strong verbs that suggest some kind of definite action. But develop? It sounds a bit vague, a euphemism for something underhand – the point when the money men change the tragic ending to a happy one, add in a shower scene and give your existential protagonist a wise-cracking talking dog.

It can be that. And worse. But that's not really development, is it? Literally speaking, that's the opposite. It's killing an idea by changing it so much that it becomes unrecognisable. Good development is about nurture and growth.

First steps

Imagine that an idea for a film is a new-born baby. It's a creature full of potential, gasping for life but at the same time it's small and delicate, nowhere near fully formed. And the parents of the baby think it's the most gorgeous baby that has ever lived, even if it's ugly.

The trick then is how to take that baby and help it grow into a fully fledged adult. A lot could go wrong during the child's growth. It could become too flabby or too slight. It could hang out with the wrong crowd. It could become seduced by money and fame rather than truth and beauty.

The parents of this child are the filmmakers – the writer, the director, the producer. It's their baby and so they bring up the idea. Who then are these script editors, readers, researchers and why do they have a say in the bringing up of someone else's baby?

Development stage

I'm one of them. At the UK Film Council's Development Fund, filmmakers apply to us with their feature film and we then select projects to fund and support through development. Sometimes people come to us with just an idea, described in a synopsis and a treatment. At other times they've got a complete draft of the script.

So the projects we fund will never be our baby. But we should still care for it and help it grow. Bad development happens when you come with your own pre-conceptions and ideas for the project. Then, either consciously or subconsciously, you impose these ideas on the filmmaker. This never ends well. We're lucky in the Development Fund because our only agenda is to help get the best film made.

On some of the projects we fund, we take an active role in the development process, but if there's already numerous voices giving their opinion, we're happy to keep quiet. Development is definitely a situation where too many cooks can spoil the broth.

Similarly, whilst it is tempting to fall back on the various theory books of scriptwriting ('your inciting incident comes too early'), it must be remembered that many great films break the rules – that voice-overs can be brilliant; that some characters do not have an arc; that some stories do not have three acts. Some of the theory books can be a great way of understanding the mechanics of scriptwriting but they should never be taken as gospel. Take them as descriptive guides rather than proscriptive laws. Don't try to fit a three-act structure onto a screenplay as if it's a straitjacket that will make the film behave.

Asking questions

The key then is to approach the project from within. What film do the filmmakers want to make? This is the first thing to find out and agree upon at the start of development. What tone? What feel? What look? What kind of world? What kind of characters? Is it a genre film? Or is it something else? One of the key questions we ask all applicants to the Development Fund is why they want to make their film. This is not only a fascinating insight into the mind of the filmmaker, it also helps us work out how we can help their project. We can then make sure that we're on the same page from the very beginning of the project.

Support

And it's at this point that a story editor becomes useful. Parents are often blind to the imperfections of their baby. So they should be. A film needs supporters who keep driving things forward. But a story editor has a distance from the project and so can be more able to see the problems. So you approach the project from within but at the same time maintain an outsider perspective. A healthy paradox.

Discussion time

On a practical level this can mean a whole variety of things. Different filmmakers discuss their projects in a variety of different ways. A lot of the time it's less about going through the minutiae of a script and more about discussing its overarching elements. A moment in the story whose set-up is too complicated; a choice by a character which seems out of step with his or her behaviour; a loss of perspective in a scene or sequence; the sense that a great confrontation feels underwhelming either in its build-up or its execution. But then for some filmmakers it's useful to go through the script, page by page, moment by moment, and discuss the whole thing. Or they want written notes to ponder over. There are no hard and fast rules and no single successful way of working. Sometimes it's vital to raise the stupid questions. If you don't understand what's going on then you have to swallow your pride and admit it. Now is the time to ask the simple questions because once the great machine of production has begun, it's much harder to change things.

Passion for the project

It's about engaging in a dialogue, a free exchange of ideas, all the while keeping at the back of the mind the original concept for the film. It might be that this original aim was not the correct one but, if you're going to diverge from it, then it's vital to understand why you're doing so, that it's a choice to diverge and not a gradual slipping away. Rather than simply bringing up problems it's about asking questions and offering suggestions. Filmmakers are understandably very sensitive about their work – they have to be – but raising questions helps everyone to keep thinking about the project.

The ultimate aim of development then is to realize the potential of a project so that it can go on to make a great film. The screenplay is never the finished article. It is a blueprint, vitally important, but not yet a fully fledged film. And you can overdevelop a script: the passion for writing can dissipate if the writer is constantly having to tinker with things. It can become lifeless. Flat. Dull. It should have energy, momentum and drama. When we look at applications to the Development Fund we are simply looking for an idea borne out of passion and told passionately.

Jon Croker is Story Editor at the UK Film Council's Development Fund.

'Play It again, Sam': Screenwriting in a Virtual World

Andrew S. Walsh

> RICK
> You know what I want to hear.
>
> SAM
> No, I don't.
>
> RICK
> You played it for her and you can play it for me.
>
> SAM
> Well, I don't think I can remember it.

AS SAM HOLDS HIS GAZE, RICK LEANS OVER THE PIANO AND SMASHES HIM IN THE FACE WITH A BOTTLE...

Writing for games is not like writing for the silver screen. The worst that cinema-goers can do to a story is to answer their phone, mutter into their popcorn, or walk out. They can't attempt to make Humphrey Bogart kiss Ingrid Bergman, or shoot Victor Lazlo. This is because the viewer is passive.

A game, however, requires a player, not a viewer; it demands interaction. A game asks the player to experience the story by walking through it, not view it from the outside. This means the player can – to an extent – dictate how the story is told and even what that story is. In a game Rick could hit Sam for refusing to play 'As Time Goes by'.

'Toto, I've a feeling we're not in Kansas anymore'

Contrary to the title, this article is not about screenwriting. Screenwriting does not exist in games. Games contain games scripts and are written by gameswriters. The evolution of gameswriting and writers is a new phenomenon and it has arisen as those working in games have realised that the comfy Kansas world of the screenplay has been picked up by a twister and dropped into a world where magic is real and nothing is what it seems.

As the stunned screenwriters gets to their feet, surrounded by munchkins and bits of relocated housing, to get their bearings they must repeat to themselves the simple mantra, 'viewers view, but players play'. This will lead to the next step in converting from screen to game, the realisation that while it is a general truth that the most important thing in a movie is the script – this is not the case in games. In games the most important thing is gameplay. Without gameplay a game is not a 'game'.

This means that any story and dialogue appearing in a game must serve the gameplay first and the narrative second. Whilst this might appear to be a terrifying thought, it does not mean that story is unimportant, many of the big-selling titles of the gameworld are story driven, or story centred. Story serving gameplay does not have to be a direct conflict, rather a happy alliance. When well combined the two elements of story and game bring out the best in each other and create an incredible experience. A story you can step into.

'Give me a camera and a place to stand and I can move the world'

When the first films to tell stories were shot they used a static camera shooting a single set, to all intents and purposes they were stageplays with a camera placed in front of them. As technology and the craft of screenwriting developed, so came the cut, then sound, colour, the steadicam; and from a medium that was seen as a novelty was born a new way of telling stories, separate from books, separate from theatre – 'the movies'.

Eighty years later along came games. Purely visual at first, games evolved to include text-based adventures, the first story-driven games. Written in the style of a novel these games were essentially books chopped into paragraphs which allowed the reader a few different ways to move through the story (turn left/turn right, shoot the girl/kiss the girl) and allowed the reader the feeling of control. Later, as the power of machines increased and game graphics improved, so story and pictures could be combined.

With the introduction of animated graphics the games industry assumed that it could tell stories in the same way as they were told in movies. Games that included stories took linear narratives, chopped them into small packets and presented them interspersed on either side of sections of gameplay. These linear narrative scenes are called cutscenes and they are defined by the fact that the player cannot control them; in essence, they are sections of a movie. The addition of cutscenes to games had the advantage of presenting players with a story, while showing off the best graphics and offering the originating company a chance to reinforce or create IP (intellectual property) which it could market and protect. However, such non-interactive linear scenes have many drawbacks. Just as shooting a single set with a static camera would not provide the world with *Gone with the Wind*, or *Citizen Kane*, neither would cutscenes bring out the true potential of gameswriting.

'It's alive! It's alive!'

By 1942, the film industry had evolved from its single shot beginnings, so when *Casablanca*, a film based upon the stage play *Everybody Comes to Rick's*, was made there was no thought of simply shooting the playscript. Instead, the play was adapted and rewritten to meet the demands of the filmmakers and their audience.

Similarly, gameswriting techniques have evolved beyond limiting a game's story to chopped up sections of movie that are separate from the gameplay. While the cutscene remains an important tool for gameswriters, gameswriting is evolving subtler ways of storytelling that allow the two elements of gameplay and narrative to intermix. Just as film moved away from shooting stage plays, so games have moved away from imposing the strictures of screenwriting upon games.

'Houston, we have a problem'

While the screenwriter may master the short film, the novella and the feature while playing with different genres and methods to explore the narrative, the way the story is delivered is always the same. Movies are watched by an audience on a screen. Games do not have this single form.

Platforms

A 'platform' is the hardware upon which games are played. This could be an arcade machine, a mobile phone, a games console, a PC, or a hand-held device, to name but a few of the game-delivery platforms. Each of these platforms has its own intrinsic strengths and weaknesses that change the way that a game and its story may be experienced. While games console such as an Xbox 360 or Playstation 3 deliver the latest stunning graphics and complex AI (artificial intelligence) capable of allowing the game to respond to the player's every move, mobile phones offer games similar to those available on personal computers twenty years ago, with dialogue appearing as text and a limited series of choices (interactions) for the player. This means that the differences between writing for one platform and another may be as broad as those between writing a novel and a screenplay, yet both fall within the sphere of gameswriting. As gameswriters are likely to write across a wide variety of platforms during their careers, they will need to learn the strengths and weaknesses of the platforms they work with if they are to write the best scripts for each one.

Games genre

Once the writer has begun to establish the technical limitations and opportunities offered by the platform for which a game is being developed, there is a second key element to take into account before commencing work. Before dealing with the narrative genre the gameswriter must first examine the game genre, for this will change not only the style of the story, but also the way in which it can be told. When conceiving a sports game the game designers (the designer is the person responsible for devising how a game works and how it interacts with the player) will put their efforts into creating an experience as close to the real sport as possible. Their time will be taken up with working out how to create a realistic, enjoyable and playable scissor kick rather than with designing a complex speech engine to deliver the writer's sparkling lines. A speech engine is a part of the game specifically tasked with delivering speech. Another thing worth noting is that sports games will generally contain no story, though they will often contain large amounts of commentary dialogue, demanding a lot of writing but no character, world or story design. Unlike the movies, games can exist with no narrative and no characters, or employ both in complex ways. An MMORPG (massively multiplayer online role-playing game) on the other hand will demand a vast amount of world and character design as the players enter a free-form storyworld in which they can easily play forty hours a week for several years. Games genre therefore effects how writers can tell their story and the tools they will have to do that.

The list below will give you an idea of the major categories of game genre:

- Role-playing games (RPGs): here the player takes on a role in order to go on a quest. The narrative genre can be fantasy, sci-fi or more down to earth. These are the most dialogue-heavy type of game (*Mass Effect* is credited with over 2.5 million words of dialogue) and they are amongst the longest in terms of their gameplay and narrative.
- Action games: these can have a tiny amount of writing involved (e.g. *Dead or Alive*), or they can have their own movie-style storyline (e.g. *Metalgear Solid* or *Bioshock*).
- Sims: short for simulation these games put the player in control of an element of life, be it controlling a city (e.g. *SimCity*), running a railway system (e.g. *Railway Tycoon*), or managing someone's life (e.g. *The Sims*).
- Strategy games: generally military in nature these games can pit the player against the platform, or against other players. Some are based on taking turns, others take place in real time (RTS: real-time strategy games).
- Platform games: this is the traditional jump-around style of game and ranges from 2D side-scrollers to newer 3D versions.
- Sports: players can play their favourite sports as a player, a team or a manager. These games often have commentaries even if there isn't interaction between the characters.
- Racing games: many racing games take an approach that is more *Fast and Furious* than Formula One sports games. These sometimes have background stories and dialogue within them (e.g. the *Driver* series).
- Flight sims: pilot a Sopwith, a 747 or a starcruiser. Often little or no dialogue and rarely story or characters.
- Children's games: niche versions of the above, generally with lower budgets and often with text rather than the spoken word.

In gameswriting, therefore, one must consider the differences that exist in a wide and varied set of platforms and games genres before considering the narrative genre, character, plot and other skills central to more established media.

The greatest event in motion picture history: starring a hundred elephants!

Why does a platform and games genre affect storytelling? Surely anything and everything can be done in a game? If a game is about sports, why can't there be complex speech? Beyond matters of choice and relevance, there is a simple fact that many non-gameswriters ignore or simply don't realise – that most games are expensive to create. Very expensive.

The reason for this is simple. As games take place in a virtual world so *everything* in that world must be created. Before a hundred elephants can be added to a game they must be designed, drawn and animated. The ground, the trees and sky surrounding the elephants must be generated. If the elephants can touch an object then each object must have a relevant animation and sound effect so as to react as if touched. If the object falls over, the game's physics engine must be able to make the object bounce, or land on the ground rather than fall through the floor, or shoot into the sky, or simply fail to react. The game's AI system must understand when to trigger these animations and sounds as well as what the consequences of this will be, e.g. a loud noise might be 'heard' by a guard and bring him and others running.

Adding an elephant to a game and making it interact with the virtual world is a complex and therefore expensive process. Expense leads to decisions about what will and will not be included in a game. This is one reason why in some games where characters can jump they cannot talk, and why in others they are verbose but unable to jump or climb.

Part of a gameswriter's job is to hide the limitations of games design within the gameworld. To do this, writers have to address two elements. Firstly, they must use their skill to disguise the gameplay by making it part of the narrative. This could be by making the information the player requires to complete the next part of the game – the need to attack a fortification, kiss the girl or save Victor Lazlo – blend into the story. In other cases it means finding story reasons to make the game's design-demand – for a level involving, say, motorbikes – fit with the characters and the narrative. Secondly, writers must service the narrative. Good designers will understand the need to reorder or rework gameplay elements if it makes the story (and therefore the game) a better experience.

When design and story work together the result is a game such as *Halflife*, where the story and gameplay cannot be extricated from the other, and they are interwoven to make one uniform experience – and one that sold a lot of copies.

'You talkin' to me?'

The most obvious way to interweave narrative and gameplay is to place story information within the levels themselves rather than limit the story to movie-style cutscenes. Such interweaving is most often achieved through the game's dialogue. Such dialogue takes many different forms, each of which changes the way a story is told and the game is experienced:

- **Cutscene dialogue.** Contained within small pre-recorded and structured sections, cutscenes remove control from the player to deliver information. These sections are the part of gameswriting that is closest to screenwriting, but they can distance the player from the story as they are separate from the gameplay.
- **Timed dialogue.** This dialogue starts following a particular event (e.g. a character opening a door) and runs through from start to finish as a recorded

piece. Gameplay continues, but the player's actions will not affect the dialogue, they merely trigger it.

- **Interactive dialogue.** In some games the player is able to hold conversations with other characters. Such dialogue often gives the player options as to how the conversation can proceed and will respond to the player's actions and decisions.
- **Barks/Emotes**. These are short lines such as orders, pain response, taunts, etc. They are triggered by the player's actions and are responsive to his or her interaction with the game.
- **Incidental dialogue.** This dialogue is placed in the world to make it feel alive. For example, as the player's character walks down a street he or she might hear the other people there holding conversations or reacting to the way the character looks or behaves.

The sorts of dialogue options available to the gameswriter are predetermined by the game's speech engine and the complexity of its AI.

'I'm sorry, Dave, I'm afraid I can't do that'

One of the crucial elements in advanced writing models is AI. A game's AI system not only allows characters to react to game events, but describes how the game understands the player's actions and how it must respond to them. Early AI models were full of holes, one oft-quoted problem were games where a player could shoot one of a pair of guards without the second guard reacting, or indeed noticing the death of his companion. Such short fallings rather burst the bubble of reality. On a positive note, more advanced and well-implemented AI models allow characters to anticipate a player's moves and to react to the player's actions. This means that game characters are imbued with intelligence and senses, e.g. guards can see characters who are in the wrong place, or hear a noise and react to it. AI is important to gameswriters, as the more complex the AI model then the greater the range of ways writers have to tell their story, as the characters and game are capable of better responding to gameplayers' actions and the world that they are creating within that game.

'Here's Johnny!'

Games characters break down into two groups, one controlled by the player, the others controlled by the game's AI system:

- **PC.** The player takes the role of a player character (PC). There may be a single PC in a game, or the player may be able to move between different characters (sometimes between the different members of the PC's team/party) or there may be a number of PCs played by different players within the same game.
- **NPC.** Characters who are not controlled by a player are called non-player characters (NPCs). These can be enemies, allies or neutrals that PCs meet

along their way. All of these will be controlled by the game itself and will be capable of reacting to the player within set parameters.

Characterisation, AI models, speech design and a knowledge of platforms is only the tip of the iceberg when learning how to write for games; but this introduction should give screenwriters an insight into the differences which exist between screenwriting and gameswriting. There is no reason why a screenwriter cannot make the transition to gameswriting, but such a shift means accepting and exploring these differences. It means playing games and learning about the ways stories are told, about how to create non-linear plots or stories that are told over much longer periods than in movies. The differences don't stop merely at the writing either. The games industry operates in a different way from other media. For instance, a gameswriter will often write more than just the game script. It is by no means unusual for the writer to be required to write everything from a concept document and story designs through to dialogue, advertising copy, the games manual and the blurb for the back of the box. Just as the writing is different, so the industry within which gameswriters conduct their business is also different from the film industry.

'Forget it, Jake. It's Chinatown'

The games industry is split into two parts, developers and publishers. Primarily, games are produced by developers. Many of these are small, single-title operations prone to financial collapse and takeover. These developers provide content to publishers who then release it to the world. There are a number of big players, such as Electronic Arts (EA) and Ubisoft, that both develop and publish games.

Unlike other media, it is extremely rare for computer-games companies to commission original ideas from writers. It is also worth noting that, with ideas and scripts, companies often work on a buyout, and don't allow residual deals. Though, with some of the big name writers beginning to demand residual rights for their creations, this is something which may change.

Show me the money!

In terms of pay, it is difficult to specify a single payscale when games projects differ wildly. Some games are as short as 100 words, while others, such as *X-Files*, need a 250-page shooting script. This said many national writers' guilds have developed (or are in the process of developing) their own pay rates, which are available on their websites or by contacting them.

To get paid, you first need to find a job, and this generally means going it alone. Unlike other media there is not a structure for agents within the games industry. Few agents understand the needs of the industry or have established contacts within it. The closest the industry gets to this is a small number of all-in-one agencies that offer developers a one-stop shop for scriptwriting through to recording. With no agents and regular routes of contacts, writers will need to find gameswriting jobs themselves. When these occur they are often advertised on companies' websites, through gaming websites, recruitment agencies, very occasionally through newspapers and magazines, but often through word of mouth.

I'm going to make him an offer he can't refuse

When approaching a games company, screenwriters should select something that shows their writing ability. This does not have to be a games script: a television, a film or even a theatre script can do. What's important is matching the sample sent to the genre of game that is being made (e.g. an action sample for an action game), where this is possible. On a technical level, gameswriters do not need to be programmers, nor are they games designers. They must, as should have become clear, understand their medium. The games industry is rife with stories of screenwriters who were hired by games companies, charged high Hollywood prices and who then delivered a script entirely unsuitable for the game. Unsurprisingly, this has made some companies wary of employing professional writers again. Such tales of woe were born from a failure of the industry and the writers to understand that gameswriting is distinct from screenwriting.

One reason some writers fail when attempting to make the move into games from other media is a lack of commitment to a project. Just as a screenplay will evolve once the film commences production, so a games script will also evolve. Changes to the gameplay necessitate changes to the script. Such alterations may be as simple as the change of a button to a lever, but they can regularly stretch to characters disappearing and whole levels being cut. If the writer is unavailable to make changes to the script, this can make the final script confusing at best and a nonsense at worst. A games script requires tender loving care from its first conception through to the day the game hits the shelves: it does not stop when a draft of the script is handed over.

This fact means that some games companies are keen to have writers working in-house and involved in the development process from the first day until the last. This is by no means the only model of employment, or necessarily the best. Many established gameswriters work as freelancers moving from contract to contract, company to company. This is because companies often don't, in reality, need the writer there every day, but just involved in the process for the duration that it runs. As such, a gameswriter is often (in-house or freelance) working on a number of titles at the same time.

You're going to need a bigger boat

For writers who feel out at sea, help is at hand for those who wish to find advice or simply to discuss their craft. In the UK, the US and beyond, the various international writers guilds are producing information to help writers break into the games industry and to find their way within it once they have made the move. The Writers' Guild of Great Britain, for instance, has published a set of guidelines for writers and people working with writers in the games industry (these are available on the Internet and also by post). The Writers' Guild of America is also publishing advice.

Separate from the guilds, The International Games Developers Association (IGDA) is an active body that connects professionals within the games industry and provides a platform for discussion of industry matters. The IGDA boasts a writers' Special Interest Group, which meets to discuss, promote and *inform* matters of gameswriting.

I think this is the beginning of a beautiful friendship

Stories have had a rough start in the world of games, but then they did at the start of the movie industry; at this point in time, though, the quality of stories is improving across the games industry. This is partly due to the development of improved AI, faster machines and more sophisticated speech engines, but it is mostly because gameswriting has evolved into its own distinct form of storytelling. The games industry is realising that it cannot force the methods used to tell story in film upon a different medium, games. As this revolution continues, so writers are being given more tools with which to tell stories. They are being given longer to do it and they are being employed earlier within the game development process. While this means that there are many more opportunities than there once were, it also means that any screenwriter hoping to make the transition between media will be expected to know more about games and gameswriting than they may have in the past. This is not something to be afraid of, but something to be embraced – after all, it's all about learning to tell stories as well as possible in a new world: one where you can step into the story and walk around.

Andrew S. Walsh has written for television, radio, theatre and film animation, and has worked on more than twenty-five games for companies such as EA, Sony, SEGA and Eidos.

Frequently Asked Questions about How to Develop a Script

Neil Peplow

Frequently Asked Questions about Time Travel is the debut film of screenwriter Jamie Mathieson. A British low-budget comedy, it follows the fortunes of three slacker friends who discover a slip in the space–time continuum in the gents toilets. It throws them back and forth in time, but only ever to the same pub. Directed by Gareth Carrivick and produced by Neil Peplow and Justin Anderson Smith, the film was financed by BBC Films and HBO Films. It is planned for a February 2009 release in the UK by Lionsgate.

How did you find the script?

The script came to us from Jamie's agent at ICM. Having strong relationships with agents is a really important part of producing, if not the most important. They are the gatekeepers to talent and as an independent producer your value depends on your talent relationships. There are two ways to develop these relationships. You can trawl the films schools, go to every film festival, hang out at the right parties and find the talent yourself. Alternatively you can ask an agent who does that for a living: aggregate talent. The best way to build your own talent network is a combination of both. You can then trade with agents: tell them about talent you've found and they'll tell you about talent they've found. Agents will be more willing to introduce you to a script first if they trust you and like you. It also helps if you have a lot of development funding.

What attracts you to an idea?

Jamie's script had already been to a number of other production companies who had turned it down. The first draft did need work but we saw a great core idea. When we like a script after a first read, we need to make sure we truly believe in it. There's nothing worse than realising six months into development that you made the wrong choice because you were blind to its flaws. To do this we put it through a ruthless examination.

Some of the decision-making process is based on a gut feeling: are we passionate enough about the idea to still be working enthusiastically on it in eighteen months' time? Does it make us laugh out loud? Would we be proud to work on it? Some of it is commercial: can we pitch it easily? What would the poster look like? What would the log line be? Can we cut a trailer from the script? Who is the audience? Who could finance the script at its projected budget level? Can we cast it to finance the budget? Some of it is practical: do we like the writer? Does the writer see the project the same way as we do? Finally, and most importantly, what is the likelihood of getting paid and how much?

How did you finance the development?

As we didn't have an overhead deal when we started our new production company, we needed to be creative in order to capture the rights. A recent experience had also convinced me that the best way to defend the integrity of a script was to keep the writer involved through the entire process. From choosing the director, the cast, sitting in rehearsals and even hanging out on set. It can be very easy for a producer to forget about the writer's original motivation behind the script when financiers are shouting for changes. To counter this and compensate for a lack of upfront money, we cut the writer in on ownership of the company that holds the rights to the script. It means that the whole process has to be transparent: from showing the accounts to what to do with the rights. Jamie gets a producer's credit and a share in all revenue. If the film is a success, he will not feel aggrieved that the producers may have done better than himself. Once we had secured the option, we got Jamie to do a redraft and polish. The writing was fantastic but it needed a more cinematic structure; the first draft felt too episodic. Once that was done, we approached BBC Films who were extremely enthusiastic about the script. They then financed subsequent drafts, budgets and even castings. Their support was fantastic.

How was the script developed before production?

I try to look at a script through two eyes. One eye is analytical, focusing on structure, subtext, character journeys and genre. The other eye is more passionate, looking at where the heart of the film lies, looking back at the history of the script and what originally inspired the writer. The conflicting focus of the two normally results in a headache and a few late night discussions. However, I believe that this approach helps shape a script that doesn't feel 'overdeveloped' and which still hits all the right story beats. I think with *FAQ* we almost got it right. You can always learn from each and every film you make and *FAQ* was no different. Once you think you know it all, you can expect a very nasty bump to be just around the corner: hubris tends to attract comeuppance.

How did you handle the relationship between the writer, the talent and the financiers?

I try to introduce into the process at a very early date a simple edict that applies to everyone on the production: to listen to all ideas with an open mind, no matter where they come from, and to examine those ideas carefully to see whether they make the film better. Everything is about improving the film. Nothing can be dismissed out of hand because of where it came from or more importantly whom it came from. I always tell writers and directors that one day even the biggest chump may come up with a great idea, and when they do you can accept that gift graciously and eventually claim it as your own. If you slap them down too often, they'll stop trying and the idea which could have made a huge difference will never be thought. Luckily, Gareth Carrivick, the director, and Jamie bought into this willingly. This really helps set the tone as the process advances. It gets more difficult to uphold when bigger egos come into play, but it's the job of the producer to handle that.

How closely involved did the writer remain in the process?

Jamie was around for casting, pre-production, rehearsal and the shoot. He was involved in a lot of decisions from casting through to helping the art department design the leading actors' t-shirts. He was also on-set for a lot of the shoot and felt comfortable enough to talk to Gareth about how scenes were being constructed. The edit was more problematic. It's always difficult for writers to see their film for the first time. It takes time for them to accept that the film very rarely turns out exactly as they saw it in their mind's eye. At this point Jamie decided to step back. It was disappointing, but I understood his decision. Sometimes a writer needs to step back in order to stay sane.

What general advice do you have for producers?

Hold onto the reason why you wanted to make the film in the first place, keep the writing talent on board as long as possible and make sure you get paid.

Neil Peplow has produced or executive produced twelve feature films, including Waking Ned, Shooting Fish, Mike Bassett: Football Manager, *and* Bright Young Things.

If my books had been any worse, I should not have been invited to Hollywood and if they had been any better, I should not have come.

Raymond Chandler

The way to make a film is to begin with an earthquake and work up to a climax.

Cecil B. DeMille

It's not necessary for a director to know how to write, but it helps if helps if he knows how to read.

Billy Wilder

Don't tell me the moon is shining; show me the glint of light on broken glass.

Anton Chekhov

Scriptwriting is the toughest part of the whole racket, the least understood and the least noticed.

Frank Capra

Production Companies

What general advice do you have for producers?
Hold onto the reason why you wanted to make the film in the first place, keep the
writing talent on board as long as possible and make sure you get paid.

      Neil Peplow in 'Frequently asked questions about how to develop a script'

Are there then any good producers? Yes, says Mamet, a few, 'but we must remember
that even Diaghilev went into the ballet because he wanted to screw Nijinsky'.

      Barry Turner on David Mamet in 'Just a Mamet'

Production Companies

Production companies sometimes fail to acknowledge, respond to or even read unsolicited submissions. It is always worth contacting a company before sending in any material to find out their current policy on new writing.

⚷ Key to Symbols

Production companies listed with symbols specialise in the following:
MOVIE Movies
TV Television
VG VideoGames

USA

1492 Pictures   MOVIE

c/o Warner Brothers, 4000 Warner Boulevard, Bldg. 3, Burbank, CA 91522
☏ 818 954 4939

Founded in 1995 by director Chris Columbus, producing films such as *Jingle All the Way*, the first three instalments in the *Harry Potter* series and *Night at the Museum*.
  Does not accept unsolicited material.

20th Century Fox   MOVIE

10201 West Pico Boulevard, Los Angeles, CA 90035
☏ 310 369 1000
www.foxmovies.com

The Fox Film Corp was founded in 1914 and merged with 20th Century Pictures in 1935 to form 20th Century Fox, one of America's leading film production companies. 20th Century Fox has produced Academy Award-winning films such as *The Sound of Music*, *The King and I*, *Alien* and *Titanic*. Recent output includes *Star Wars Episode III*, *Mr. & Mrs. Smith*, *Walk the Line*, *Ice Age: The Meltdown*, *X-Men: The Last Stand*, *The Devil Wears Prada*, *Eragon* and *Night at the Museum*.

20th Century Fox Television   TV

10201 W Pico Boulevard, Bldg 88, Los Angeles, CA 90035
☏ 310 369 1000
www.foxhome.com

Founded in 1949, 20th Century Fox Television is the television production division of the 20th Century Fox movie studio, a subsidiary of News Corporation. Its shows include *The Simpsons*, *Prison Break* and *24*.

2929 Entertainment   MOVIE

9100 Wilshire Boulevard, Suite 500, West, Beverly Hills, CA 90212
☏ 310 309 5200 🖷 310 309 5716
www.2929entertainment.com

Founded in 2005, producing independent feature films such as *Good Night and Good Luck* (nominated for 6 Academy Awards including Best Picture) and *Akeelah and the Bee* (starring Academy Award-nominees Laurence Fishburne and Angela Basset).
  Submit a brief 2–3 paragraph synopsis of project via website.

3 Arts Entertainment   MOVIE

9460 Wilshire Boulevard, 7th Floor, Beverly Hills, CA 90212
☏ 310 888 3200

Produced films such as *I am Legend*, *Carpoolers*, *Bio-Dome* and *A Scanner Darkly*, which won the Austin Film Critics Award and was nominated for the Saturn award for Best Animated Film.

44 Blue Productions   TV

4040 Vineland Avenue, Suite 105, Studio City, CA 91604
☏ 818 760 4442 🖷 818 760 1509
reception@44blue.com
www.44blue.com

Established in 1984, producing reality, lifestyle, action-adventure and documentary programming for many leading broadcasters.

4th Row Films (MOVIE) (TV)

27 West, 20th Street, Suite 1006, New York, NY 10011
℡ 212 974 0082 🖷 212 627 3090
miguel@4throwfilms.com
www.4throwfilms.com

Produces feature films, television programmes and documentaries, working across genres. Output includes *Pills, Anytown, USA, The X-Effect, Funny Peculiar* and *Owning the Weather*.

Will consider unsolicited submissions of screenplays, as well as documentary treatments, television pitches and writing samples. Email query letter in first instance with a logline and synopsis. Do not mail hard copies unless specifically requested. Generally takes up to four weeks for decision. For return of hard copies, include a pre-paid and addressed envelope.

50 Cannon Entertainment (MOVIE) (TV)

1950 Sawtelle Boulevard, Suite 333, Los Angeles, CA 90025
℡ 310 244 4622

Director Mike Newell's production company, producing the *Huff* series for television. Other titles include *I Capture the Castle, The Ice at the Bottom of the World* and *The Elfstones of Shannara*.

A-V-A Productions (TV)

4760 E. 65th Street, Indianapolis, IN 46220
℡ 317 255 6457 🖷 317 253 6448
avaprods@comcast.net
www.avavideoproductions.com
Owner and Executive Producer *Bud Osborne*

Established in 1980, an award-winning television production company specializing in documentaries. Past productions include documentary segments for The History Channel, The Discovery Channel and E! Entertainment Network.

Accepts unsolicited manuscripts by mail. Welcomes the opportunity to see new writers' work and 'will give feedback about what we like and don't like'.

Activision, Inc. (VG)

Corporate Headquarters, 3100 Ocean Park Boulevard, Santa Monica, CA 90405
℡ 310 255 2000
www.activision.com

A leading international publisher of interactive entertainment software products. Produces games across genres for an array of markets and platforms. Has development agreements with the likes of DreamWorks, Hasbro Properties and Harrah's Entertainment, Inc. and has created successful brands such as *True Crime* and *Call of Duty*.

Actual Reality Pictures (MOVIE) (TV)

6725 West Sunset Boulevard, Suite 350, Los Angeles, CA 90028
℡ 310 202 1272 🖷 310 202 1502
questions@arp.tv
www.actualreality.tv

Works across range of platforms and genres including series, features, documentaries and animation. Output includes *30 Days, Black.White, Flip That House* and *Thin*.

Does not accept unsolicited show ideas.

AEI (MOVIE)

c/o AEI Submissions, 518 South Fairfax Avenue, Los Angeles, CA 90036
℡ 323 932 0407
submissions@aeionline.com
www.aeionline.com

A one-stop full-service literary management and motion picture production company. Offers development, sales, marketing, production and licensing services.

Send query letters only. Additional material will be discarded unread. No calls. Allow 8–10 weeks for a response on manuscripts, non-fiction proposals and screenplays. Send sae for reply and return of material. Full submission guidelines available on website.

Agamemnon Films Inc. (MOVIE)

650 N. Bronson Avenue, Suite B-225, Los Angeles, CA 90004
℡ 323 960 4066
www.agamemnon.com

Founded in 1981 by Fraser and Charlton Heston (*Ben Hur*). Produced films such as *The Bible* and the animated version of *Ben Hur*.

Alchemy (TV)

8530 Wilshire Boulevard, Suite 400, Beverly Hills, CA 90211
℡ 310 289 7766 🖷 310 289 7833
sales@alchemy.tv
www.alchemy.tv

Produces and distributes TV event drama. Output includes the mini series, *Coco Chanel* and *Diamonds*. Also has an office in London.

Alpine Pictures, Inc. (MOVIE)

3500 W. Magnolia Boulevard, Burbank, CA 91505
℡ 818 333 2600 Ext. 208 🖷 818 333 2662
scottv@alpinepix.com
www.alpinepix.com

Founded in 1995. Credits include *Dorothy of Oz, Dark Honeymoon* and *Love is the Drug*.

Does not accept unsolicited manuscripts. To submit an idea or synopsis, approach via email.

Ambush Entertainment ⬚MOVIE

7364 1/2 Melrose Avenue, Los Angeles, CA 90046
☎ 323 951 9197 🖷 323 951 9998
info@ambushentertainment.com
www.ambushentertainment.com

Independent feature film production company. Aims to produce/finance around four pictures a year. Credits include *The Oh in Ohio*, *Unearthed*, *Wonderful World*, *Hindsight* and *Against the Current*.

Amen Ra Films ⬚MOVIE

9460 Wilshire Boulevard, Suite 400, Beverly Hills, CA 90212
☎ 310 246 6510

Wesley Snipe's production company, founded in 1991 and responsible for the *Blade* trilogy. Other credits include *The Big Hit*, *Down in the Delta*, *Nature Boy* and *Undisputed*.

American Cinema International ⬚MOVIE

15363 Victory Boulevard, Van Nuys, CA 91406
☎ 818 907 8700 🖷 818 907 8719
www.aci-americancinema.com

Founded in 2000 to produce and distribute filmed entertainment internationally. Aims to work on up to six titles per year. Output includes *Hatchetman*, *By Appointment Only* and *The Lost Samaritan*.

American Empirical Pictures ⬚MOVIE

36 East 23rd Street, 6th Floor, New York, NY 10010
☎ 212 475 1771

Company formed by director Wes Anderson. Its debut production was 1998's *Rushmore*. Subsequent features include *The Royal Tenenbaums*, *The Life Aquatic with Steve Zissou*, *The Squid and the Whale*, *Hotel Chevalier* and *The Darjeeling Limited*.

American Zoetrope ⬚MOVIE⬚ ⬚TV

916 Kearny Street, San Francisco, CA 94133
☎ 415 788 7500 🖷 415 989 7910
www.zoetrope.com

Founded by Francis Ford Coppola and George Lucas in the late 1960s. Has received no less than 15 Academy Awards and 68 nominations. Output includes *Lost in Translation*, *Kinsey* and *The Godfather* trilogy.

Craig Anderson Productions ⬚MOVIE

444 N. Larchmont Boulevard, Suite 109, Los Angeles, CA 90004

☎ 323 463 2000 🖷 323 463 2022
info@cappix.com
www.cappix.com

Founded in 1990, producing films such as *The Christmas Blessing*, *American Meltdown* and *Wilder Days*.
 In first instance submit a 2–5 page synopsis of the project.

Anonymous Content ⬚MOVIE⬚ ⬚TV

8522 National Boulevard, Suite 101, Culver City, CA 90232
☎ 310 558 6031
info@anonymouscontent.com
www.anonymouscontent.com

Founded in 1999 and working across film, integrated, commercials, music video and television. Works closely with talent agencies, distributors and broadcasters in order to assemble the right team for any project. Output includes *Babel* and *Eternal Sunshine of the Spotless Mind*.

Antidote International Films, Inc. ⬚MOVIE

200 Varick Street, Suite 502, New York, NY 10014-4810
☎ 646 486 4344 🖷 646 486 5885
info@antidotefilms.com
www.antidotefilms.com

Founded in 2000 by producer Jeffrey Levy-Hinte. Has produced such titles as *The Hawk is Dying*, *Mysterious Skin* (nominated for IFP Gotham and Independent Spirit Awards), *Thirteen* (starring Holly Hunter and Evan Rachel Wood) and *Chain*.
 Does not accept any unsolicited submissions.

Apertura ⬚MOVIE⬚ ⬚TV

535 Main Street, Orford, NH 03777
☎ 603 353 9067 🖷 603 353 4646
karol@apertura.org
www.apertura.org
Producer/Filmmaker *John Karol*

Founded in 1969, producing documentary features and television programmes. Nominated for an Academy Award for Best Documentary Feature.
 Welcomes unsolicited manuscripts by mail or email.

Appian Way ⬚MOVIE

9255 Sunset Boulevard, Suite 615, West Hollywood, CA 90069
☎ 310 300 1390

Leonardo DiCaprio's production company whose first credit was *The Assassination of Richard Nixon* in 2004. Subsequent films include

The Aviator, The 11th Hour, Gardener of Eden, In Dark Woods and *The Rise of Theodore Roosevelt.*

Arden Entertainment `TV`

12034 Riverside Drive, Suite 200, North Hollywood, CA 91607
T 818 985 4600 F 818 985 3021
dan@ardenentertainment.com
www.ardenentertainment.com

Founded in 2000, producing for television. Output includes the documentary series *Growing Up*, the studio-based reality series *Dream Decoders* and the docu-soap *K9 Karma*.

Does not accept unsolicited materials or programme proposals.

Arenas Entertainment `MOVIE`

3375 Barham Boulevard, Los Angeles, CA 90068
T 323 785 5555 F 323 785 5560
info@arenasgroup.com
www.arenasgroup.com

Founded in 1988. Produces, acquires, markets and distributes films in all media—including theatrical, video and television—that target the US Latino market. Produced the award-winning film *Nicotina* starring Diego Luna, and co-produced *Imagining Argentina*, starring Antonio Banderas and Emma Thompson.

Arjay Entertainment `TV`

1627 Pontius Avenue, Los Angeles, CA 90025
T 310 481 2282
info@arjayentertainment.com
www.arjayentertainment.com

Formed in 2003, developing, producing and distributing celebrity-based programming. Production credits include *The Red Carpet, Above the Line Beauty, VIP Access* and *Premiere Beauty.*

The Artists' Colony `MOVIE` `TV`

256 S. Robertson Boulevard, Suite 1500, Beverley Hills, CA 90211
T 310 720 8300
theartistscolony@sbcglobal.net
www.theartistscolony.com
President/CEO/Producer *Lloyd A. Silverman*
V.P., Production *Rick Husong*
V.P., Creative *Michale Fiore*
V.P., Development *P. J. Koll*
V.P., NY *Steve Grill*

Founded in 1997, creating content for film, TV and the web. Specializes in drama, thrillers and documentaries. Credits include *Snow Falling on Cedars, Shattered Image, 12* and *A Girl, Three Guys and a Gun*. Has recently launched a corporate identity/branding and communications

firm in conjunction with other film and TV companies, called communic8.

Only packaged projects will be considered. Approach by email.

Ascendant Pictures `MOVIE`

406 Wilshire Boulevard, Santa Monica, CA 90401
T 310 288 4600 F 310 288 4601
info@ascendantpictures.com
www.ascendantpictures.com

Founded in 2002. Produced the film *The Big White* and has co-produced titles including *Lord of War, Ask the Dust* and *Lucky Number Slevin.*

The Asylum `MOVIE`

1012 North Sycamore, Los Angeles, CA 90038
T 323 850 1214 F 323 850 1218
www.theasylum.cc

Founded in 1997 to produce, finance and distribute motion pictures. Fully finances and produces 10–15 titles per year. Examples of output include *666: The Beast, Alien Abduction, King of the Lost World, Shapeshifter, Snakes on a Train, The Hitchhiker, Universal Soldiers* and *When a Killer Calls.*

Avatar Productions `TV`

329 270 Sparta Avenue, Ste 104, Sparta, NJ 07871
T 973 486 2875
contactavatar@avatarproductions.com
www.avatarproductions.com

Has been providing production services for broadcast, industrial and corporate television for over 20 years. Has received several awards for their work, including 2 Emmy awards.

Axial Entertainment `MOVIE`

20 West 21st Street, 8th Floor, New York, NY 10010
submissions@axialentertainment.com
www.axialentertainment.com

Production and management company, with a model loosely based on the 'hot house' studio environment of the 1940s. Brings writers and film executives together to encourage the development process before studios, networks or producers become involved. Allows for a chance to be taken on a new writer or original idea.

It is company policy not to accept unsolicited materials. If interested in submitting work, email a 1-page brief summary of material.

Baldwin Entertainment Group `MOVIE`

3000 West Olympic Boulevard, Suite 200, Santa Monica, CA 90404

Ⓣ 310 453 9277
erin@baldwinent.com
www.baldwinent.com

Able to see a project through from concept to theatrical release. Credits include *Ray*, *Sahara*, *Death Sentence*, *Mandrake* and *Atlas Shrugged*.

Will not accept or consider creative ideas, suggestions or materials other than those specifically requested.

Barcelona Films (MOVIE) (TV)

4220 Duval Street, Austin, TX 78751
Ⓣ 512 320 0743
info@barcelonafilms.com
www.barcelonafilms.com

Founded in 2001, developing film, TV and commercial projects, either independently, as co-producers or as commissions. Has worked across genres including comedy and documentary. Output includes *Brayton Field*, *Chisholm Trail* and *Two Queens Cooking*.

Barnstorm Films (MOVIE) (TV)

73 Market Street, Venice, CA 90291
Ⓣ 310 396 5937 Ⓕ 310 450 4988
tony@barnstormfilms.com
President *Tony Bill*

Film production company founded in 1975. Recent credits include *Flyboys*, *North Country* and *Untamed Heart*. Company president Tony Bill won a Best Picture Oscar for *The Sting*.

Welcomes new writing. Send a one-page synopsis.

Bauer Martinez Studios (MOVIE)

801 West Bay Drive, Suite 800, Largo, FL 33770
Ⓣ 727 852 9939
bmstudios1@aol.com
www.bauermartinezstudios.com

Feature film producer and distributor. Production credits include *Dot.Kill*, *I Could Never Be Your Woman*, *The Number One Girl*, *The Piano Player*, *Wake of Death* and *Welcome to the Jungle*.

Beacon Pictures (MOVIE)

120 Broadway, Suite 200, Santa Monica, CA 90401
Ⓣ 310 260 7000
www.beaconpictures.com

Producing films since 1990. Credits include *A Lot Like Love*, *Air Force One*, *Ladder 49*, *The Commitments*, *The Family Man* and *Thirteen Days*.

Will not accept unsolicited material.

Belladonna Productions (MOVIE) (TV)

118 West 22nd Street, Suite 3, New York, NY 10011
Ⓣ 212 807 0108 Ⓕ 212 807 6263
cordelia@belladonna.bz
www.belladonna.bz
Head of Development *Cordelia Stephens*

Produces a range of media content, including feature films, documentaries and TV shows. Output includes: *A Guide to Recognizing Your Saints* and *Transamerica*.

Sometimes welcomes unsolicited manuscripts.

Benderspink (MOVIE) (TV)

110 S. Fairfax Avenue, Suite 350, Los Angeles, CA 90036
Ⓣ 323 904 1800 Ⓕ 323 904 1802
info@benderspink.com
www.benderspink.com

Founded in 1998, this management/production company has sold over 20 client scripts and co-produced *American Pie* with Universal Pictures in its first year. Also co-produced *American Pie 2*, *Final Destination*, and *Cats & Dogs*. Produced the television series *Kyle XY*.

Always accepting query letters for screenplays and short or feature films of all genres. Submit a brief synopsis via email.

Beyond Pix (TV)

950 Battery Street, San Francisco, CA 94111
Ⓣ 415 434 1027 Ⓕ 415 434 1032
www.beyondpix.com

Production company offering broadcasting, production, editorial and internet media services. It produces the weekly programme *For the Record* for Bloomberg Television.

Black Sheep Entertainment (MOVIE) (TV)

The Lot, Formosa Building, Suite 100, West Hollywood, CA 90046
Ⓣ 323 850 3999 Ⓕ 310 424 7117
www.blacksheepent.com

An independent film and television production company founded in 1990. Examples of output include *The Big Gig*, *Cottonwood*, *It Had To Be You* and *Hooking Up*.

Most projects generated internally. May occasionally partner a screenwriter but generally not able to consider uncompleted scripts or scripts with an unknown director or non-recognizable talent attached to any of the lead roles. Before submitting a screenplay, forward a one page synopsis of the story including a logline (one sentence description) and a page count. Company will then make contact if interested in seeing full screenplay (which should be in industry accepted format and accompanied by a signed copy of release form provided by the company). Submission of a project is voluntary and not in confidence.

Black Watch Productions, Inc. (MOVIE) (TV)

Black Watch Stages, 49 Murray Street, Suite 1, New York, NY 10007
(T) 212 349 0369 (F) 212 349 1335
jfxa@blackwatchproductions.com
www.blackwatchproductions.com
Executive Producer *John FX Anderson*

Founded in 1989. Can provide all aspects of commercial production under one roof, from concept to post-production. Works across genres: documentaries, TV specials, children's programming, commercials and corporate films.

Bleiberg Entertainment, Inc. (MOVIE)

9454 Wilshire Boulevard, Suite 200, Beverly Hills, CA 90212
(T) 310 273 0003 (F) 310 273 0007
info@bleibergent.com

Established in 2005, distributing and producing feature films. Output includes *Adam Resurrected*, *Dance of the Dead* and *The Band's Visit*.

Blizzard Entertainment (VG)

PO Box 18979, Irvine, CA 92623
www.blizzard.com

A major developer and publisher of entertainment software. Founded in 1991 as Silicon & Synapse and relaunched in 1994 as Blizzard. Examples of games include *World of Warcraft*, *Diablo* and *StarCraft*. The company's research and development group includes over 250 designers, producers, programmers, artists, and sound engineers.

Blowback Productions (MOVIE)

601 West 26 Street, Suite 1776, New York, NY 10001
(T) 212 352 3007 (F) 212 352 3015
kara@601nw.com
www.blowbackproductions.com

Formed by Marc Levin in 1988. Output includes *Mr. Untouchable*, *Heir to an Execution* and *Gladiator Days: Anatomy of a Prison Murder*.

Blue Rider Pictures (MOVIE) (TV)

2801 Ocean Park Boulevard, Suite 193, Santa Monica, CA 90405
(T) 310 314 8405 (F) 310 314 8402
info@blueriderpictures.com
www.blueriderpictures.com

Founded in 1991. Has produced over 100 movies and TV projects. Involved in the production of *Around The World In 80 Days*, *The Call of the Wild* and *The Incredible Mrs. Ritchie*, starring James Caan and Gena Rowlands, which won an Emmy Award.

For submissions, send an email in first instance.

Blue Sky Studios (MOVIE)

44 South Broadway, White Plains, New York, NY 10601
(T) 914 259 6500 (F) 914 259 6499
info@blueskystudios.com
www.blueskystudios.com

A division of Fox Filmed Entertainment since 1997. Produces pioneering animated content for film and television. Credits include *Bunny* and *Ice Age*.

Boca Productions LLC (MOVIE) (TV)

44314 Date Avenue, Lancaster, CA 93534
(T) 661 949 2848
bocaproductions@yahoo.com
bocaproductions.scriptmania.com
Executive Producer *Gary L. Myers*

Founded in 2001, producing feature films and made-for-TV features. Production credits include *BOCA*, *Emissary* and *Evan's Heaven*.

Welcomes original screenplays 'with financing in place' or 'adaptations of successful novels where rights have been acquired.'

Braun Entertainment Group, Inc. (MOVIE) (TV)

280 S. Beverly Drive, Suite 500, Beverly Hills, CA 90212
(T) 310 888 7727 (F) 310 888 7726
jplace@braunentertainmentgroup.com
www.braunentertainmentgroup.com

Produces for film and TV, including series, television movies and feature films. Output includes *A Girl Like Me: The Gwen Araujo Story*, *Edges of the Lord*, *Lethal Vows* and *Witness for the Prosecution: Amber Frey*.

Does not accept unsolicited submissions of screenplays, treatments or similar material. Will be returned unread. Happy to receive queries (consisting of a log line and brief synopsis) via email or fax. Will then request further material if interested in project (when you will be required to include a release form). No calls.

Braverman Productions, Inc. (MOVIE) (TV)

3000 Olympic Boulevard, Santa Monica, CA 90404

☏ 310 264 4184 ☏ 310 388 5885
chuck@bravermanproductions.com
www.bravermanproductions.com

With a long history of producing both for TV and cinema, but now concentrating on documentaries for channels including A&E, History Channel and Discovery Channel.

Brillstein-Grey Entertainment   MOVIE TV

9150 Wilshire Boulevard, Suite 350, Beverly Hills, CA 90212
☏ 310 275 6135 ☏ 310 275 6180

Brillstein-Grey Entertainment is a Hollywood agency that includes Brad Pitt and Courteney Cox among its clientele. The company has also produced movies including *The Wedding Singer*, and television shows including *Just Shoot Me*.

Bristol Bay Productions   MOVIE

1888 Century Park East, 14th Floor, Los Angeles, CA 90067
☏ 310 887 1000 ☏ 310 887 1001
info@bristolbayproductions.com
www.bristolbayproductions.com

Specializes in major motion pictures. Credits include *Ray, Sahara, The Game of Their Lives, Amazing Grace* and *The Great Buck Howard*.

Brooklyn Films   MOVIE TV

PO Box 20412, New York, NY 10021
☏ 212 744 2845 ☏ 718 421 2702
inquiries@brooklynfilms.com
www.brooklynfilms.com

Founded in 1988, producing feature films, television series and HDTV productions. Past titles include *Flesh-Eating Mothers, The Suckling, The Appartment* and *Man Date*.

Brookwell/McNamara Entertainment   MOVIE TV

1600 Rosecrans Boulevard, Raleigh Studios Bldg. 6A, 3rd Floor, North Wing, Manhattan Beach, CA 90266
☏ 310 727 3353 ☏ 310 727 3354
laurie@bmetvfilm.com
www.bmetvfilm.com

Has produced films including *Cake, Beyond the Break* and *Dance Revolution*. Other credits include the award-winning television series *That's So Raven* from 2002 to 2006.

Jerry Bruckheimer Films   MOVIE TV

1631 10th Street, Santa Monica, CA 90404
☏ 310 664 6250
www.jbfilms.com

Produces films and television series in the genre of action, drama and science fiction. Most of these big-budget productions feature trademark elaborate special effects. Credits include *Top Gun, The Rock, Armageddon* and *Pirates of the Caribbean*. Also produces television shows including *CSI: Crime Scene Investigation, E-ring* and *The Amazing Race*.

Bunim/Murray Productions   MOVIE TV

PO Box 10421, Van Nuys, CA 91410
www.bunim-murray.com

Founded in the early 1990s, a leading independent producer of TV content, ranging from reality programmes to scripted dramas. Includes the BMP Films division.

Butchers Run Films   MOVIE TV

1041 N. Formosa Avenue, Santa Monica Building E-200, West Hollywood, CA 90046
☏ 323 850 2703

Production company formed by Robert Duvall. Recent credits include *Broken Trail*, made for television.

C2 Pictures   MOVIE

2308 Broadway, Santa Monica, CA 90404
☏ 310 828 2765

Cinergi Pictures Entertainment was a small production company formed in 1992 and produced films such as *Tombstone, Nixon* and *Evita*. In 2002, it merged with Carolco Pictures to form C2 Pictures. Output includes *Terminator 3: Rise of the Machines* and *Basic Instinct 2*.

C3 Entertainment Inc.   MOVIE TV

1415 Gardena Avenue, Glendale, CA 91204
☏ 818 956 1337
query@c3entertainment.com
www.c3entertainment.com

Founded in 1959 as Comedy III Productions by the Three Stooges. Produced the television movie *The Three Stooges* in association with Mel Gibson, and also co-produced *The Three Stooges Greatest Hits* with Sony Pictures Television.

See website for information on screenplay submissions.

Camelot Entertainment Group, Inc.   MOVIE TV

130 Vantis, Suite 140, Aliso Viejo, CA 92656
☏ 949 334 2950
www.camelotfilms.com

Develops, produces and distributes output for film, TV and new media. Seeks to acquire 'writing talent off of the Hollywood radar'.

Capcom USA, Inc. [VG]

Capcom Entertainment Inc., 800 Concar Drive, Suite 300, San Mateo, CA 94402-2649
T 650 350 6500
www.capcom.com

Founded in Japan in 1979, manufacturing and distributing electronic game machines. In 1983 Capcom Co., Ltd was established and over 25 years has produced some of the most successful games franchises in the world. Titles include *Resident Evil*, *Street Fighter*, *Breath of Fire* and *Mega Man*. Offices in Tokyo, Osaka, California, England, Germany and Hong Kong.

Capital Arts Entertainment [MOVIE] [TV]

17941 Ventura Boulevard, Suite 205, Encino, CA 91316
T 818 343 8950 F 818 343 8962
info@capitalarts.com
members.aol.com/capartsent

Formed in 1995, producing for cinema and TV. Early successes were *Casper: A Spirited Beginning* and *Casper Meets Wendy*. Other output includes *Addams Family Reunion*, *After the Storm*, *Au Pair*, *Happy Endings*, *Oh Baby*, *Richie Rich's Christmas Wish*, *Route 9*, *The Michael Jordan Story* and *The Prince and Me*.

Carsey-Werner, LLC. [TV]

12001 Ventura Place, 6th Floor, Studio City, CA 91604
T 818 299 9600 F 818 299 9650
www.cwm.com

Established by Marcy Carsey and Tom Werner in 1981, incorporating production and distribution arms. Particular reputation for its hit TV comedy series. Output includes *3rd Rock from the Sun*, *A Different World*, *Cybill*, *Grace Under Fire*, *Roseanne*, *That '70s Show* and *The Cosby Show*.

Castle Rock Entertainment [MOVIE] [TV]

335 N. Maple Drive, Suite 135, Beverly Hills, CA 90210
T 310 285 2300 F 310 285 2345
www2.warnerbros.com

Founded in 1987 by Martin Shafer, director Rob Reiner, Andy Scheinman, Glenn Padnick and Alan Horn, with Columbia Pictures as a strategic partner. Has produced a number of award-winning feature films and television shows. Titles include *When Harry Met Sally*, *A Few Good Men*, *The Polar Express*, *Music and Lyrics*, and the television show *Seinfeld*.

CatchLight Films [MOVIE]

T 310 295 0071 F 310 341 3806
mailbox@catchlightfilms.com
www.catchlightfilms.com

Produces features, short films and documentaries, as well as commercials, corporate spots and music videos. Launched in 1999 with *It Can Be Done* at the Venice Film Festival. Further output includes *Amy's Orgasm*, *Break a Leg*, *Rings*, *Man of God*, *Heart of the Beholder*, *Red Tide Rising* and *Lesser of Three Evils*.

Will not accept unsolicited material. In first instance email or fax a one page introduction and a summery of screenplay. Submission agreement available from website.

Centropolis Entertainment [MOVIE]

www.centropolis.com

Formed in 1985 by director Roland Emmerich and producer Ute Emmerich. Trackrecord of blockbuster credits including *Independence Day*, *What Women Want*, *The Patriot*, *Eight Legged Freaks*, *The Day After Tomorrow*, *Trade* and *10,000 B.C.*

Contact via email form on website. Does not accept unsolicited material. Material must be submitted to Centropolis by an established talent or literary agency.

CFP Productions [MOVIE]

Paramount Pictures, 5555 Melrose Avenue, Lucy Bungalow 105, Hollywood, CA 90038
T 310 470 0845 F 310 470 0842
www.cfpprod.com

Partnered with Paramount Pictures and founded by Christine Forsyth-Peters. Production credits include *How to Lose a Guy in 10 Days* and *The Thing About Jane Spring*. (See also www.CPmotionpictures.com; email: info@ CPmotionPictures.com)

Charnjit Films [MOVIE]

1209 Liberty Avenue, Union, NJ 07083
T 212 302 1906 F 212 302 0500
charnjitfilms@aol.com
Contact *Chani Singh*

Founded in 2001, producing Bollywood films. Recent productions include *Bollywood Rocks*.

Does not welcome unsolicited manuscripts.

Cherry Road Films [MOVIE]

10960 Wilshire Boulevard, Suite 980, Los Angeles, CA 90024
T 310 458 6550

info@cherryroadfilms.com
www.cherryroadfilms.com

Founded in 2002, with credits including *Spring Break in Bosnia*, *White Jazz* and *Illegal Superman*.

Company policy not to accept unsolicited manuscripts.

Cine Excel Entertainment MOVIE TV

1219 West El Segundo Boulevard, Gardena, CA 90247-1521
T 323 754 5500 F 818 848 1590
cineexcel@msn.com
www.cineexcel.com

Operating since 1989 and incorporated in 1991 A distribution company that also develops, finances and produces its own productions. Credits include *Internetrix*, *The Abominable*, *Reptilicant*, *WildCat* and *Carnival Evil*. Also has an office in San Francisco.

Does not accept any unsolicited scripts or submissions.

Cinema Libre Studio MOVIE

8328 De Soto Avenue, Canoga Park, CA 91304
T 818 349 8822 F 818 349 9922
info@CinemaLibreStudio.com
www.CinemaLibreStudio.com

Founded in 2003. Aims to produce and distribute films with global appeal. Production emphasis is on narrative features and documentary films that explore cultural, political or global topics. Examples of output include *Tre*, *Conventioneers*, *The Empire in Africa* and *Heads N Tailz*.

Cinetel Films MOVIE

8255 Sunset Boulevard, Los Angeles, CA 90046
T 323 654 4000 F 323 650 6400
info@cinetelfilms.com
www.cinetelfilms.com

Founded in 1980. Has produced films including *Deadly Surveillance* , *Solar Attack* and *The Bone Eater*.

Company policy not to accept unsolicited manuscripts.

Cineville International MOVIE

3400 Airport Avenue, Santa Monica, CA 90405
T 310 397 7150 F 310 397 7155
www.cineville.com

Founded in 1990, and has produced over 40 movies including *Hurly Burly*, *Mi Vida Loca* and *Swimming With Sharks*.

Circe's World Films, LLC MOVIE TV

149 South Barrington Avenue, Suite 190, Los Angeles, CA 90049-3310
T 310 229 5388
info@circesworldfilms.com
www.circesworldfilms.com

Committed to the advancement and encouragement of independent films and their makers. Interested in innovative scripted and improvisational material for TV and film. Aims to provide minor equity financing but does not take on the role of primary financial provider.

Two open submission periods run in January and June each year. Send (via mail or email) a brief synopsis and a rough estimate of funding requirements. Interested in all genres, shorts, documentaries and features.

City Lights Pictures MOVIE TV

6 East 39th Street, New York, NY 10016
T 646 519 5200 F 212 679 4481
www.citylightsmedia.com

Produces for television and cinema. Also has an office in Los Angeles. Film credits include *A History of Sex*, *A Dirty Shame*, *Tamara*, *Descent*, *The Ten*, *Harold* and *Interrupted*. City Lights Television has signed deals with 27 different networks.

Clarity Pictures MOVIE TV VG

1107 Fair Oaks Avenue, Suite 155, South Pasadena, CA 91030
T 877 868 8298
submissions@claritypictures.net
www.claritypictures.net

Production company working across formats including film, television, video game development, music video and commercials. Films have appeared on HBO, Lifetime Television, at The Toronto Film Festival and at the California Independent Film Festival. Examples of output include *The Fiesta Grand*, *The Clique* and *Hiding in Plain Sight: Tales of an American Predator*.

Always looking for new writers and good stories. Submit a query letter in first instance. If company asks to see further material, you will need to download submission form from the website. Does not return submissions.

Dick Clark Productions TV

2900 Olympic Boulevard, Santa Monica, CA 90404
T 310 255 4600
www.dickclarkproductions.com

Produces wide range of television content including comedy, children's programming, TV movies and dramatic series.

Columbia Tristar Motion Picture Group / Sony Pictures MOVIE TV VG

10202 W Washington Boulevard, Culver City, CA 90232

T 310 244 4000 F 310 244 2626
www.sonypictures.com

Columbia TriStar Motion Pictures is part of Sony Pictures Entertainment Inc. It releases about 25 films per year and has 12 Academy Award Best Picture films in its library. Recent output includes *The Da Vinci Code*, *Talladega Nights*, *Casino Royale*, *Rocky Balboa*, *The Benchwarmers*, *Ghost Rider* and *Spider-Man 3*.

Connection III Entertainment Corp. MOVIE TV

8489 W. Third Street, Los Angeles, CA 90048
T 323 653 3400
info@connection3.com
www.connection3.com

A multimedia producer-syndicator with output including *Made in Hollywood* and *Christmas at Water's Edge*.

Cooper's Town Productions MOVIE TV

302A W. 12th Street, Suite 214, New York, NY 10014
T 212 255 7566 F 212 555 0211
info@cooperstownproductions.com
www.cooperstownproductions.com

Philip Seymour Hoffman's production company, seeking to produce for cinema and TV. Output includes *Capote* and *The Savages*.

In first instance, email with a logline and brief synopsis. Only send a full script if requested. Non-agency submission must be accompanied by a signed release form, available from the website.

Crane Wexelblatt Entertainment Ltd MOVIE TV

6061 Galahad Drive, Malibu, CA 90265
T 310 457 4821 F 310 457 3888
twomoguls@aol.com
Producer *Peter Crane*
Producer *Linda Wexelblatt*

Established in 1978, producing drama and comedy for film and television.

Welcomes unsolicited ideas; send a synopsis by email.

The Crosley Company MOVIE TV

PO Box 1545, Edmond, OK 73083
T 405 348 1628
Producer *Richard Crosley*

Over 30-years experience in commercial and educational television producing, directing and writing. Working mostly for corporate clients.

Welcomes new writing.

Crystal Sky Pictures MOVIE

10203 Santa Monica Boulevard, 5th Floor, Los Angeles, CA 90067
T 310 843 0223 F 310 553 9895
info@crystalsky.com
www.crystalsky.com

Involved in film production, sales and financing since 1977. Able to finance or co-finance projects with budgets ranging from $5m to $125m. Its *Baby Geniuses* franchise has generated over $100m worldwide in revenues. Other production credits include *Bratz*, *Ghost Rider*, *Doomsday* and *Losing My Virginity*.

Dakota Films MOVIE TV

4133 Lankershim Boulevard, North Hollywood, CA 91602
T 818 760 0099 F 818 760 1070
info@dakotafilms.com
www.dakotafilms.com

Dakota Films specializes in reality based programming as well as scripted series and specials. Notable productions include Billy Crystal's opening films for the Oscars, HBO's *Tenacious D* and MTV's *Viva La Bam*.

Does not accept unsolicited material.

Dalaklis McKeown Entertainment, Inc. TV

1750 Berkeley Street, Santa Monica, CA 90401
T 310 460 0200 F 310 460 0202
production@dmetv.net
www.dmetv.net

Specializes in creating, developing, and producing programming for cable and network broadcasters. Production credits include *Rob and Amber Get Married*, *Insider's List*, *I Do Diaries: Wedding Videos Revealed*, several episodes of *Intimate Portrait* and *Extreme Beaches*.

Lee Daniels Entertainment MOVIE

315 West, 36th Street, Suite 1002, New York, NY 10018
T 212 334 8110 F 212 334 8290
info@leedanielsentertainment.com
www.leedanielsentertainment.com

Produces feature films, with credits including *Monster's Ball*, *The Woodsman*, *Shadowboxer*, *Tennessee* and *Push*.

To submit ideas, send a logline and one paragraph synopsis via email. A response will take 2 to 4 weeks but if interested, a copy of the script will be requested at this stage. Will only accept non-agency scripts if accompanied by a signed release form (available from the website).

Dark Horse Entertainment   [MOVIE]

421 S. Beverly Drive, Beverly Hills, CA 90212
T 310 789 4751
www.dhentertainment.com

Founded in 1992, producers of *The Mask*, *TimeCop*, *Barb Wire*, *The Mystery Men* and *Hellboy*.

Darkwoods Productions   [MOVIE]

1041 N. Formosa Avenue, SME #108,
West Hollywood, CA 90046
T 323 850 2497
www.darkwoodsproductions.com

Film production company with credits including *The Green Mile*, *The Majestic*, *The Salton Sea*, *The Mist* and *Foreign Babes in Beijing*.

Dino De Laurentiis Company   [MOVIE]

100 Universal City Plaza Bungalow 5195,
Universal City, CA 91608
T 818 777 2111 F 818 866 5566
ddlcoffice@ddlc.net
www.ddlc.net
Producer *Dino De Laurentiis*
Producer/CEO *Martha De Laurentiis*
Co-Producer *Lorenzo De Maio*
Head of Development *Ed Wacek*

Film production company producing narrative feature films. Established in 1983, past credits include *Hannibal Rising*, *U-571* and *Breakdown*.
   Does not accept unsolicited manuscripts.

DeeGee Entertainment   [MOVIE]

366 N. La Cienega Boulevard, Los Angeles,
CA 90048
T 310 652 9955 F 310 652 9050
info@coronet-theatrela.com
www.coronet-theatrela.com

Founded in 1986, producing films including *Ricochet River* (starring Kate Hudson and Jason James Richter), *Hostile* (starring Rob Lowe) and *The Wedding Planner* (starring Matthew McConaughey and Jennifer Lopez). Also run a writer's lab once a week.

Dimension Films   [MOVIE]

375 Greenwich Street, 4th Floor, New York,
NY 10013
www.weinsteinco.com

Founded in 1995, Dimension Films is a subsidiary of the Weinstein Company. Its credits include the *Scream* horror trilogy, *Spy Kids* and the *Scary Movie* franchise.

Dingo Production LLC   [MOVIE]

12 Havemeyer Street, Suite 3L, Brooklyn, NY 11211
T 718 554 3558 F 510 856 3973
enquiries@dingoproduction.com
www.dingoproduction.com
Producer *Nan Sandle*
Producer *Mike Urdoneta*

Founded in 2006, producing a range of media content including feature films, podcasts, documentaries and advertising.
   Welcomes unsolicited manuscripts by email.

Disney Interactive Studios, Inc.   [VG]

500 South Buena Vista Street, Burbank, CA 91521
T 818 553 5000
www.disney.go.com/disneyinteractivestudios

The interactive entertainment arm of the Walt Disney Company. Self-publishes and licences a range of games for many platforms and all markets.

Distant Horizon   [MOVIE]

Suite A, 8282 Sunset Boulevard, Los Angeles,
CA 90046
T 323 848 4140 F 323 848 4144
la@distant-horizon.com
www.distant-horizon.com

Has produced in the US, Canada, United Kingdom and Africa. Credits include *Chain of Desire*, *Dead Beat*, *Scorpion Spring* and *I Capture the Castle*. Also has an office in London.

Dreamworks SKG   [MOVIE] [TV] [VG]

650 Madison Avenue, 22nd Floor, New York,
NY 10022
T 212 588 6229 F 212 588 6233
www.dreamworks.com

Founded in 1994. Major American film studios. Develops, produces and distributes films, video games and television programming. Won three consecutive Best Picture Academy Awards in 1999, 2000 and 2001 for *American Beauty*, *Gladiator* and *A Beautiful Mind* (the latter two with Universal). Also produces the television series *Into the West*, *Las Vegas*, *Rescue Me* and *The Contender*. Sold to Viacom (parent company of Paramount Pictures) in 2006. Independent animation branch DreamWorks Animation SKG spun off in 2004 and now produces two CG animated family features per year. Recent credits include *Shark Tale*, *Shrek*, *Shrek 2* and *Madagascar*.

Dreyfuss/James Productions (MOVIE)

1041 N. Formosa Avenue, Pickford Building,
Rm. 110, West Hollywood, CA 90046
☏ 323 850 3140

The production company of Richard Dreyfuss and
Judy James. Its first production in 1991 was *Once
Around*. More recent credits include *The Forest*
and *Road to Ladakh*.

Echo Lake Productions, LLC (MOVIE)

421 South Beverly Drive, 8th Floor, Beverly Hills,
CA 90212
☏ 310 789 4790 ☏ 310 789 4791
contact@echolakeproductions.com
www.echolakeproductions.com

An independent film production company with
access to a private investment fund. Develops
many projects in-house and has a particular
interest in working with international directors.
Generally seeks projects costing under $5 million.
Also offers some bridge loans, sales agency
advances, gap loans and finishing funds. Credits
include *Twelve and Holding*, *Water*, *Away from
Her*, *Dreamland*, *Before the Rains*, *You Kill Me*
and *Devil to Pay*.
　Does not accept unsolicited submissions.

Edmonds Entertainment (MOVIE) (TV)

1635 N. Cahuenga Boulevard, Los Angeles,
CA 90028
☏ 323 860 1550 ☏ 323 860 1554
www.edmondsent.com

Founded by Grammy-winning singer/producer
Kenneth "Babyface" Edmonds and his wife Tracy.
It has produced films including *Light it Up*, *Hav
Plenty*, *College Hill* and *Soul Food*, which spawned
the award-winning television drama series of the
same name.

El Dorado Pictures (MOVIE)

725 Arizona Avenue, Suite 100, Santa Monica,
CA 90401
☏ 310 458 4800 ☏ 310 458 4802
www.alecbaldwin.com/eldorado

Founded by actor Alec Baldwin, with credits
including *State and Main* (winner of the Jury
Award at the Ft. Lauderdale International Film
Festival), *Second Nature* and *The Devil and Daniel
Webster*.

Electric Entertainment (MOVIE) (TV)

1438 N. Gower Street, Suite 24, Los Angeles,
CA 90028
☏ 323 817 1308
www.electric-entertainment.com

Founded in 2001, and repsonsible for films
such as *Flyboys*, *Eight Legged Freaks*, *Cellular*
(starring Kim Basinger) and *Who Killed the
Electric Car?* (nominated by the Writer's Guild
of America for Best Documentary Screenplay
and by the Broadcast Film Critics Association for
Best Documentary Feature). Also produced the
made-for-television film *The Librarian*, starring
Noah Wyle.

Electro-Fish Media LLC (MOVIE) (TV)

PO Box 301348, Austin, TX 78703
☏ 512 351 7133 ☏ 512 524 0011
info@electro-fish.com
www.electro-fish.com
Director *Chris Elley*

Founded in 2004, film and television
production companies primarily specializing in
documentaries. Recent titles include *Barbecue:
A Texas Love Story* and *Ghost Town: 24 Hours in
Terlihgua*.
　Prefers introductory contact before
submissions are sent.

Electronic Arts (EA) (VG)

209 Redwood Shores Parkway, Redwood City,
CA 94065
☏ 650 628 1500 ☏ 650 628 1414
www.ea.com

Founded in 1982, one of the giants of the gaming
world, employing over 7,000 people worldwide.
An independent developer and publisher of
interactive entertainment software for advanced
entertainment systems. Works with storywriters
and film directors. Among its properties are *Need
for Speed*, *The Sims*, *FIFA Soccer*, *Fight Night*, *Tiger
Woods PGA Tour* and *The Lord of the Rings*. Its
five hub studios are in Redwood Shores (US), Los
Angeles (US), Orlando (US), Vancouver (Canada)
and Surrey (UK).

Element Films (MOVIE)

8000 Beverly Boulevard, Los Angeles, CA 90048
☏ 323 655 8000 ☏ 323 655 8001
info@sbe.com
www.sbeent.com/elementfilms/elementfilms.html

Came into being in 2004. A fully integrated film
company with ability to develop, finance, produce,
distribute and market titles. Examples of output
include *Waiting*, *The Last Time*, *Five Fingers*, *Mr.
Brooks*, *Pride* and *Kill Theory*.

Elixir Films (MOVIE)

8033 W. Sunset Boulevard, West Hollywood,
CA 90046
☏ 323 848 9867 ☏ 323 848 5945

info@elixirfilms.com
www.elixirfilms.com

Made its first film in 2001 with the screen adaptation of Eric Bogosian's *Wake Up and Smell the Coffee*. Also produced the films *Where the Red Fern Grows*, *The Good Thief*, and the documentary *Long Way Round* (starring Ewan McGregor and Charley Boorman).

Emerging Pictures Corporation MOVIE TV

245 West 55th Street, 4th Floor, New York, NY 10019
T 212 245 6767 F 212 202 4984
inquiries@emergingpictures.com
www.emergingpictures.com

Supplies original content to both traditional media outlets and new platforms.

Only accepts material from literary agents and entertainment lawyers with whom a relationship already exists. Unrepresented writers can mail, fax or email a one-page project brief, including a synopsis, details of actors, directors or producers attached to the project. State whether the project is to be considered for development, financing or marketing. If suitably interested to see the complete manuscript, the company will require the writer to complete a release form (available from website).

Energy Entertainment MOVIE

999 N. Doheny Drive, Suite 711, Los Angeles, CA 90069
T 310 274 3440
info@energyentertainment.net
www.energyentertainment.net

Founded in 2001, its film credits include *The Number 23*, starring Jim Carrey.

Unsolicited calls or mail will not be returned.

EnterAktion Studios, Inc. MOVIE TV

c/o Sound City Center, Stage 15466, Cabrito Road, Van Nuys, CA 91406
T 818 994 5494 F 818 994 5794
www.enteraktion.com

Working across motion pictures, television, animation, interactive and web entertainment. Includes a team of in-house writers. Credits include *Los Opolis: The Movie* and *Life with Dwegons*.

Epiphany Pictures, Inc. MOVIE TV

10625 Esther Avenue, Los Angeles, CA 90064
T 310 815 1266
epiphanysubmissions@gmail.com
www.epiphanypictures.com

A full service film and TV production company. Works across features, TV series, cable, documentaries, reality, home video, promo and online productions. Also has an office in Chicago. Output includes *Road Kings*, *Picture Windows*, *Phenomenon* and *Rosemary*.

Do not send script in first instance. If you receive a response requesting script, download release form from the website.

Evamere Entertainment LLC MOVIE

575 Broadway, 6th Floor, New York, NY 10012
T 212 475 7555 F 212 475 1717
contact@evamere.com
www.evamere.com

Founded as Hart Sharp Entertainment, partner John Hart launched Evamere in February 2007 after the team decided to part ways. This new production company will focus on feature films and Broadway theatrical productions. Forthcoming features include *Life During Wartime* and *Alice Goes To Harlem*.

Does not accept unsolicited material.

Evergreen Films LLC TV

1515 Palisades Drive, Suite N, Pacific Palisades, CA 90272
T 319 573 9978 F 310 573 1137
info@evergreenfilms.com
www.evergreenfilms.com

Has produced films including *The Secret Adventures of Jules Verne*, *When Dinosaurs Roamed America*, *Red Flag*, and the mini-series *Alien Planet*.

Evolution Media TV

3310 W. Vanowen Street, Burbank, CA 91505-1239
T 818 260 0300 F 818 260 1333
www.evolutionusa.com

Operating since 1987, producing network and cable programming across genres including documentaries, children's and educational programming. Production credits include *The Real Housewives of Orange County*, *Ocean Force*, *Desperate Spaces*, *Now You See It*, *Clean Sweep*, *Switched!* and *House Rules*.

Will not accept unsolicited submissions.

Exodus Film Group MOVIE TV

1201 Electric Avenue, Venice, CA 90291
T 310 392 7778
info@exodusfilmgroup.com
www.exodusfilmgroup.com

Independent production company. Also has offices in Paris and New York. Produces both live action and animated work for cinema and TV.

Has created a private equity animation film fund. Credits include *Amarillo Armadillo*, *Bunyan and Babe*, *EcoOdyssey*, *Igor* and *The Hero of Color City*.

Will not accept unsolicited materials, ideas, scripts or treatments.

Fabrication Films `MOVIE`

8701 West Olympic Boulevard, Los Angeles, CA 90035
T 310 289 1232 F 310 289 1292
chris@fabricationfilms.com
www.fabricationfilms.com

Founded in 2003, producing and distributing commercially-viable feature films.

FatKid Films `MOVIE`

2372 Veteran Avenue, Suite 201–202, Los Angeles, CA 90064
T 310 234 0942 F 310 601 1866
info@fatkidfilms.com
www.fatkidfilms.com

Established in 2001, producing feature films. First movie was *June*, a romantic comedy. Other titles include *A Darker Reality*, *Bled*, *Butcher House* and *Dark Reality*.

Filbert Steps Productions `MOVIE`

270 Lafayette Street, Suite 409, New York, NY 10012
T 212 246 2301 F 212 246 2285
info@filbertsteps.com
www.filbertsteps.com

Produces features and documentaries. Examples of output include *Two Family House*, *Runaway* and *Trumbo*.

Film Garden Entertainment `TV`

6727 Odessa Avenue, Van Nuys, CA 91406
T 818 301 4500
www.filmgarden.tv

Founded in 1994, with a track record of over three hundred hours of programming, including prime-time network specials, prime-time reality series, daytime strip series, and documentaries. Credits include *Baby Panda's First Year*, *Platinum Weddings*, *Crash Test Dummies*, *Carpocalypse* and *Taste of America*.

Film Police! `MOVIE`

4310 North Mozart Street, Chicago, IL 60618
T 773 463 4010
info@filmpolice.com
www.filmpolice.com
Producer *Phillip Koch*
Producer *Sally Marshall*

Founded in 1980, producing feature films and documentaries. Output includes *Medua Challenger*, *Pink Nights* and *The Ameican Flag* (a documentary for PBS).

No unsolicited manuscripts. In first instance, send an inquiry.

Firebrand Productions `MOVIE` `TV`

1524 Riverside Drive, Burbank, CA 91506
T 818 955 5711 F 818 955 5158
salli@firebrandproductions.com
www.firebrandproductions.com

Develops and produces feature films plus television movies, mini-series and episodic series for domestic and international markets. Examples of output include *Crazy Horse*, *Four Eyes and Six-Guns*, *Just Ask My Children*, *Slight Case of Murder*, *The Dead Will Tell*, *The Good Old Boys*, *When a Man Loves a Woman* and *Widow on the Hill*.

Firelight Media `MOVIE`

1201 Martin Luther King Jr. Way, 2nd Floor, Oakland, CA 94612
T 510 587 0000 F 510 587 0002
info@firelightmedia.org
www.firelightmedia.org

An independent non-profit production company dedicated to telling stories of people, places, cultures and issues that are underrepresented in the mainstream media. Most Firelight Media productions are created for national broadcast on PBS. They have produced films including *Jonestown: The Life and Death of Peoples Temple*, and the documentary *Beyond Beats and Rhymes* (which examines representations of manhood, sexism and homophobia in hip-hop culture).

First Look Studios, Inc. `MOVIE`

2000 Avenue of the Stars, Suite 410, Century City, CA 90067
T 424 202 5000
info@firstlookmedia.com
www.firstlookmedia.com

Major independent studio. Production credits include *A Map of the World*, *After.Life*, *An American Crime*, *Black Supaman*, *Dedication*, *Evelyn*, *Kicked, Bitten and Scratched*, *Skins* and *Ted Bundy*.

Express policy not to accept unsolicited concepts, creative ideas, suggestions, stories, artwork, or other content or potential content.

First Street Films `MOVIE`

T 202 518 6007 F 202 518 6009
info@1streetfilms.com
www.1streetfilms.com

Production credits include *High School Musical 1* and *2*.

Five Sisters Productions (MOVIE) (TV)

171 Pier Avenue, Suite 207, Santa Monica, CA 90405
Ⓣ 310 712 5443
info@fivesistersproductions.com
www.fivesistersproductions.com
Director/Actor/Producer *Gabrielle Burton*
Producer/Writer *Jennifer Burton*
Director/Actor/Producer *Maria Burton*
Actor/Director *Ursula Burton*
Producer *Charity Burton*

Founded in 1996. Run by five real-life sisters, Maria, Jennifer, Ursula, Gabrielle and Charity Burton. Aims to make high quality independent films 'that are entertaining yet also contribute to a sense of hope on a personal or social level'. Credits include *Manna from Heaven, Temps, Just Friends, Sign My Snarling Movie, Letting Go of God* and *The Happiest Day of His Life*.

Currently not considering new projects without production financing in place. Will not read unsolicited approaches. Email only if enquiring after producers-for-hire.

Flexitoon (MOVIE) (TV)

46 West 73 Street, Suite 3A, NY 10023
Ⓣ 212 877 2757
craigmarin@flexitoon.com
www.flexitoon.com
Co-owner *Craig Marin*
Co-owner *Olga Felgemacher*

Founded in 1979 by puppeteers Olga Felgemacher and Craig Marin, specializing in family entertainment and commercials across a wide range of media. Past children's programmes include *Hamlin, Loopy Sloop* and *Pinwheel*.

Does not welcome unsolicited manuscripts.

Florentine Films (MOVIE)

PO Box 613, Walpole, NH 03608
Ⓣ 603 756 3038 Ⓕ 603 756 4389
www.florentinefilms.com

Founded in 1976, specializing in documentaries. Credits include *The Civil War, The West, Thomas Jefferson, Frank Lloyd Wright, Jazz, Mark Twain, Unforgivable Blackness: The Rise and Fall of Jack Johnson, Niagara Falls, The War, The American Brew* and *Through Deaf Eyes*.

Flower Films (MOVIE)

9220 Sunset Boulevard, Suite 309, Los Angeles, CA 90069
Ⓣ 310 285 0200

Drew Barrymore's film company, producing since 1999. Output includes the *Charlie's Angels* franchise, *Donnie Darko, 50 First Dates, Fever Pitch* and *Music and Lyrics*.

Fobia Films, LLC (MOVIE)

6230 Wilshire Boulevard, Suite 2065, Los Angeles, CA 90048
Ⓣ 310 494 0183 Ⓕ 866 848 4626
www.fobiafilms.com

Production company whose credits include *America 101* and *The Final Season*.

Focus Features (MOVIE)

65 Bleecker Street, 3rd Floor, New York, NY 10012
Ⓣ 212 539 4000
www.focusfeatures.com

Formed in 2002, the art house film division of NBC Universal's Universal Studios. Also serves as a producer and distributor of low-budget action/horror films through its Rogue Pictures. Focus has produced Academy Awarding-winning films such as *Gosford Park, Traffic, The Constant Gardner, Pride and Prejudice, Brokeback Mountain* and *Atonement*.

Fortis Films (MOVIE) (TV)

8581 Santa Monica Boulevard, Suite 1, West Hollywood, CA 90069
Ⓣ 310 659 4533

Production company established by Sandra Bullock. Responsible for the *Miss Congeniality* series of movies and the *George Lopez Show* for television. Other film credits include *Gun Shy, Sudbury, All About Steve* and *Grace*.

Fortress Entertainment (MOVIE) (TV)

6725 W. Sunset Boulevard, Los Angeles, CA 90028
Ⓣ 323 467 4700
scripts@fortress-ent.com
www.fortress-ent.com

Established in 2004, a fully self-financed feature film and television development and production company. Credits include *Pride, The Joy of Funerals, The Midnight Man* and *The Secret Diary of Adrian Mole*.

Submit a logline (50 words or less) and synopsis (250 words or less). The company will then request a copy of the script if interested in the project. At this stage, you will be required to sign a submission form, available from the website.

Forty Acres & A Mule Filmworks [MOVIE]

124 Dekalb Avenue, Brooklyn, NY 11217

T 718 624 3703

Production company founded by director Spike Lee, beginning production in 1983. Output includes *She's Gotta Have It, Do the Right Thing, Mo' Better Blues, Malcolm X, Tales from the Hood, Clockers, Summer of Sam, The Original Kings of Comedy, She Hate Me, Jesus Children of America, Inside Man* and *Lovers & Haters*.

FourBoys Films [MOVIE] [TV]

T 323 656 2710 F 323 656 2712
www.fourboysfilms.com

Producing for film, TV and theatre. Output includes *Amazing Grace, The Engagement Ring* and *The Bituminous Coal Queens of Pennsylvania*.

Company policy not to accept or consider unsolicited creative ideas, suggestions or materials. Any unsolicited submissions via email, facsimile, mail or otherwise will be treated as non-confidential and shall be deemed, and shall remain, the property of FourBoys Films.

Fox Atomic [MOVIE]

10201 West Pico Boulevard, Bldg. 38, Los Angeles, CA 90035

T 310 369 1000 F 310 369 2000
www.foxatomic.com

A division of 20th Century Fox, founded in 2006 and producing films, comics and digital content for young adults. Output includes *Turistas, The Hills Have Eyes II* and *28 Weeks Later*.

Fox Searchlight Pictures [MOVIE]

10201 W Pico Boulevard, Bldg 38, Los Angeles, CA 90035

T 310 369 4402
www.foxsearchlight.com

Fox Searchlight Pictures was established in 1994 as a speciality film division of 20th Century Fox to produce and distribute indie-styled films. It has produced Academy Award-winning movies such as *Boys Don't Cry, Sideways, The Last King of Scotland* and *Little Miss Sunshine*.

Fries Film Group [MOVIE] [TV]

22817 Ventura Boulevard, Suite 909, Woodland Hills, CA 91364

T 818 888 3052 F 818 888 3042

Production/distribution company founded in 1994. Output includes *Treasure Island, Wildflowers* and *LAPD: To Protect and Serve*.

Furst Films [MOVIE] [TV]

8954 West Pico Boulevard, 2nd Floor, Los Angeles, CA 90035

T 310 278 6468 F 310 278 7401
info@furstfilms.com
www.furstfilms.com

Founded in 1999. Film output includes *The Matador* (starring Pierce Brosnan, Greg Kinnear, and Hope Davis), *The Cooler* (starring Alec Baldwin and William H. Macy) and *Owning Mahowny* (starring Philip Seymour Hoffman, Minnie Driver and John Hurt). Also produced the television show *Tough Trade*.

Furthur Films [MOVIE]

100 Universal City Plaza Building 507-4G, Universal City, CA 91608

T 818 777 6700

Michael Douglas's film company with credits including *Don't Say a Word, The In-Laws, The Sentinel, The Mechanic* and *Racing the Monsoon*.

Future Films USA, LLC [MOVIE]

1531 14th Street, Santa Monica, CA 90404

T 310 393 7124 F 310 393 7251
info@futurefilmgroup.com
www.futurefilmgroup.com

Formed in 2000, providing a full range of services from pre-financing to post-production. Highly experienced as both a co-producer on international productions and as executive producer on other productions.

FXF Productions Inc. [MOVIE] [TV]

2554 Lincoln Boulevard, Suite 1062, Venice Beach, CA 90291

T 310 577 5007 F 310 577 1960
info@fxfproductions.com
www.fxfproductions.com
Director *L. Lonnie Peralta*
Executive Producer *Eric Alan Donaldson*

Founded in 1997, producing for television and film. Works across genres and formats: commercials, music video, sports, reality shows, documentary.

Will accept unsolicited approaches by mail and email. No returns.

Ghost House Pictures [MOVIE]

315 South Beverly Drive, Suite 216, Beverly Hills, CA 90212

T 310 785 3900
info@ghosthousepictures.com
www.ghosthousepictures.com

Founded in 2002 by filmmakers Sam Raimi and Rob Tapert, specializing in horror films. Works with Mandate Pictures to finance, develop and distribute high-concept genre films. Production credits include *30 Days of Night, Rise, Siren, The Evil Dead, The Grudge, The Grudge 2* and *The Messengers*.

Cannot accept unsolicited work at this time.

Gigantic Pictures `MOVIE`

59 Franklin Street, Ground Floor, New York, NY 10013
T 212 925 5075 F 212 925 5061
info@giganticpictures.com
www.giganticpictures.com

Producing low and mid-level budgeted feature films and documentaries. Examples of output include *3 Weeks After Paradise, Cosmopolitan, Flannel Pajamas, Goodbye Baby, Satellite, The Doorman, The First Seven Years, The Suitor* and *The Toe Tactic*.

Go Girl Media `TV`

9062 Hayvenhurst Avenue, North Hills, CA 91343
T 310 472 8910
www.gogirlmedia.com

Founded in 2004, a television development and production company working in both scripted and non-scripted entertainment properties.

Goff-Kellam Productions `MOVIE`

8491 Sunset Boulevard, Suite 1000, West Hollywood, CA 90069
T 323 656 2001 F 323 656 1002
goffkellam@aol.com
www.goffkellam.com

Established in 1998, producing movies and documentaries. Production credits include *Dropped, Girl Play, Heavy Put-Away, Out at the Wedding* and *Seventy*.

Gold Circle Films `MOVIE`

2000 Avenue of the Stars, Suite 600-N, Los Angeles, CA 90067
T 310 278 4800 F 310 278 0885
info@goldcirclefilms.com
www.goldcirclefilms.com

Founded in 2000 and has produced films such as *My Big Fat Greek Wedding, White Noise* (and the sequel *White Noise 2*) and *Rock Paper Scissors*.

Has a policy not to accept or consider unsolicited submissions of concepts, creative ideas, suggestions, stories, artwork or other potential content.

Goldcrest Films `MOVIE`

1240 N. Olive Drive, Los Angeles, CA 90069
T 323 650 4551 F 323 650 3581
mail@goldcrestfilms.com
www.goldcrestfilms.com

Founded in 1977. Produces, finances and distributes motion pictures. Titles in the library include *Chariots of Fire, Ghandi*, and *The Killing Fields*.

Samuel Goldwyn Films `MOVIE`

9570 W Pico Boulevard, Suite 400, Los Angeles, CA 90035
T 310 860 3100 F 310 860 3198
info@samuelgoldwyn.com
www.samuelgoldwynfilms.com

Develops, produces and distributes feature films and documentaries. Titles include *The Squid and the Whale* (nominated for an Academy award), *Raising Victor Vargas, Japanese Story* and *Super Size Me*.

Gracie Films `MOVIE` `TV`

10202 W. Washington Boulevard, Poitier 2nd Floor, Culver City, CA 90232
T 310 244 4000
www.graciefilms.com

Founded in 1986, producing award-winning films and television series. Output includes *Broadcast News, Jerry Maguire*, and *The Simpsons*.

Grammnet Productions `TV`

5555 Melrose Avenue, Los Angeles, CA 90038
T 323 956 5000

A TV production company run by Kelsey Grammar. Titles include *Gary the Rat, Medium, The Game* and *Dash 4 Cash*.

GRB Entertainment `TV`

13400 Riverside Drive, Sherman Oaks, CA 91423
T 818 728 7600 F 818 728 7601
www.grbtv.com

Specializes in unscripted alternative programming. Has produced television shows such as *True Caribbean Pirates, Princes of Malibu, Growing Up Gotti, Simply Irresistible* and *Beauty Shop Secrets*.

Green Dog Films `MOVIE`

2030 S. Sherbourne Drive, Suite 16, Los Angeles, CA 90034
T 310 287 0198
info@greendogfilm.com
www.greendogfilm.com
Contact *Jason Gurvitz*

Established in 1999, film production company specializing in drama and documentaries. Credits include *The Ungodly*.

Welcomes unsolicited manuscripts by email.

Green Moon Productions [MOVIE]

11718 Barrington Court, Suite 827, Los Angeles, CA 90049

Ⓣ 310 471 8800

Founded by Melanie Griffith and Antonio Banderas. Output includes *Crazy in Alabama*, *Forever Lulu*, *Tart*, *The Body*, *Imagining Argentina*, *El Camino de los ingleses*, *Have Mercy* and *Tres días*.

Greenestreet Films Inc [MOVIE]

9 Desbrosses Street, 2nd Floor, New York, NY 10013

Ⓣ 212 609 9000 Ⓕ 212 609 9099
general@gstreet.com
www.greenestreetfilms.com

Has produced movies including *In the Bedroom*, *Swimfan*, *Uptown Girls* and *Yes*.

Robert Greenwald Productions [MOVIE] [TV]

10510 Culver Boulevard, Culver City, CA 90232
info@rgpinc.com
www.rgpinc.com

Producing feature films, television programmes and documentaries for over 20 years. Credits include *Every Breath You Take*, *The Crooked E: The Unshredded Truth About Enron*, *The Book of Ruth*, *The Dead Will Tell*, *Beach Girls*, *Augusta*, *Gone* and *A Christmas Wedding*.

The Greif Company [TV]

9233 W Pico Boulevard, Suite 218, Los Angeles, CA 90035

Ⓣ 310 385 1200 Ⓕ 310 385 1207
www.greifcompany.com

Founded in 1990 by Leslie Greif, working across a range of genres including non-fiction TV programmes and TV movies. It has produced numerous episodes of the series *Intimate Portrait and Biography*. Other credits include *Lovebites*, *Steve McQueen: The Essence of Cool*, *Gene Simmons: Family Jewels* and *Brando*.

Grey Line Entertainment, Inc. [MOVIE]

115 W. California Boulevard, Suite 310, Pasadena, CA 91105-3005

Ⓣ 626 943 0950
submissions@greyline.net
www.greyline.net
Contact *Sara Miller*

A literary management and motion picture production company. Works with clients to develop, edit, promote, sell and market material.

Interested in hearing from prospective writers. Initial queries should be sent via email for the attention of Sara Miller. Any queries including an attached file will be deleted unread. Do not send multiple submissions. For screenplay, fiction and non-fiction projects, send a cover letter (under one page in length), including a logline, brief project synopsis, any relevant publication and/or production history, and current contact information. Allow two to three weeks for a response. If interested, further material will be requested at this stage. Any unrequested material will be discarded unread. See website for further submissions guidelines.

Gryphon Films [MOVIE]

13042 Rose Avenue, Los Angeles, CA 90066
Ⓣ 310 861 5383 Ⓕ 310 388 3012
info@gryphonfilms.com
www.gryphonfilms.com

Develops and produces its own motion pictures as well as investing in selected projects. Seeks projects with mass market appeal and that are "purposeful, but not 'artsy'". Production credits include *Driver's Ed*, *Five Generations of Jerks*, *Lady Jayne Killer*, *Steele's Island*, *The Cooler* and *Thunderhead*.

Will not accept unsolicited submissions.

Guy Walks into a Bar [TV]

7421 Beverly Boulevard, No. 4, Los Angeles, CA 90036
Ⓣ 323 930 9935
info@guywalks.com
www.guywalks.com

Television production company, founded in 1998. Output includes *Compton Cricket*, *Number One Girl*, *The Retreat* and *Elf*.

Hallmark Entertainment [TV]

1325 Avenue of the Americas, 21st Floor, New York, NY 10019
Ⓣ 212 977 9001 Ⓕ 212 977 9049
www.hallmarkent.com

Since 1994 Hallmark Entertainment (the successor to RHI entertainment) has been among the largest supplier of movies and mini-series in the television industry. Its productions have received 448 Emmy Nominations, 104 Emmy Awards and 15 Golden Globe Awards. Its critically-acclaimed productions include *Animal Farm*, *Arabian Nights*, *Don Quixote*, *Merlin* and *Human Trafficking*.

Hallmark Hall of Fame   TV

12001 Ventura Place, Suite 300, Studio City,
Los Angeles, CA 91604
T 818 505 8461
Manager of Development *J. Callahan*

Founded in 1951, specialists in television movies
and character dramas. Winner of over 75 Emmy
awards. Production credits include *Magic of
Ordinary Days* and *The Russell Girl*.
    Does not accept unsolicited manuscripts.

Hamzeh Mystique Films   MOVIE

61 Blaney Street, Swampscott, MA 01907
info@HamzehMystiqueFilms.com
www.hamzehmystiquefilms.com

Has produced films such as *Blood of Eden* and
The Letter, which won best documentary at the
Boston International Film Festival.
    Accepts submissions for feature films, MOWs,
and television pilots for consideration as future
projects. To submit material, complete a release
form available on website and post the first 10
pages of material to the address provided.

Hannibal Pictures   MOVIE

8265 Sunset Boulevard, Suite 107,
West Hollywood, CA 90046
T 323 848 2945 F 323 848 2946
contactus@hannibalpictures.com
www.hannibalpictures.com

Founded in 1999, specializing in international
co-productions of English language motion
pictures. Finance, produces and distributes 3–6
films annually, with budgets ranging from $3–40
million. Credits include *Crime Spree*, *Irish Jam*
and *Escobar*.

Happy Madison Productions   MOVIE

10202 West Washington Boulevard, Judy Garland
Bldg., Culver City, CA 90232
T 310 244 3100
www.adamsandler.com

Founded by Adam Sandler and has produced
films such as *Deuce Bigalow*, *The Animal* and *The
Benchwarmers*.

Harbor Lights Entertainment   MOVIE

1438 North Gower Street, Box 4, Building 2,
Third Floor, Hollywood, CA 90028
T 323 462 3887
www.harborlightentertainment.com

An independent motion picture development
and production company with credits including
Mindscan (starring Morgan Freeman), *Manjiro*
and *28/6*.

HBO   MOVIE TV VG

1100 Avenue of the Americas, New York,
NY 10036
www.hbo.com

HBO Films (part of the HBO cable television
network) has its main focus on the television
market, with successful series such as *Band of
Brothers*, *Angels in America*, *Entourage*, *Sex
and the City*, *The Sopranos* and *The Wire*. It
has also had success in the film industry with
critically-acclaimed films such as *American
Splendour*, *Elephant* (winner of the Palme D'Or at
the Cannes Film Festival) and *Elizabeth I*.

HDNet   TV

320 S. Walton Street, Dallas, TX 75226
T 214 672 1740
www.hd.net

Network producing wide range of original sports,
music, news and entertainment programming.
Also has office in Denver.

HDNet Films   MOVIE

122 Hudson Street, 5th Floor, New York, NY 10013
hdinquiries@hdnetfilms.com
www.hdnetfilms.com

Founded in 2003, financing and producing
narrative and documentary features shot on
high-definition video. Examples of output include
Gonzo: The Life and Work of Hunter S. Thompson,
Broken English, *Bubble*, *Diggers*, *Hunter*,
Mr. Untouchable, *One Last Thing*, *Quid Pro Quo*,
The Night Fisherman and *The War Within*.
    Projects for consideration should have a
finished screenplay, a director and a producer
attached. Budget cap is $5m per film. In first
instance, send only a synopsis and a brief
biography. If of interest, you will be asked you
to sign a standard release agreement before
submitting complete screenplay.

Jim Henson Co.   MOVIE TV

1416 N La Brea Avenue, Hollywood, CA 90028
T 323 802 1500 F 323 802 1825
www.henson.com

Founded by Jim Henson, the creator of *The
Muppets*, in 1958. Other television credits
include *Farscape* as well as the film *MirrorMask*.
Also responsible for the creation of the Emmy
Award-winning *Bear in the Big Blue House*.

Hit Entertainment   TV

1133 Broadway, New York, NY 10010
T 212 645 3555
www.hitentertainment.com

Launched in 1989, one of the world leading children's entertainment producers. Included in its portfolio are internationally-renowned children's properties such as *Bob the Builder*, *Thomas & Friends* and *Barney*.

Company policy not to accept unsolicited submissions or new programme ideas. Submissions will only be accepted through agents, publishers and content providers for the entertainment industry. A submissions pack is available for downloading from their website.

Hollywood Pictures · MOVIE

500 South Buena Vista Street, Burbank, CA 91521
T 818 560 1000
www.disney.go.com
Contact *Jo Johnson*

Part of the Walt Disney Corporation, producing films for a more mature, adult audience than Walt Disney Pictures. Has produced films such as *Arachnophobia*, *The Sixth Sense* and *The Invisible*.

Hollywood Studios International · MOVIE

9595 Wilshire Boulevard, Suite 900, Beverly Hills, CA 90212
T 310 358 9007
losangeles@hsifilms.com
www.hsifilms.com

Produces feature films through a diversified group of wholly-owned production companies, joint venture holdings, and strategic co-production agreements. Also offices in New York, Miami, London, Dubai, Mumbai and Hong Kong.

Iceman Productions · MOVIE

EBS World Entertainment, 3000 West Olympic Boulevard, Santa Monica, CA 90404
T 310 449 4065 F 310 449 4061
www.ebsla.com

The in-house production company of EBS World Entertainment. Output includes *The Remnant*, *A Conspiracy* and *Stage Fright*.

Icon Productions · MOVIE TV

808 Wilshire Boulevard, 4th Floor, Santa Monica, CA 90401
T 310 434 7300
www.iconmovies.com

Icon Productions was founded by actor/director Mel Gibson in 1989. Has produced award-winning films such as *Braveheart* and *The Passion of the Christ*. Also produces television movies and series such as *The Three Stooges*, *Complete Savages* and *Kevin Hill*.

Illuminati Entertainment · MOVIE

11901 Santa Monica Boulevard, Suite 494, Los Angeles, CA 90025
info@illuminatientertainment.com
www.illuminatientertainment.com

A management and production company focusing on clients with backgrounds in comic books and video games. Interested in creator-owned material and also building new brands and franchises.

Unable to accept unsolicited submissions.

Imagine Entertainment · MOVIE TV

9465 Wilshire Boulevard, 7th Floor, Beverly Hills, CA 90212
T 310 858 2000
www.imagine-entertainment.com

A film and television production company founded in 1986 by director Ron Howard and producer Brian Grazer. Output includes the Golden Globe- and Emmy Award-winning television series *24* and *Arrested Development*, as well as the Academy Award-winning films *Apollo 13* and *A Beautiful Mind*.

Impact Pictures LLC · MOVIE VG

9200 Sunset Boulevard, Suite 800, Los Angeles, CA 90069
T 310 247 1803
production@impactpix.com

Major feature-film production company specializing in adapting videogames for the screen. Past projects include *Resident Evil*, *Resident Evil: Apocalypse* and *DOA: Dead or Alive*.

No unsolicited scripts. Material accepted only via an agent.

Infinitum Nihil · MOVIE

c/o The Management Group, 9100 Wilshire Boulevard, Suite 725-E, Beverly Hills, CA 90212
T 310 271 0300
www.infinitumnihil.com

Formed by Johnny Depp in 2004. Its first film, *Shantaram*, is scheduled to appear in 2009.

Intermedia Films · MOVIE

9242 Beverley Boulevard, Suite 201, Beverly Hills, Los Angeles, CA 90210
T 310 777 0007 F 310 777 0008
info@intermediafilm.com
www.intermediafilm.com
CEO *Martin Schuermann*
President of Production *Scott Kroopf*

Founded in 1995, producing feature films with budgets between US$4m. and US$30m.

Examples of output include *Breach*, *RV*, *Alexander* and *Terminator 3*.

Does not welcome unsolicited approaches.

Jersey Films   MOVIE TV

PO Box 491246, Los Angeles, CA 90049
T 310 550 3200 F 310 550 3210

Founded by Danny DeVito and Rhea Perlman. Credits include *Matilda*, *Pulp Fiction*, *Erin Brockovich* and *Freedom Writers*. Also produced the television sitcom *Reno 911*.

Kaplan/Perrone Entertainment   MOVIE TV

10202 West Washington Boulevard, Astaire Building, Suite 3024, Culver City, CA 90232
submission@kaplanperrone.com
Partner *Aaron Kaplan*

Founded in 2000, producing for TV and film as well as computer games. Credits include *You, Me and Dupree*, *Made of Honor*, *Knowing* and *Pet*.

David E. Kelley Productions   TV

1600 Rosecrans Avenue, Building 4B, Manhattan Beach, CA 90266
T 310 727 2200
www.youstinka.com

Specializing in television drama series. Credits include *Ally McBeal*, *Boston Legal*, *Chicago Hope*, *Halley's Comet*, *Picket Fences*, *Snoops* and *The Practice*.

Kettledrum Films   MOVIE

4961 Agnes Avenue, Valley Village, CA 91607
T 818 506 7525
kettledrum@worldnet.att.net
Partner *Judd Bernard*
Partner *Patricia Casey*

Founded in 1967, producing films across a wide range of genres, from *Monty Python* to thrillers and westerns. Has worked with actors including Michael Caine, Lee Marvin and Glenda Jackson.

Welcomes new writing, 'depending on subject matter'.

Killer Films Inc.   MOVIE

380 Lafayette Street, Suite 202, New York, NY 10003
T 212 473 3950 F 212 473 6152
www.killerfilms.com

Has produced films such as *I'm Not There*, *An American Crime*, *Mrs Harris* (Primetime Emmy nominee), *Infamous* and, most recently, *Then She Found Me* (directed by Helen Hunt).

Will not accept unsolicited submissions.

Kismet Entertainment Group   MOVIE

8350 Wilshire Boulevard, Suite 200, Beverly Hills, CA 90211
T 323 556 0748
www.kismetent.com

Independent film production company since 2000. Examples of output include the *Dog Soldiers* films, *Cemetery Gates*, *Neo Ned*, *Mojave* and *The House on Turk Street*.

Has policy not to accept unsolicited scripts.

Kopelson Entertainment   MOVIE

1900 Avenue of the Stars, Suite 500, Los Angeles, CA 90067
T 310 407 1540
www.kopelson.com

Specializing in movies, with credits including *The Devil's Advocate*, *A Perfect Murder*, *Joe Somebody*, *Don't Say a Word*, *Twisted*, *In the Navy* and *A Killing on Carnival Row*.

Kultur International Films, Inc.   MOVIE

195 Highway 36, West Long Branch, NJ 07764
T 732 229 2343 F 732 229 0066
info@kultur.com
www.kultur.com
Chairman *Dennis Hedlund*
Vice President *Pearl Lee*
Managing Director *Ronald Davis*
Acquisitions Director *Rupert Hull*

Founded in 1980, independent company specializing in performing arts documentaries, including ballet, opera and music.

Accepts unsolicited proposals.

LAIKA, Inc.   MOVIE

1400 NW 22nd Avenue, Portland, OR 97210
T 503 225 1130 F 503 226 3746
ask_us@laika.com
www.laika.com

Established in 2005, specializing in animated features and short films. Examples of projects include *Moongirl*, *Jack and Ben's Animated Adventure* and *Coraline*.

To submit an unsolicited feature idea (in the form of script, treatment, outline, artwork or any combination of the above), send a copy of materials with a signed literary submission agreement by post. Do not send originals. Email submissions will not be opened or acknowledged.

Lakeshore Entertainment Corp   MOVIE

9268 W 3rd Street, Beverly Hills, CA 90210
☎ 310 867 8000 ⊕ 310 300 3051
info@lakeshoreentertainment.com
www.lakeshoreentertainment.com

Founded by Tom Rosenberg in 1994, Lakeshore has produced over 40 films including *Runaway Bride*, *Underworld* and *The Mothman Prophecies*. Also produced the multi-award-winning *Million Dollar Baby*.

Leaudouce Films   MOVIE

1626 Wilcox Avenue, Suite 424, Los Angeles, CA 90028
☎ 323 469 3546 ⊕ 323 417 4710
info@leaudouce.com
www.leaudouce.com
President *Ana Marie Laperal*
Creative Executive *Maria Pagan*

Founded in 2003, film production company specializing in features and international co-productions. Credits include *Seed of Contention*.
    Does not accept unsolicited manuscripts.

Levinson/Fontana Company   TV

185 Broome Street, New York, NY 10002
☎ 212 206 3585 ⊕ 212 206 3581
www.levinson.com

Television production company of Barry Levinson and Tom Fontana. Output includes *Hudson's Law*, *The Jury*, *The Bedford Diaries*, *3 lbs.* and *M.O.N.Y.*

Lifetime Entertainment   MOVIE TV

309 West 49th Street, New York, NY 10019
☎ 212 424 7000
www.lifetimetv.com

Specializing in women's television, a major producer of original drama and made-for-TV movies.

Light Renegade Entertainment, Inc.   MOVIE

8383 Wilshire Boulevard, Suite 510, Beverly Hills, CA 90211
☎ 323 653 0076 ⊕ 323 653 6005
info@lightrenegade.com
www.lightrenegade.com

Committed to creating commercially viable films that 'celebrate the human spirit'. Credits include *A Tryst in Time*, *Aloha, Where Are You?*, *Dark House*, *Finding Gardel* and *Rockin' the Suburbs*.

Lightstorm Entertainment   MOVIE

919 Santa Monica Boulevard, Santa Monica, CA 90401
☎ 310 656 6100

The production company of James Cameron. Output includes *The Abyss*, *Terminator 2: Judgment Day*, *True Lies*, *Titanic*, *Solaris* and *The Dive*.

Lion Rock Productions   MOVIE

2120 Colorado Avenue, Suite 225, Santa Monica, CA 90404
☎ 310 309 2980

The company of producer John Woo. Films include *The Big Hit*, *Bulletproof Monk*, *Paycheck*, *Tian tang kou* and *Second Sight*.

Lionsgate Films   MOVIE TV

2700 Colorado Avenue, Santa Monica, CA 90404
☎ 310 449 9200
www.lionsgatefilms.com

Founded in 1976, focusing on foreign and independent films. Known for distributing controversial films such as *Fahrenheit 9/11* and *American Psycho*. Received an Academy Award for Best Picture for *Crash*. Also the producer of the television series the *Dead Zone* and *Weeds*.

LivePlanet   MOVIE TV

11150 Santa Monica Boulevard, Suite 1200, Los Angeles, CA 90025
☎ 310 664 2400 ⊕ 310 664 2401
info@liveplanet.com
www.liveplanet.com

Co-founded by Ben Affleck, Matt Damon and Sean Bailey (chairman). Creates and develops for feature films and television, as well as new media. Credits include *American Wedding*, *Battle of Shaker Heights*, *Feast*, *First Descent*, *JoyRide*, *Matchstick Men*, *Running the Sahara*, *Stolen Summer*, *The Core* and *The Emperors Club*.

LMNO   TV

15821 Ventura Boulevard, Suite 320, Encino, CA 91436
☎ 818 380 8000
jmayer@lmnotv.com
www.lmnotv.com

Independent television production company, established in 1989. Creates reality, documentary, informational and entertainment programming for network, cable and syndicated television. Recent credits include *A Face for Yulce*, *Anatomy of a Giant*, *Fire Me Please*, *Iraq: Frontline ER*, *Lance Armstrong: Running for Life*, *My War Diary*, *Secrets of Arlington National Cemetery* and *Secrets of Hoover Dam*.

LucasArts   VG

PO Box 29908, San Francisco, CA 94129-0908
T 415 746 8000
www.lucasarts.com

Established in 1982 by George Lucas, a
leading publisher and developer of interactive
entertainment software for videogame console
systems and personal computers. Works closely
with sister company Industrial Light & Magic.
Among its properties are *Star Wars*, *Indiana
Jones* and *Mercenaries*.

Lucasfilm Ltd   MOVIE TV VG

PO Box 29919, San Francisco, CA 94129-0919
T 415 623 1000
www.lucasfilm.com

Founded in 1971 by George Lucas, the architect
behind the *Starwars* and *Indiana Jones* films
series. The production company has won a total
of 19 Academy Awards and produced the Emmy
Award-winning television series *The Young
Indiana Jones Chronicles*.

Mace Neufeld Productions   MOVIE TV

9100 Wilshire Boulevard, Suite S17 - East Tower,
Beverly Hills, CA 90212
T 310 401 6868
President *Mace Neufeld*
Vice President, Productions *Kel Symons*
Creative Executive *Ryan Patterson*
Office Manager *Kathy Day*

Film and television production company with
credits including *The Hunt for Red October*,
Patriot Games, *The Saint* and *Sahara*.
    Does not welcome unsolicited manuscripts.

MacGillivray Freeman Films, Inc.   MOVIE

PO Box 205, Laguna Beach, CA 92652
T 949 494 1055 F 949 494 2079
www.macfreefilms.com
Manager, Sponsorship and Development *Patty
Collins*
General Manager *Harrison Smith*

Established in 1963, producers with particular
reputation for large-format films for IMAX
cinemas. Past credits include *The Living Sea*
and *Dolphins*, both nominated for Academy
Awards in the Best Documentary/Short Subject
category.

Mad Chance   MOVIE

9021 Melrose Avenue Suite 202, West Hollywood,
CA 90069,
T 310 285 2077
www.madchance.com

With production credits including *10 Things I
Hate About You*, *Space Cowboys*, *Cats & Dogs*,
Confessions of a Dangerous Mind, *Catch That Kid*
and *Get Smart*.

MakeMagic Productions   MOVIE

8489 West 3rd Street, Suite 1044, Los Angeles,
CA 90048
T 323 653 3108
ddavid@makemagicproductions.com
www.makemagicproductions.com
President/Producer *Denise David Williams*
Reader *Chris Duarte*

Film company founded in 1998. Examples of
output include *My Dinner with Ovitz*.
    Approach via email. Offers script consulting
service to new and unproduced writers.

Malpaso Productions   MOVIE

4000 Warner Blvd, Building 81, Burbank,
CA 91522
T 818 954 3367

Clint Eastwood's production company, making
films since 1977. Credits include *Pink Cadillac*,
Unforgiven, *The Bridges of Madison County*,
Midnight in the Garden of Good and Evil, *Million
Dollar Baby*, *Flags of Our Fathers* and *Letters from
Iwo Jima*.

Mandalay Pictures   MOVIE

4751 Wilshire Boulevard, 3rd Floor, Los Angeles,
CA 90010
T 323 549 4300
info@mandalay.com
www.mandalay.com

Film studio founded in 1995 as part of the
Mandalay Entertainment Group. Output includes
Seven Years in Tibet, *Donnie Brasco*, *I Know What
You Did Last Summer* and the sequel *I Still Know
What You Did Last Summer*.
    Has a company policy not to accept unsolicited
materials.

Mandate Pictures   MOVIE

8750 Wilshire Boulevard, Suite 300 East, Beverly
Hills, CA 90211
T 310 360 1441 F 310 360 1447
info@mandatepictures.com
www.mandatepictures.com

Film production and distribution entity founded
in 2005. Releases include *Harold & Kumar Escape
from Guantanamo Bay*, *Juno*, *Mr. Magorium's
Wonder Emporium*, *Stranger Than Fiction* and
The Strangers.

Matador Pictures LLC   `MOVIE`

12021 Wilshire Boulevard, Suite 117, Los Angeles,
CA 90025
Ⓣ 310 472 6220
admin-la@matadorpictures.com
www.matadorpictures.com

Founded in 1999. Has produced award-winning
films across genres including *Ae Fond Kiss*
(winner of two awards at the Berlin Film Festival
and nominated for The Golden Bear), *Ten
Minutes Older* and *The Wind that Shakes the
Barley* (winner of the Palme D'Or and numerous
other awards). Also has offices in London.

Matthau Company   `MOVIE` `TV`

11661 San Vicente Boulevard, Suite 609,
Los Angeles, CA 90049
Ⓣ 310 454 3300
info@matthau.com
www.matthau.com

Founded in 1990, producing films including
Dennis the Menace, Grumpy Old Men and *The
Grass Harp*. The company focuses on stories with
'humanity, heart and humour'.

M8 Entertainment   `MOVIE`

15260 Ventura Boulevard, Suite 710,
Sherman Oaks, CA 91403
Ⓣ 818 325 8000 Ⓕ 818 325 8020
info@media8ent.com
www.media8ent.com

Media 8 was formed in 1993 and subsequently
changed its name to M8 Entertainment in 2004.
Finances, develops, produces and distributes
feature films in all media. Also has an office in
Montreal. Credits include *Lovewrecked, The
Upside of Anger* and the Academy Award-winning
Monster.

Merchant Ivory Productions   `MOVIE`

250 W 57th Street, Suite 1825, New York,
NY 10107
Ⓣ 212 582 8049 Ⓕ 212 459 9201
contact@merchantivory.com
www.merchantivory.com

Founded in 1961 and best known for producing
period pieces such as *A Room With A View* and
Howards End, each of which won three Academy
Awards. Also received eight Academy Award
nominations for *The Remains of the Day*.

Microsoft Game Studios (MGS)   `VG`

Microsoft Corporation, One Microsoft Way,
Redmond, WA 98052-6399
www.microsoft.com/games

The game creation wing of the Microsoft
Corp. Develops and publishes games for
Windows-based PCs and Xbox consoles.
Examples of output include *Combat Flight
Simulator, Dungeon Siege, Halo* and *Kingdom
Under Fire*.

Mikemon International, LLC   `MOVIE` `TV`

5225 Pooks Hill Road, Suite 1817-N, Bethesda,
MD 20814
Ⓣ 302 256 7073
mm@mikemoninternational.com
www.mikemoninternational.com
President *Michael Monroe*

Established in 2002, producing documentaries
on historical figures, travel and human interest
stories.
   Welcomes unsolicited manuscripts, with a
letter of introduction.

Mindfire Entertainment   `MOVIE`

3740 Overland Avenue, Suite E, Los Angeles,
CA 90034
Ⓣ 310 204 4481 Ⓕ 310 204 5882
www.mindfireentertainment.com

Founded in 1997 by Mark Gottwald and
filmmaker Mark A. Altman. Produced the film
Free Enterprise, winner of the Best Feature Film
and Best Screenplay at the AFI International Film
Festival. Also produced *The Specials*, starring
Rob Lowe, Jamie Kennedy and Thomas Haden
Church.

Mirage Productions Inc   `TV`

111 Spring Street, Newton, NJ 07860
Ⓣ 973 300 9477 Ⓕ 973 300 9467
pm@mirageproductions.com
www.mirageproductions.com

Mirage is a broadcast television and corporate
multimedia video production company founded
in 1994.

Miramax Films   `MOVIE`

8439 Sunset Boulevard, West Hollywood,
CA 90069
Ⓣ 323 822 4100 Ⓕ 323 822 4216
www.miramax.com

Miramax Films was founded in 1979 by brothers
Harvey and Bob Weinstein. The company
has been bought out by the Walt Disney
Company, and has Dimension Films as one
of its subsidiaries. Has produced Academy
Award-winning films such as *Pulp Fiction,
Shakespeare in Love, The English Patient, Chicago*
and *Cold Mountain*.

Montivagus Productions   MOVIE TV

13930 Burbank Boulevard, Suite 100, Sherman Oaks, CA 91401-5046

T 818 782 1212 F 818 782 1931
mail@montivagus.com
www.montivagus.com

Founded in 1989, producing for TV and cinema. Have developed and produced novels, plays, screenplays and short stories by new and emerging writers. Examples of output include *I'm No Dummy*, *Along for the Ride* and *The Second Room*.

Moonstone Entertainment   MOVIE

PO Box 7400, Studio City, CA 91614
T 818 985 3003 F 818 985 3009
etchie@moonstonefilms.com
www.moonstonefilms.com

Founded in 1993, Moonstone has produced films including *Digging to China*, *Miss Julie* and *Hotel*. The company also produced *Wu Ji* (nominated for a Golden Globe award for Best Foreign Language Film) and *Cookies Fortune* (winner of the Prize of the Guild of German Art House Cinemas at the Berlin International Film Festival).

Morgan Creek Productions   MOVIE

10351 Santa Monica Boulevard, Suite 200, Los Angeles, CA 90025
T 310 432 4848 F 310 432 4844
www.morgancreek.com

Founded in 1988, producing a number of box office successes including *Young Guns*, *True Romance*, *Ace Ventura: Pet Detective* and *Robin Hood: Prince of Thieves*. Also produced the Academy Award-winning *Last of the Mohicans* and the Academy-nominated *The Good Shepherd*.

Mosaic Media Group   MOVIE TV

9200 Sunset Boulevard, 10th Floor, Los Angeles, CA 90069
T 310 786 4900 F 310 777 2185

Mosaic Media Group is a talent management firm that represents comedy stars, including Jim Carrey, Will Ferrell, and the Wayans brothers. The company has also produced films such as *Elf*, the *Scooby Doo* series, *The Brothers Grimm* and *Talladega Nights: The Ballad of Ricky Bobby*.

Motion Picture Corporation of America   MOVIE TV

10635 Santa Monica Boulevard, Suite 180, Los Angeles, CA 90025
T 310 319 9500 F 310 319 9501

info@mpcafilm.com
www.mpcafilm.com

Founded in 1986 to produce, acquire and distribute commercial motion picture and television productions. Has subsequently produced 21 major studio theatrical releases and 40 television and made-for-video movies. Credits include *A Love Song for Bobby Long*, *Beverly Hills Ninja*, *Dumb and Dumber*, *Kingpin*, *Pumpkinhead* and *Riding the Bullet*.

MPH Entertainment, Inc.   MOVIE TV

1033 North Hollywood Way, Suite C, Burbank, CA 91505
T 818 441 5040 F 818 441 5050
info@mphent.com
www.mphent.com

Established in 1996, focusing on the development and production of feature films, telefilms, television series and specials. Credits include the series *The Dog Whisperer* and *History's Mysteries*, *Strangers Online*, *Egypt: Land of Mummies*, *Men Seeking Women* and *My Big Fat Greek Wedding*.

Mr Mudd   MOVIE TV

5225 Wilshire Boulevard, Suite 604, Los Angeles, CA 90036
www.mrmudd.com

Founded in 1998, producing feature films, documentaries and original TV series. Established by Lianne Halfon, John Malkovich and Russell Smith. Credits include *Ghost World*, *The Dancer Upstairs*, *Art School Confidential*, *The Libertine* and *Juno*.

MTV Films   MOVIE

5555 Melrose Avenue, Los Angeles, CA 90038
T 323 956 8023 F 323 862 1386
www.mtvfilms.com

Founded in 1995, producing films based on MTV programmes such as *Beavis and Butt-head Do America* and *Jackass: The Movie*. Also produced *Freedom Writers*, starring Academy Award-winner Hillary Swank.

Company does not accept unsolicited submissions.

MWG Productions   TV VG

8075 West 3rd Street, Suite 304, Los Angeles, CA 90048
T 323 937 8313 F 323 937 5239
wynne9@aol.com
President/Producer *Max Goldenson*
Associate Producer *Anna Tkatch*

Established in 1978, producing TV drama, miniseries and documentaries, as well as computer games. Winner of two Emmy Awards.

Welcomes unsolicited manuscripts by email or fax.

Myriad Pictures (MOVIE) (TV)

3015 Mainstreet, Suite 400, Santa Monica, CA 90405

T 310 279 4000 F 310 279 4001
info@myriadpictures.com
www.myriadpictures.com

Founded in 1998, Myriad has produced films including *The Good Girl*, *Jeepers Creepers*, *Imagining Argentina* and *Van Wilder*. It received an Oscar nomination for *Being Julia*.

Namco Bandai Games America Inc. (VG)

4555 Great America Parkway, Suite 201, Santa Clara, CA 95054

support@namcobandaigames.com
www.namcobandaigames.com

Video games development company, with global headquarters in Japan. Among its properties are *Galaxian*, *Pac-Man*, *Time Crisis*, *Tekken*, *Ace Combat* and *Ridge Racer*.

Namesake Entertainment (MOVIE) (TV)

7608 West Highway, 146 English Manor II, Suite 100, Pewee Valley, KY 40056

T 502 243 3185 F 502 243 3187
info@namesakeentertainment.com
www.namesakeentertainment.com

Established in 1996, specializing in 'life-affirming family entertainment for values–seeking general audiences'. Production credits include *Left Behind*, *Hangman's Curse*, *The Visitation*, *Thr3e* and *House*.

National Geographic Feature Films (MOVIE) (TV)

9100 Wilshire Boulevard, Suite 401E, Beverly Hills, CA 90212

T 310 858 5800 F 310 858 5801
www.nationalgeographic.com

Output includes *God Grew Tired of Us: The Story of Lost Boys of Sudan*, which won the Grand Jury Prize at the Sundance Film Festival. Examples of other titles include *The Gospels of Judas* and *The Mysteries of Egypt*.

NBC Entertainment (TV)

100 Universal City Plaza, Universal City, CA 91608
www.nbc.com

NBC Entertainment creates and produces programming for the network's primetime, late-night, daytime and Saturday-morning schedules. Over the years NBC has produced a number of top-rated award-winning television shows. Its popular comedy series include *Frasier*, *Friends* and *Scrubs*, and its signature dramas include *ER*, *Law and Order* and *The West Wing*.

Neo Art and Logic (MOVIE) (TV)

8315 Beverly Boulevard, Los Angeles, CA 90048

T 323 653 6007 F 323 653 0409
www.neoartandlogic.com
Creative Executive *Courtney Balaker*
Creative Executive *Aaron Ockman*

Established in 1990, producing drama, action and documentaries, but specializing particularly in horror films and TV shows. Production credits include *Feast*, *Pulse*, *The Prophecy* and *Trekkies*.

Not able to accept unsolicited manuscripts.

Neverland Films, Inc. (MOVIE)

9229 Sunset Boulevard, Suite 615, Los Angeles, CA 90069

T 310 772 0008 F 310 772 0006
contact@neverlandfilms.com
www.neverlandfilms.com

Has produced films including *Scorched* (starring Woody Harrelson, Alicia Silverstone, Rachael Leigh Cook and John Cleese), *Drowning Mona* (starring Danny DeVito, Bette Midler, Jamie Lee Curtis and Neve Campbell) and *Noel* (directed by Chazz Palminteri, starring Susan Sarandon, Penelope Cruz, Paul Walker and Alan Arkin). Also co-produced a full-length docu-drama about Studio 54 entitled *The Last Dance*.

New Amsterdam Entertainment, Inc. (MOVIE)

1133 Avenue of the Americas, Suite 1621, New York, NY 10036

T 212 922 1930 F 212 922 0674
mail@newamsterdamnyc.com
www.newamsterdamnyc.com

Particular focus on adapting novels to make movies and other forms of entertainment. Production credits include the *Creepshow* series, *Dawn of the Dead* and several Stephen King adaptions (*The Night Flier*, *Thinner*, *Pet Sematary*).

Will not accept unsolicited material. In first instance, mail (not email) to request permission to submit a script for evaluation and ask for a writer's release form. Indicate why your script is a good basis for a commercial movie and include a brief description of the story. Provide some information about yourself include contact information. If interested in reading the script, the company will send out a standard release form. Unsuccessful applicants will not receive notification.

New Crime Productions (MOVIE)

555 Rose Avenue, Venice, CA 90291
Ⓣ 310 396 2199
contact@newcrime.com
www.newcrime.com

Production company of John Cusack. Output includes *Grosse Pointe Blank*, *High Fidelity*, *Grace Is Gone* and *War, Inc.*

New Line Productions Inc. (MOVIE) (TV)

116 N. Robertson Boulevard, Suite 200,
Los Angeles, CA 90048
Ⓣ 310 854 5811 Ⓕ 310 854 1824
www.newline.com

New Line Cinema was founded in 1967 and is a subsidiary of Time Warner. It is responsible for producing the *Austin Powers* trilogy and the award-winning *Lord of the Rings* trilogy. Among its other titles is the Oscar-nominated *A History of Violence*.

New Regency Productions (MOVIE)

10201 W Pico Boulevard, Bldg 12, Los Angeles,
CA 90035
Ⓣ 310 369 8300 Ⓕ 310 969 0470
www.newregency.com

New regency Productions was founded in 1991 and is a subsidiary of Regency Enterprises. It has produced over 50 movies including *Mr & Mrs Smith*, *Fight Club*, *Natural Born Killers* (Golden Globe nominee) and *Big Momma's House*.

New Root Films (MOVIE) (TV)

39 East 12th Street, Suite 709, New York,
NY 10003
contact@newrootfilms.com
www.newrootfilms.com

Aims to make commercially-viable and credible material for cinema and TV at any budget level. Output includes *Livingston Avenue*, *The Cake Eaters* and the TV series, *Gnome*.

Currently not accepting any unsolicited material (whether received by mail or electronically).

New Wave Entertainment (MOVIE) (TV)

2660 West Olive Avenue, Burbank, CA 91505
Ⓣ 818 295 5000
www.nwe.com

The television and feature film development and production department has produced hundreds of hours of programming for clients including NBC, ABC, HBO, TBS, Logo, Oxygen, UPN, History Channel, Arts & Entertainment and SciFi.

Newmarket Films (MOVIE)

202 North Canon Drive, Beverly Hills,
Los Angeles, CA 90210
Ⓣ 310 858 7472
info@newmarketcap.com
www.newmarketfilms.com
President *Bob Berney*

Creative production arm of the Newmarket Entertainment Group founded in 1994 by William Tyrer and Chris Ball. Grew from a private film finance company into major producer and distributor of independent film. Distribution credits include *Donnie Darko*, *Monster* and *The Passion of the Christ*. In-house production began in 1998 with Christopher Nolan's *Memento*. Follow-ups include *The Mexican*, *Cruel Intentions* and *The Prestige*. 'Defiantly maintains independence from major corporations.'

Nickelodeon Movies (MOVIE) (VG)

5555 Melrose Avenue, Lubitsch Annex,
Hollywood, CA 90038
www.nickelodeonmovies.com

Founded in 1996, Nickelodeon Movies is the motion picture production arm of the children's cable channel Nickelodeon. It has produced popular films including *Harriet the Spy*, *Rugrats the Movie*, *Charlotte's Web* and *Lemony Snicket's A Series of Unfortunate Events* (nominated for four Academy Awards).

Does not accept unsolicited material.

Nintendo of America (VG)

PO Box 957, Redmond, WA 98073
www.nintendo.com

A giant among the creators of interactive entertainment, having sold over one billion video games worldwide. Nintendo was founded in Japan and Nintendo of America is the focus of its Western hemisphere activities. Responsible for household names such as *Mario*, *Donkey Kong*, *The Legend of Zelda* and *Pokémon*.

Nite Owl Productions, Inc. (MOVIE) (TV)

126 Hall Road, Aliquippa, PA 15001
Ⓣ 724 775 1993 Ⓕ 801 881 3017
niteowlprods@aol.com
www.niteowlproductionsltd.com

Founded in 2001, a privately funded motion picture and TV production company. Specializes in moderately budgeted films with known stars. Aims to produce 5–10 films per year.

Happy to evaluate the work of unknown writers. In first instance, send a one page query letter via mail or email. You will then be contacted if the company is interested in seeing more of your work. All script submissions must be accompanied by a dated and signed submission release form. See website for more details.

Nova Pictures `MOVIE`

6496 Ivarene Avenue, Los Angeles, CA 90068
T 323 462 5502 F 323 463 8903
pbarnett@novapictures.com
www.novapictures.com

A development, production and distribution company. Output includes the award-winning short, *The Yellow Badge of Courage*.
To submit a screenplay, send a brief pitch or synopsis via email.

Nu Image `MOVIE`

6423 Wilshire Boulevard, Los Angeles, CA 90048
T 310 388 6900 F 310 388 6901
production@nuimage.net
www.nuimage.net

Founded in 1992, a film producer whose credits include *16 Blocks*, *The Beat*, *Black Dahlia* and *Hard Cash*.

Lynda Obst Productions `MOVIE`

Paramount Pictures, 5555 Melrose Avenue, Milland Building 210, Hollywood, CA 90038
T 323 956 8744 F 323 862 2287
www.lyndaobst.com

Founded in 1989, specializing in feature films. Credits include *How to Lose a Guy in 10 Days*, *How to Lose It All*, *Interstellar* and *This Side of the Truth*.

Odd Lot Entertainment `MOVIE`

9601 Jefferson Boulevard, Suite A, Culver City, CA 90232
T 310 652 0999 F 310 652 0718
info@oddlotent.com
www.oddlotentertainment.com

Founded in 2001, producing films including *Suburban Girl* (starring Sarah Michelle Gellar and Alec Baldwin) and *Green Street Hooligans* (starring Elijah Wood).

Oddworld Inhabitants `VG`

info@oddworld.com
www.oddworld.com

Founded in 1994 by the experienced special effects and computer animation duo Sherry McKenna and Lorne Lanning. Based in San Luis Obispo, California. Aims 'to produce experiences rich in emotionality, empathy and entertainment value'. Titles include *Oddworld: Abe's Oddysee* and *Oddworld: Abe's Exoddus*.

Orchard Films `MOVIE` `TV`

68 Jay Street, Suite 218, Brooklyn, NY 11201
T 718 923 1950
info@orchardfilms.com
www.orchardfilms.com

Founded in 2000, producing documentaries and other non-fiction programming for broadcast. Examples of output include the *Indie Sex* series, *Fabulous! The Story of Queer Cinema*, *Chasing the Crown*, *In the Company of Women*, *Beauty in a Jar* and *Who Is Alan Smithee?*

Outlaw Productions `MOVIE`

10202 Washington Boulevard, Poitier Building 3214, Culver City, CA 90232
T 310 244 3445 F 310 244 1139
james@outlawfilm.com
www.outlawfilm.com

Has co-produced several films including *Phat Girlz*, *National Security*, *Training Day* and *Three to Tango*.

Overbrook Entertainment `MOVIE`

450 North Roxbury Drive, 4th Floor, Beverly Hills, CA 90210
T 310 432 2400

Will Smith's production company, which debuted in 2001 with the Oscar-nominated Muhammad Ali biopic, *Ali*. Other credits include *I, Robot*, *Hitch*, *The Pursuit of Happyness*, *I Am Legend*, *Hancock* and *The Secret Life of Bees*.

Palm Pictures `MOVIE`

76 Ninth Avenue, Suite 1110, New York, NY 10011
T 212 320 3600 F 212 320 3609
www.palmpictures.com

Founded in 1998, with credits including *The Basketball Diaries*, *Sex and Lucia*, *The Believer*, *Fulltime Killer*, *Gunner Palace*, *Cronicas*, *The Eye* and *DIG!*

Paramount Motion Pictures Group `MOVIE` `TV`

5555 Melrose Avenue, Hollywood, CA 90038-3197
T 323 956 5000 F 323 862 1204
www.paramount.com

Founded in 1912 and one of the giants of the movie world. Output in more recent times encompasses (either producing or co-producing) includes *Forrest Gump*, *Mission Impossible*, *Ghost*, *Titanic*, *Vanilla Sky*, *Changing Lanes*, *Team*

America: World Police, Dreamgirls and *Zodiac*. Their television series include *The Brady Bunch, MacGyver*, the *Star Trek* franchise and *Sabrina the Teenage Witch*.

Participant Productions (MOVIE)

info@participantproductions.com
www.participantproductions.com

Founded in 2004. Production credits include *Arna's Children, North Country* and *Syriana*. Won the Oscar for best documentary for *An Inconvenient Truth*.

Has a company policy not to accept unsolicited manuscripts.

Pathfinder Pictures LLC (MOVIE)

Ocean Front Walk, Suite 7, Venice, CA 90291
(T) 310 664 1500
info@pathfinderpictures.com
www.pathfinderpictures.com

Founded in 1998 and has produced *Double Deception* (starring Louis Mandylor), the sci-fi thriller *Shadow Fury* and the drama *Until The Night*.

Does not accept unsolicited submissions, concepts, creative ideas, stories or content of any sort.

Perdido Productions (MOVIE)

140 W. 57th Street, Suite 4-B, New York, NY 10019
(T) 212 582 9062

Production company formed by Woody Allen. Credits include *Hollywood Ending, Anything Else, Melinda and Melinda* and *Scoop*.

Persistent Entertainment (MOVIE) (TV)

9107 Wilshire Boulevard, Suite 500, Beverly Hills, CA 90210
(T) 310 777 0126 (F) 310 777 5259
info@persistent-ent.com
www.persistent-ent.com

Founded in 1997, producing films including *The Beautiful Ordinary, Walker Payne* and *Southland Tales* (starring Dwayne 'The Rock' Johnson, Sean William Scott and Sarah Michelle Gellar).

Company policy not to accept unsolicited manuscripts. Email a brief synopsis of project ideas for consideration.

Phoenix Pictures (MOVIE)

10202 W. Washington Boulevard, Frankovich Building, Rm. 204, Culver City, CA 90232
www.phoenixpictures.com

Founded in 1995. Output includes *The Mirror Has Two Faces, The People vs. Larry Flynt, Urban Legends: Final Cut, Whatever It Takes, Miss Potter, All the King's Men, Resurrecting The Champ, Pathfinder* and *Zodiac*.

Pixar Animation Studios (MOVIE)

1200 Park Avenue, Emeryville, CA 94608
(T) 510 922 3000 (F) 510 922 3151
www.pixar.com

Pixar Animation Studios, a subsidiary of Disney (since 2006) is a seven-time Academy Award-winning computer animation studio. Credits include *Toy Story, Finding Nemo, Cars* and *The Incredibles*.

Plan B (MOVIE)

9150 Wilshire Boulevard, Suite 350, Beverly Hills, CA 90210
(T) 310 275 6135

Production company, owned by Brad Pitt. First film was *Troy*. Subsequent titles include *Charlie and the Chocolate Factory* (starring Johnny Depp), *The Departed, Year of the Dog, A Mighty Heart* and *The Assassination of Jesse James by the Coward Robert Ford*.

Plum Pictures (MOVIE)

636 Broadway, Suite 814, New York, NY 10012
(T) 212 529 5820 (F) 212 529 5824
plumpic@plumpic.com
www.plumpic.com

Established in 2003, making features (both studio and independent films). Examples of output include *Dedication, Grace is Gone, Great World of Sound, Lonesome Jim, Pack Strap Swallow, The Baxter* and *The Ground Truth: After the Killing Ends*.

Point Made (MOVIE)

40 West 83rd Street, New York, NY 10024
(T) 212 724 6534
info@pointmade.com
www.pointmade.com

Founded in 1997, specializing in documentaries on American life. Output includes *Adopted: The New American Family* and *In 500 Words or Less*.

PorchLight Entertainment (MOVIE) (TV)

11050 Santa Monica Boulevard, Los Angeles, CA 90025
(T) 310 477 8400 (F) 310 477 5555
info@porchlight.com
www.porchlight.com

Makes and distribute television programmes and movies, with an emphasis on family entertainment. Recent production credits include the *Leap Frog* series, *Animalia*, *Meltdown*, *Personal Effects* and *I Downloaded a Ghost*.

Edward R. Pressman Film Corporation (MOVIE)

1648 N. Wilcox Avenue, Hollywood, CA 90028
T 323 871 8383 F 323 871 1870
www.pressman.com

Producing movies since 1969. Also has an office in New York. Film credits include *American Psycho*, *The Crow* trilogy, *Das Boot*, *The King*, *Reversal of Fortune*, *Rhapsody*, *Two Girls and a Guy* and *Wall Street*.

Runs policy not to accept unsolicited materials.

Production Partners, Inc. (MOVIE) (TV)

4421 Riverside Drive, Suite 206, Burbank, CA 91505
T 818 556 5065 F 818 556 5069
contact@productionpartners.com
www.productionpartners.com

Founded in 1990, producing in a variety of media. Particular strength in comedy. Production credits include *Curb Your Enthusiasm* and *Fat Actress*.

Approach via agent. Will not accept unsolicited material.

Protozoa Pictures (MOVIE)

104 North 7th Street, Brooklyn, NY 11211
T 718 388 5280 F 718 388 5425

Movie production company with credits including *The Wrestler*, *Black Swan*, *Flicker*, *The Fountain*, *Below*, *Requiem for a Dream* and *Pi*.

Punch Productions (MOVIE) (TV)

11661 San Vincente Boulevard, Suite 222, Los Angeles, CA 90049
T 310 442 4888

Founded by double Oscar-winner, Dustin Hoffman. Production credits include *Tootsie*, *Death of a Salesman* (for TV), *Outbreak*, *American Buffalo*, *The Devil's Arithmetic* and *A Separate Peace* (for TV).

Radar Pictures (MOVIE)

10900 Wilshire Boulevard, Los Angeles, CA 90024
T 310 208 8525
www.radarpictures.com

Movie production company with an output including *They*, *Le Divorce*, *The Texas Chainsaw Massacre*, *The Last Samurai*, *The Chronicles of Riddick*, *The Amityville Horror*, *The Heartbreak Kid*, *Emerald City*, *All About Steve* and *The Box*.

Raecom Productions (MOVIE) (TV)

5333 S. Genoa Way, Centennial, CO 80015
T 303 699 8110 F 303 317 3022
customerservice@RaecomProductions.com
www.RaecomProductions.com
Executive Producer *Villa Rae McClure*
Associate Producer *Chris Horton*

Founded in 2003, producing across a wide range of genres, including documentaries, music videos, short films and feature films.

Welcomes unsolicited manuscripts.

Rainbow Media Holdings LLC (TV)

11 Penn Plaza, New York, NY 10001
T 212 324 8500
www.rainbow-media.com

Produces targeted, multi-platform content for global distribution. Output includes *Wife, Mom, Bounty Hunter*, *North Face Expedition: Legends of the Lost World*, *Without Prejudice?*, *Behind the Label* and *My Funny Valentine*.

Rainstorm Entertainment Inc. (MOVIE)

15821 Ventura Boulevard, Suite 515, Encino, CA 91436
T 818 784 7500 F 818 981 4618
steve@rainstormentertainment.com
www.rainstormentertainment.com

Founded in 2002. Privately-financed, developing, producing, financing and distributing feature films. Credits include *The Big Empty*, *Red White Black and Blue* and *Big Bad Wolf*.

Happy to encourage new writers. In first instance send a one- or two-page treatment for consideration. Screenplays will only be accepted via a known agent or company associate. All submissions must be accompanied by a signed release form, available from the website.

Red Hour Films (MOVIE)

193 N. Robertson Boulevard, Beverly Hills, CA 90211
T 310 289 2565
www.redhourfilms.com

Company run by Ben Stiller and Stuart Cornfeld, specializing in comedy features. Examples of output include *Zoolander*, *Starsky & Hutch*, *Dodgeball: A True Underdog Story*, *Tenacious D in The Pick of Destiny*, *Blades of Glory* and *Tropic Thunder*.

Red Om Films (MOVIE)

16 W. 19th Street, 12th Floor, New York, NY 10011
T 212 243 2900

Julia Roberts's production company, formerly called Shoelace Productions. Film credits include *Maid in Manhattan*, *Mona Lisa Smile*, *The Friday Night Knitting Club* and the *American Girl* series.

Red Storm   [VG]

3200 Gateway Ctr. Boulevard, Suite 100, Morrisville, NC 27560
T 919 460 1776 F 919 460 1502
www.redstorm.com

Founded in 1996, the brainchild of Doug Littlejohns and novelist Tom Clancy. Now a leader in content development for multiple-media.

Regent Entertainment   [MOVIE] [TV]

10990 Wilshire Boulevard, Penthouse 1800, Los Angeles, CA 90024
T 310 806 4288 F 310 806 4268
info@regententertainment.com
www.regententertainment.com

Production credits include *Tom and Viv*, *One False Move* and *Gods and Monsters* (winner of the 1999 Academy Award for Best Adapted Screenplay).

Renegade Animation   [MOVIE] [TV] [VG]

116 North Maryland Avenue, Lower Level, Glendale, CA 91206
T 818 551 2351 F 818 551 2350
www.renegadeanimation.com

Founded in 1992, producing animation over a range of formats including episodic television, feature films, games, and web and mobile content. Examples of output include *Christmas Is Here Again*, *The Mr. Men Show*, *Hi Hi Puffy Ami Yumi* and *Captain Sturdy*.

Interested in proposals for new projects. It is recommended that submissions be made through an agent or entertainment attorney, accompanied by a signed submission agreement (available from the website).

Revolution Studios   [MOVIE]

2900 W. Olympic Boulevard, Santa Monica, CA 90404
T 310 255 7000 F 310 255 7001
info@revolutionstudios.com
www.revolutionstudios.com

Founded in 2000. Credits include *Punch-Drunk Love*, *13 Going on 30*, *The Animal* and *Daddy Day Care*.

River Road Entertainment   [MOVIE]

2000 Avenue of the Stars, Los Angeles, CA 90067

T 310 860 9470 F 310 461 1490
www.riverroadentertainment.com

Founded in 1987. Produces films and documentaries. Recent credits include *Brokeback Mountain*, *A Prairie Home Companion* and *Chicago 10*.

Roberts/David Films, Inc.   [MOVIE]

100 Universal City Plaza, T6139, Universal City, CA 91608
T 818 480 3139 F 818 480 3292
info@robertsdavid.com
www.robertsdavid.com

Established in 1994, utilizing up-and-coming writing, directing and acting talents. Output include *Danika*, *Dark Ride*, *Eastside*, *Extreme Dating*, *Outta Time*, *Plump Fiction*, *Strangers With Candy* and *Trash*.

Rockstar Games   [VG]

622 Broadway, Suite 4, New York, NY 10012-2600
T 212 334 6633
www.rockstargames.com

Founded in 1998, publishers of titles including *Grand Theft Auto*, *Midnight Club*, *Max Payne*, *Smuggler's Run*, *Manhunt* and *Red Dead Revolver*.

Welcomes input from the gaming community but any submissions of any nature whatsoever become the sole and exclusive property of Rockstar Games, which shall have full right, title and interest thereto, including under copyright, in all media now existing or hereafter created, and without any obligation to account or make any payment to the submitter for any use thereof.

Rogue Pictures   [MOVIE]

65 Bleecker Street, New York, NY 10012
T 212 539 4000
www.roguepictures.com

Launched in 2004, Rogue Pictures is a division of Focus Features, the speciality film division of Universal Studios. Mainly produces and distributes low-budget action/horror films. Output includes *Shaun of the Dead*, *Seed of Chucky*, *Assualt on Precinct 13* and *Hot Fuzz*.

Alex Rose Productions   [MOVIE]

8291 Presson Place, Los Angeles, CA 90069
T 323 654 8662 F 323 654 0196
alexroseproductions@hotmail.com
Producer *Alexandra Rose*
Assistant *Joe Pezzula*

Established by Oscar nominee Alex Rose, film production company producing drama, comedy and romance. Output includes *The Other Sister* and *Exit to Eden*.

Seeking new writing that they consider 'original, gratifying and uplifting'. Must sign release form.

Scott Rudin Productions `MOVIE`

560 South Buena Vista Street, Los Angeles, CA 91521-1759
⊤ 818 560 4600

An independent film and theatre production house at Paramount Pictures. Has produced films including *The Truman Show*, *I Heart Huckabees*, *Zoolander*, the Oscar-nominated *Notes on a Scandal* and the Oscar-winning *The Queen*.

Salty Features `MOVIE`

104 West 14th Street, 4th Floor, New York, NY 10011
⊤ 212 924 1601 Ⓕ 212 924 2306
info@saltyfeatures.com
www.saltyfeatures.com

Film production company founded in 2003, committed 'to films of the best quality made at modest budgets'. Output includes *The Inner Life of Martin Frost*, *Brief Interviews with Hideous Men*, *Bam Bam and Celeste* and *Evergreen*.

Not able to accept unsolicited material.

Saturn Films `MOVIE`

9000 Sunset Boulevard, Suite 911, Los Angeles, CA 90069
www.saturnfilms.com

Film production company, with credits including *The Family Man*, *The Life of David Gale*, *The Wicker Man*, *Next*, *National Treasure: Book of Secrets*, *The Dance* and *Bangkok Dangerous*.

Joel Schumacher Productions `MOVIE`

1149 N Gower Street, Suite 247, Los Angeles, CA 90038
⊤ 323 785 2274 Ⓕ 323 785 2275

Production company whose output includes *Phantom of the Opera* (based on the Andrew Lloyd Webber musical). Other titles include *Centricity* and *1.30 Train*.

Scott Free Productions `MOVIE` `TV`

614 N La Peer Drive, Los Angeles, CA 90069
⊤ 310 360 2250 Ⓕ 310 360 2251

Film and television production company founded in 1995 by Ridley Scott and his brother Tony. Has produced films such as *Black Hawk Down*, *In Her Shoes*, *Matchstick Men*, *Kingdom of Heaven* and the multi-Academy Award-winning *Gladiator*.

Has been producing the CBS crime series *Numb3rs* since 2005.

Screen Door Entertainment `TV`

15223 Burbank Boulevard, Sherman Oaks, CA 91411
⊤ 818 781 5600 Ⓕ 818 781 5601
generalinfo@sdetv.com
www.sdetv.com

Founded in 2001. Specializes in non-fiction programming including reality, travel, food, documentary, lifestyle, entertainment and how-to.

Will not view show ideas unless accompanied by a release form, available from website.

Section Eight `MOVIE`

4000 Warner Boulevard, Building 15, Burbank, CA 91522
⊤ 818 954 4860

Production company of George Clooney and Steven Soderbergh founded in 2001. Credits include the *Ocean's* franchise, *Insomnia*, *Confessions of a Dangerous Mind*, *Syriana*, *The Good German*, *A Scanner Darkly* and *Michael Clayton*.

SEGA of America `VG`

650 Townsend Street, Suite 650, San Francisco, CA 94103-4908
⊤ 415 701 6000
www.sega.com
Contact *Tony Harnett*

Sega of America was founded in 1986 by the Sega Corporation of Japan but can trace its roots back to the 1940s. A giant of the gaming world, its franchises include *Phantasy Star*, *Sonic the Hedgehog*, *Shining Force*, *Total War* and *Sega Rally*.

Seven Arts Pictures `MOVIE`

6310 San Vicente Boulevard, Suite 510, Los Angeles, CA 90048
⊤ 323 634 0990
www.7artspictures.com

Has produced films such as *An American Rhapsody* (winner of the Hollywood Discovery Award for best feature at the Hollywood Film Awards) and *Interstate 60*.

ShineBox Motion Pictures `MOVIE` `TV`

288 E. 1090 N., Orem, UT 84057
⊤ 801 427 3638
contact@shineboxmp.com
Producer *Bryan Young*

Producer *Steven Greenstreet*
Producer *Elias Pate*

Established in 1999, producing documentaries, narrative features and shorts. Recent credits include *This Divided State*, winner of Best Documentary at the 2005 Santa Cruz International Film Festival.

Welcomes unsolicited manuscripts.

Shoreline Entertainment, Inc.   [MOVIE]

1875 Century Park East, Suite 600, Los Angeles, CA 90067
T 310 551 2060 F 310 201 0729
info@shorelineentertainment.com
www.shorelineentertainment.com

Founded in 1992. Has produced titles including *The Secret*, *Sidekick* and *Marilyn Hotchkiss Ballroom Dancing and Charm School*, *Dark Corners* and *The Fifth Patient*.

Sierra Entertainment   [VG]

6060 Center Drive, 5th Floor, Los Angeles, CA 90045
www.sierra.com

Video games and interactive entertainment developer and publisher. Its studios include Massive Entertainment (founded in 1997; sample games: *Ground Control*); Radical Entertainment (sample games: *The Simpsons: Hit & Run*, *The Incredible Hulk: Ultimate Destruction* and *Scarface*); Swordfish Studios (founded in 2002; sample games: *Brian Lara International Cricket*); and High Moon Studios (sample games: *Darkwatch*).

Signature Pictures   [MOVIE]

725 Arizona Avenue, Suite 202, Santa Monica, CA 90401
T 310 394 1000 F 310 394 1001
frontdesk@signaturepictures.com
www.signaturepictures.com

Producing feature films, including *Spartan*, *Imaginary Heroes*, *A Sound of Thunder*, *Circadian Rhythm*, *The Black Dahlia* and *Amazon*.

Silver Lion Films   [MOVIE]

701 Santa Monica Boulevard, Suite 240, Santa Monica, CA 90401
T 310 393 9177 F 310 458 9372
slf@silverlionfilms.com
www.silverlionfilms.com

Established in 1987, an independent feature film production company which self-finances its projects. Past credits include *Man on Fire*, *Flipper* and *Crocodile Dundee in Los Angeles*.

Email synopsis in first instance before sending complete manuscript.

Silvers/Koster Productions   [MOVIE]

353 S. Reeves Drive, Penthouse, Beverly Hills, CA 90212
T 310 991 4736 F 310 284 5797
skfilmco@aol.com
www.silvers-koster.com
Owner *Iren Koster*
CEO *Tracey Silvers*
Vice President, Development *Karen Corcoran*

Established in 1985 by the daughter of comedian Phil Silvers and the nephew of director Henry Koster. Film production company producing drama, comedy and family films.

Does not welcome unsolicited manuscripts. Prefers first contact by email.

Sirk Productions, LLC   [MOVIE] [TV]

12 West 31st Street, 5th Floor, New York, NY 10001
T 212 244 4934 F 212 244 5429
sirk@sirkproductions.com
www.sirkproductions.com

Founded in 1997, producing feature films and television programming. Credits include *Severe Clear*, *Anytown*, *USA* and *The Inside Reel*.

Sneak Preview Entertainment   [MOVIE]

6705 Sunset Boulevard, 2nd Floor, Hollywood, CA 90028
T 323 962 0295 F 323 962 0372
indiefilm@sneakpreviewentertain.com
www.sneakpreviewentertain.com

Produces independent films, including *Hellbent*, *When do we Eat?* and *The Civilization of Maxwell Bright*. Some titles focus on adult content.

Snowfall Films   [MOVIE]

2321 W. Olive Avenue, Suite A, Burbank, CA 91506
T 818 558 5917 F 818 842 4112
www.snowfallfilms.com

Founded by producers Suzanne Lyons and Kate Robbins. Film credits include *Bailey's Billion$*, *Jericho Mansions* and *Plots With A View*. Also operates WindChill Films, specializing in genre films.

Sobini Films   [MOVIE]

c/o Lionsgate Entertainment, 2700 Colorado Avenue, Santa Monica, CA 90404
www.sobini.com

Established in 2001, producing specialized and mainstream features. Aims to produce 2–4 projects a year, with budgets ranging from US$10–40 million. Does not accept unsolicited manuscripts. Examples of output include *Streets of Legend*, *The Prince & Me*, *Prince & Me II: The Royal Wedding*, *Peaceful Warrior* and *Nellie Bly*.

Sommers Company · MOVIE

204 Santa Monica Boulevard, Suite A, Santa Monica, CA 90401
info@sommerscompany.com
www.sommerscompany.com

Founded in 2004. Produced *Van Helsing*, (starring Kate Beckinsale). Other projects include *Jason and the Argonauts*, *Flame over India* and *When Worlds Collide*.
Does not accept unsolicited manuscripts.

Sony Pictures Entertainment Inc. · MOVIE TV VG

10202 W Washington Boulevard, Culver City, CA 90232
T 310 244 4000 F 310 244 2626
www.sonypictures.com

Sony Pictures Entertainment is the television and film production/distribution unit of the Sony media conglomerate. It comprises various studios and entertainment brands including Columbia Pictures, TriStar Pictures, MGM and United Artists. In 1992 Sony created Sony Pictures Classics for art-house fare and its output includes the Academy Award-winning *Crouching Tiger, Hidden Dragon* and the Academy Award-nominee *House of Flying Daggers*.

Sony Pictures Imageworks · MOVIE

9050 W. Washington Boulevard, Culver City, CA 90232
T 310 840 8000 F 310 840 8100
www.imageworks.com

An Academy Award-winning, state-of-the-art digital production studio dedicated to the art of visual effects production and character animation. Film credits include *Spider-man*, *The Chronicles of Narnia: The Lion, The Witch and the Wardrobe* and *Stuart Little*.

Southern Skies/Great Norther Film and Music · MOVIE

1104 South Holt Avenue, Suite 302, Los Angeles, CA 90035
T 310 855 9833 F 310 855 0220
gr8norther@hotmail.com
President/Director *Edward D. Markley*

Founded in 1980, feature film production company with past credits including *City Slickers*, *Alien 3* and *Major League*. Specializes in drama and comedy features.
Welcomes unsolicited manuscripts, but 'only if they're great!'

Spirit Dance Entertainment · MOVIE TV

1023 North Orange Drive, Los Angeles, CA 90038
T 323 512 7988

Forest Whitaker's production company. Having decided to concentrate on his acting career since 2005, the company was involved in producing *Powder Blue* for release in 2008. Other titles include *American Gun* and *First Daughter*, as well as episodes of *The Twilight Zone* for television.

Spitfire Pictures · MOVIE TV

9348 Civic Center Drive, Mezzanine, Beverly Hills, CA 90210
T 310 300 9000 F 310 300 9001
info@spitfirepix.com
www.spitfirepix.com
Creative Executive *Anna Bocchi*

Founded in early 2003 by Guy East and Nigel Sinclair. A specialized development, production and finance company for motion pictures and television, with access to substantial development funds and other equity. Established a documentary division in 2005. Has a First Look development and production deal with Hammer Films. Output includes the movies *Possession* (with Sarah Michelle Gellar) and *Masked and Anonymous*, and the documentaries *No Direction Home: Bob Dylan* (directed by Martin Scorsese) and *Amazing Journey: The Story of The Who*.

Spyglass Entertainment · MOVIE TV

10900 Wilshire Boulevard, 10th Floor, Los Angeles, CA 90024
T 310 443 5800 F 310 443 5912
www.spyglassentertainment.com

Spyglass Entertainment was founded in 1998 and has produced Academy Award-nominated films such as *The Sixth Sense*, *The Insider* and *Seabiscuit*. It also produced the Academy Award-winning *Memoirs of a Geisha* and the television drama series *Miracles*.
Company policy not to accept unsolicited materials.

Square Enix Co., Ltd. · VG

999 N. Sepulveda Boulevard, Third Floor, El Segundo, CA 90245
T 310 846 0345
support@square-enix.com
www.square-enix.com

Founded in Tokyo, Japan in 1975. Creating, publishing and distributing entertainment content including interactive entertainment software and publications in Asia, North America and Europe. Properties include *Final Fantasy*, *Dragon Quest* and *Kingdom Hearts*.

Star Entertainment Group, Inc. [MOVIE]

13547 Ventura Boulevard, Suite 140,
Sherman Oaks, CA 91423
T 818 988 2200 F 818 988 2202
starentgrp@earthlink.net
www.findinghomemovie.com
Chairman *Lawrence David Foldes*
President *Victoria Paige Meyerink*

Independent production company specializing in films 'with substance'. Recent feature film *Finding Home* won best picture awards at a number of international film festivals.

Welcomes unsolicited approaches by mail (no faxes) but prefers submission of character descriptions and a synopsis only. 'No mindless comedies, please!'

Andrew Stevens Entertainment [MOVIE]

468 North Camden Drive, Suite 213, Beverly Hills, CA 90210
T 310 285 5358 F 310 285 5359
info@astevensent.com
www.astevensent.com

Founded in 2003, developing, producing and/or financing feature films. Also has an office in Dallas. Output includes *7 Seconds*, *Black Dawn*, *Fire from Below*, *Missionary Man*, *The Detonator*, *The Marksman* and *Walking Tall*.

Stonebridge Communications [MOVIE] [TV]

3027 Abell Avenue, Baltimore, MD 21218
T 410 467 3570 F 410 467 2281
jkahoe@stonebridgecommunicationsinc.com
www.stonebridgecommunicationsinc.com
President/Producer *Janet Kahoe*
Director/DP *Timothy Kahoe*

Founded in 2001, though owners previously ran Big Shot Productions from 1988–2001. Specializes in corporate programmes with a strong story-telling stance. Credits include *Sudan - Remembering, Responding, Rebuilding*, a fund-raising documentary for Darfur.

Approach by email. Most writing done in-house but does make use of freelance writers.

Storm Entertainment [MOVIE]

127 Broadway, Suite 222, Santa Monica, CA 90401
T 310 656 2500 F 310 656 2510
storment95@aol.com
www.stormentertainment.com

Founded in 1995, Storm Entertainment has produced films including *Hell's Kitchen* (starring Angelina Jolie), *Being Considered*, *The Criminal* and *The Shiner* (starring Michael Caine).

Storyline Entertainment [TV]

500 S. Buena Vista Old Animation 3D, Burbank, CA 91521-1840
T 818 560 2928 F 818 560 5145
info@storyline-entertainment.com
www.storyline-entertainment.com

Particular emphasis on producing content for television. Examples of output include *A Raisin in the Sun*, *Fahrenheit 451*, *Family Man*, *Hairspray*, *Lucy*, *The Music Man*, *The Reagans* and *Wedding Wars*.

Strand Releasing [MOVIE]

6140 Washington Boulevard, Culver City, CA 90232
T 310 836 7500 F 310 836 7510
strand@strandreleasing.com
www.strandreleasing.com

Although Strand Releasing started off as a film distribution company in 1989, it has ventured into film production since 2000. Produces art-house films such as *Split*, *Wo de mei li xiang chou* and *Psycho Beach Party*.

Streiffschuss Films [MOVIE] [TV]

124 East Broadway, New York, NY 10002
T 212 349 8747
tomi@streiffschuss.com
Director *Tomi Streif*

Founded in 1992, and the recipient of 20 festival awards. Produces for film and television across genres including drama, documentary, comedy, romance and thrillers. Output includes *The Wedding Cow*. Also has offices in Switzerland and Argentina.

Open to new writing. Send in synopsis via email.

Summit Entertainment [MOVIE]

1630 Stewart Street, Suite 120, Santa Monica, CA 90404
T 310 309 8400
www.summit-ent.com

Major motion picture distributor and producer. Credits include *The Hottie and the Nottie*, *The Alibi*, *Mr. & Mrs. Smith*, *Insomnia*, *Memento*, *American Pie*, *The Loss of Sexual Innocence*, *Lock, Stock and Two Smoking Barrels* and *Evita*.

Telling Pictures MOVIE TV

10 Arkansas Street, Suite F, San Francisco,
CA 94107
info@tellingpictures.com
www.tellingpictures.com

Founded in 1987 by filmmakers Rob Epstein
and Jeffrey Friedman. Produces non-fiction
feature films, and documentaries for television
and corporate clients. Credits include
*Celluloid Closet, Common Threads, Crime
and Punishment, Isaiah's Rap, Paragraph
175* and *Where Are We? (Our Trip Through
America).*

Thinkfilm MOVIE

72 Madison Avenue, 6th Floor, New York,
NY 10016
T 212 444 7900 F 212 444 7901
info@thinkfilmcompany.com
www.thinkfilmcompany.com

Founded in 2001 by four former Lionsgate
Films executives, producing films including
*Going to Pieces: The Rise and Fall of the
Slasher Film, When Stand Up Stood Out* and
Bloodlines.

Threshold Entertainment MOVIE TV

1649 11th Street, Santa Monica,
CA 90404
T 310 452 8899 F 310 452 0736
sales@thethreshold.com
www.thethreshold.com

Develops, manages, produces, and publishes
intellectual properties in all media including
feature films and television shows.

TIG Productions, Inc. MOVIE

100 Universal City Plaza, Universal City,
CA 91608
T 818 777 2737

The production company of Kevin Costner, which
debuted with the Oscar-winning *Dances with
Wolves* in 1990. Subsequent credits include *The
Bodyguard, Wyatt Earp, Thirteen Days, Open
Range, Mr. Brooks, The Explorers Club* and *Swing
Vote.*

Time River Productions MOVIE TV

Box 70615, Pasadena, CA 91117
T 626 304 0080
timerivr@aol.com
Founded in 1986, producing motion pictures and
television programmes.
  Develop screenplays in-house so does not
welcome any unsolicited manuscripts.

Timeline Films MOVIE

9725 Culver Boulevard, Culver City,
CA 90232-2739
T 310 287 3702 F 310 287 1370
info@timelinefilms.com

Particular emphasis on documentaries. Output
includes *Olive Thomas: The Most Beautiful Girl in
the World, Masters of Production: The Hidden Art
of Hollywood, The Woman with the Hungry Eyes*
and *Iraq: A Tale of Censorship.*

Tollin/Robbins Productions MOVIE TV

4130 Cahuenga Boulevard, Unit 305, Toluca Lake,
CA 91602
T 818 755 3000 F 818 766 8488
info@tollinrobbins.com
www.tollinrobbins.com

Founded in 1993. Television output includes
Smallville and *One Tree Hill.* Feature film
productions include *Wild Hogs* (starring Tim
Allen, John Travolta and Martin Lawrence)
and *Norbit* (starring Eddie Murphy). Also the
award-winning documentary *Hardwood Dreams,*
narrated by Wesley Snipes.

Touchstone Pictures MOVIE VG

500 S Buena Vista Street, Burbank, CA 91521
touchstone.movies.go.com

Established in 1984, a film division of the Walt
Disney Company. It releases features with
themes of a more mature nature than those that
get released under the Walt Disney Pictures
banner. Output includes *Pretty Woman,
Sister Act, Armageddon, Hidalgo* and the
Academy-nominated *Apocalypto.*

Trancas International MOVIE TV

1875 Century Park East, Suite 1145, Los Angeles,
CA 90067
T 310 553 5599 F 310 553 0536
info@trancasfilms.com
www.trancasfilms.com

Founded in 1977. The TV division produces
reality shows, one-hour dramas, sitcoms, movies
of the week, game shows and documentaries.
Film credits include *The Message, Lion of the
Desert* and the *Halloween* series. Also has an
office in London.

Treasure Entertainment, Inc. MOVIE

468 N. Camden Drive, Suite 200, Beverly Hills,
CA 90210
T 310 860 7490 F 310 943 1488
info@treasureentertainment.net
www.treasureentertainment.net

Co-Chairman/President *Jesse Felsot*
CFO *Ian Vishnevsky*
COO *Kevin Asbell*
Co-Chairman/CEO *Mark Heidelberger*

Founded in 2000, working across film, music videos and commercials. Credits include *Harsh Times*, *You've Got a Friend*, *Flintown Kids* and *Man Overboard*. Interested in films with budgets of $10 million and under.
New writers by referral only.

Tribeca Productions [MOVIE]

375 Greenwich Street, 8th Floor, New York, NY 10013
T 212 941 4000 F 212 941 4044

Motion picture production company co-founded in 1989 by Robert de Niro. Output includes *36*, *The Good Shepherd*, *Meet the Fockers*, *About a Boy*, *Meet the Parents*, *Analyze This*, *A Bronx Tale* and *Cape Fear*.

Trigger Street Productions [MOVIE]

www.triggerstreet.com

Founded in 1997 by Kevin Spacey. Credits include *Beyond the Sea*, *The United States of Leland* and *The Big Kahuna*. In 2002, TriggerStreet. com was established as a web-based filmmaker and screenwriter's community. Aims to assist the careers of emerging filmmakers and screenwriters.

Trilogy Entertainment Group [MOVIE] [TV]

325 Wilshire Boulevard, Suite 203, Santa Monica, CA 90401
T 310 656 9733
www.trilogyent.com

Founded by John Watson and Pen Densham, producing motion pictures, and movies, series and mini-series for TV. Film output includes *Tank Girl*, *Moll Flanders*, *The Dangerous Lives of Altar Boys*, *Backdraft* and *Pushing Up Daisies*. For TV it has produced *The Outer Limits*, *The Twilight Zone*, *Houdini*, *Poltergeist: The Legacy*, *Carrie* and *The Taking Of Pelham 1-2-3*.

TSG Productions [MOVIE] [TV]

718 S. 22nd Street, Philadelphia, PA 19146
T 218 546 1448
wd24p@hotmail.com
Producer *Robert M. Goodman*

Established in 1989, independent film and television production company producing drama, comedy and documentaries. Recent features include *Stone Reader*.
Welcomes unsolicited manuscripts, but prefers a query letter first. Pleased to receive new

writing: 'if it's good we don't care'. High budget or effects-based ideas are unsuitable.

Twilight Pictures & Production Services Inc. [MOVIE] [TV]

12400 Ventura Boulevard, Suite 735, Studio City, CA 91604
T 403 286 3248 F 403 286 3240
www.twilight-pictures.com

Develops, produces and co-produces across genres including dramatic series, feature films and movies of the week, children's series, documentaries, lifestyle programming, educational and corporate videos. Output includes *The Doctor Rode Side Saddle*, *All That Glitters - The BreX Story* and *Natural Born Killers* (a docu-drama). Also has an office in Calgary.

Type A Films [MOVIE]

100 Universal City Plaza, Building 1320, Rm. 2-E, Universal City, CA 91608
T 818 777 6222

The production company of Reese Witherspoon, with credits including *Legally Blonde 2: Red, White & Blonde*, *Penelope* and *Sammy*.

Ubisoft Entertainment [VG]

625 Third Street, San Francisco, CA 94107
T 415 547 4000 F 415 547 4001
www.ubisoftgroup.com

Has 15 in-house production studios across 11 countries. Distributes games in over 50 countries. Examples of output include *Blazing Angels*,*Brothers in Arms* and the *Chessmaster*, *The Settlers*, *Prince of Persia*, *Rayman* and *Tom Clancy's Ghost Recon* series.

Ugly Betty Productions, Inc. [MOVIE]

618 Carroll Street, 2nd Flr., Brooklyn, NY 11215
T 718 399 7544 F 718 399 7415
ubp@uglybetty.com
www.uglybetty.com

Has produced the short films *Thirsty*, *Face Value*, *Lucky* and *Ping Pong Love*.

Underground Films [MOVIE] [TV]

447 S. Highland Avenue, Los Angeles, CA 90036
T 323 930 2588 F 323 930 2334
submissions@undergroundfilms.net
www.undergroundfilms.net

Produces feature films and content for TV, working across genres. Production credits include *A Call for Help*, *Zoom*, *License to Wed*, *Class Act*, *Drive Sensitive* and *Venus Kincaid*.

In first instance, send an email including logline and brief synopsis. Will only respond to those emails it finds especially interesting and will then request a copy of the screenplay, which must be accompanied by a signed release form.

Union Square Entertainment (MOVIE)

9595 Wilshire Boulevard, Suite 900, Beverly Hills, CA 90212
Ⓣ 310 300 8410
info@unionsquareent.com
www.unionsquareent.com

Finances and produces feature films with budgets between $1–15 million.
Will not accept unsolicited material. Contact should be via an accredited representative.

United Artists Entertainment LLC (MOVIE)

10250 Constellation Boulevard, Los Angeles, CA 90067
Ⓣ 310 449 3000
www.unitedartists.com

Founded in 1919 by Mary Pickford, Charles Chaplin, Douglas Fairbanks and D. W. Griffith. A subsidiary of MGM Studios. In 2006, United Artists came under the control of Tom Cruise and Paula Wagner. It is responsible for the *Rocky*, *Pink Panther* and *James Bond* franchises and has produced a number of Academy Award-winning films such as *Rebecca*, *Rain Man* and *Capote*. Other recent titles include *Coffee and Cigarettes*, *Hotel Rwanda*, *Romance & Cigarettes* and *The Woods*.

Universal Pictures (MOVIE)

100 Universal City Plaza, Universal City, CA 91608
www.universalpictures.com

A subsidiary of NBC Universal and one of the major American film studios. The second longest-lived studio in Hollywood, producing films for over 80 years. Has won numerous awards for its films over the years. A taster of its recent credits includes *Babe*, *American Pie*, *Children of Men*, *The Bourne Supremacy*, *Van Helsing*, *The 40-Year-Old Virgin*, *The Interpreter*, *King Kong*, *Serenity*, *Miami Vice*, *United 93*, *The Bourne Ultimatum* and *American Gangster*.
Does not accept unsolicited material.

Vanguard Animation (MOVIE)

8703 W. Olympic Boulevard, Los Angeles, CA 90035
Ⓣ 310 360 8039 Ⓕ 310 360 8059
mail@vanguardfilms.com
www.vanguardanimation.com

Production and intellectual property development company specializing in computer generated animation feature films. Aims to serve a broad family audience. Examples of output include *Space Chimps* and *Happily Never After*.

Vanguard Documentaries (MOVIE)

PO Box 26635, Brooklyn, NY 11202-6624
Ⓣ 212 517 4333
charleshobson@yahoo.com
www.vanguarddocumentaries.com

Specializes in films based on arts, cultural affairs and politics. Co-produced the film *Harlem in Montmartre: Paris Jazz*.

View Askew Productions, Inc. (MOVIE)

116 Broad Street, Red Bank, NJ 07701
Ⓣ 732 842 6933 Ⓕ 732 842 3772
viewaskew@viewaskew.com
www.viewaskew.com

Founded in 1994 by Kevin Smith and Scott Mosier. Has produced cult films such as *Clerks*, *Chasing Amy*, *Dogma* and *Jersey Girl*.
Does not accept submissions of material.

Village Roadshow Pictures (MOVIE)

3400 Riverside Drive, Suite 900, Burbank, CA 91505
www.villageroadshowpictures.com

Village Roadshow Pictures is the film production division of Australia's Village Roadshow Limited (a leading media and entertainment company). It has been co-producing films with its principal partner Warner Bros since the late 1990s. Credits include *Charlie and the Chocolate Factory*, the *Matrix* triliogy, *Miss Congeniality* and the Oscar-winning *Training Day*.

Vulcan Productions (MOVIE)

www.vulcanproductions.com

Produces feature films with budgets under $1 million. Credits include *Hard Candy*, *Far From Heaven*, *The Safety of Objects* and *Where God Left His Shoes*. Has also produced documentaries such as *Cracking the Code of Life*, *Evolution* and *Strange Days on Planet Earth*.
Company policy not to accept unsolicited manuscripts. Screenplays, treatments, and/or pitches must be submitted through established entertainment agents and/or attorneys.

Walden Media (MOVIE)

1888 Century Park East, 14th Floor, Los Angeles, CA 90067
Ⓣ 310 887 1000 Ⓕ 310 887 1001

info@walden.com
www.walden.com

Producing films with a strong educational slant, often based on established literary works. Also has an office in Boston. Production credits include the *Chronicles of Narnia* series, *Journey to the Center of the Earth 3D*, *Mr. Magorium's Wonder Emporium*, *Bridge to Terabithia*, *Charlotte's Web* and *Around the World in 80 Days*.

Walt Disney Animation Studios MOVIE

500 South Buena Vista Street, Burbank, CA 91521
T 818 560 1000
www.disney.go.com

Founded in 1934. Has produced Academy Award-winning features such as *Beauty and the Beast*, *The Little Mermaid* and *The Lion King*.

Walt Disney Pictures MOVIE TV VG

500 South Buena Vista Street, Burbank, CA 91521
T 818 560 1000
www.disney.go.com

Established in 1983 as part of Walt Disney Productions, it has produced several award-winning films. Output includes *The Pirates of the Caribbean* trilogy, *Meet the Robinsons*, *Aladdin* and *Finding Nemo*. Films produced by the company are aimed at children. Also produces television shows such as *That's So Raven* and *The Replacements*.

Warner Bros Entertainment, Inc. MOVIE TV VG

4000 Warner Boulevard, Bldg 5, Burbank, CA 91522
T 818 954 6000
www2.warnerbros.com

Warner Bros (founded in 1923), a subsidiary of Time Warner, is one of the world's largest film and television producers. Has won numerous awards over the years and is responsible for the *Harry Potter*, *Lethal Weapon* and *Batman* franchises. Television output includes *Smallville* and *ER*.

Washington Square Films MOVIE TV

310 Bowery, 2nd Floor, New York, NY 10012
T 212 253 0333 F 212 253 0330
production@wsfilms.com
www.wsfilms.com

Has produced films including *Sweet Flame*, *Adrift in Manhattan* and *Old Joy*, plus the television programme *Words in Your Face*.
Does not accept unsolicited scripts.

Watershed Films MOVIE

833 Moraga Drive, Suite 1, Los Angeles, CA 90049
T 310 472 8750 F 310 472 2526
info@watershedfilms.com
www.watershedfilms.com

Founded in 1999. Has produced the films *The Dinosaur Hunter*, *Plato's Run*, *Wind In The Wire* and *Tall Tales*.

Jim Wedaa Productions MOVIE TV

3631 Laurel Canyon Boulevard, Studio City, CA 91604
T 818 508 4630
info@jimwedaaproductions.com
www.jimwedaaproductions.com

A 'writer-centric production company' focusing on developing and producing for cinema and TV. Credits include *Big Trouble*, *Black Dog*, *Mission to Mars*, *Nine Lives* and *Red Team*.
Will not accept unsolicited material. Send a query letter and short synopsis in the first instance.

The Weinstein Company MOVIE

375 Greenwich Street, Tribeca Film Center, New York, NY 10013
T 212 941 3800
www.weinsteinco.com

Founded in 2005 and includes the Dimension Films label. Has produced award-winning films including *TransAmerica* and *Mrs Henderson Presents*.

Jerry Weintraub Productions MOVIE

c/o Warner Bros, 4000 Warner Boulevard, Burbank, CA 91522
T 818 954 2500 F 818 954 1399

Based at Warner Bros Studios. Has produced movies including *The Specialist*, *The Avengers* and *Ocean's 11* (as well as *Ocean's 12* and *13*).

Weintraub/Kuhn Productions MOVIE

1452 2nd Street, Santa Monica, CA 90401
T 310 458 3300

Has Weintraub/Kuhn productions have produced films including *Amazons and Gladiators*, *Endangered Species* and *Warrior Angels*.

Weller Grossman TV

5200 Lankershim Boulevard, 5th Floor, North Hollywood, CA 91601
T 818 755 4800
contact@wellergrossman.com
www.wellergrossman.com

Has produced over 3,500 shows across genres including reality, documentaries, informational and entertainment.

While encouraging new ideas, unsolicited materials (including treatments/ideas) will not be accepted. Approaches should be made directly to the office via a representative.

WhiteLight Entertainment `MOVIE`

5200 Lankershim Boulevard, Suite 350,
North Hollywood, CA 91601
T 818 655 9747 F 818 763 8121
www.whitelight-entertainment.com

Aims to produce 'content responsible films that entertain, enlighten and educate'.

Winkler Films `MOVIE`

211 S Beverly Drive, Suite 200, Beverly Hills,
CA 90212
T 310 858 5780 F 310 858 5799

Has produced films including *The Net*, *Home of the Brave* and the Golden Globe-nominated *De-Lovely*.

World of Wonder `MOVIE`

6650 Hollywood Boulevard, Suite 400,
Hollywood, CA 90028
T 323 603 6300
wow@worldofwonder.net
www.worldofwonder.net

Founded in 1991, specializing in documentary films, often based on the notion that 'today's marginal is tomorrow's mainstream'. Also has a London office. Output includes *A Transamerican Love Story*, *Party Monster* and *The Eyes Of Tammy Faye*.

Worldwide Pants Inc. `TV`

1697 Broadway, Suite 805, New York, NY 10019
T 212 975 5300 F 212 975 4780

Founded in 1993 by talk show host David Letterman. Output includes *The Late Late Show*, *The Knights of Prosperity* and *Everybody Loves Raymond*.

Yari Film Group `MOVIE`

10850 Wilshire Blvd, 6th Floor, Los Angeles,
CA 90024
T 310 689 1450 F 310 234 8975
online@yarifilmgroup.com
www.yarifilmgroup.com

Finances, produces and distributes feature films, working with first time writers and directors to experienced industry professionals. Production credits include *The Matador*, *The Hoax*,

Resurrecting the Champ, *Assassination of a High School President* and *Possession*.

The Saul Zaentz Company `MOVIE`

2600 Tenth Street, Berkeley, CA 94710
T 510 549 1528
info@zaentz.com
www.zaentz.com

Founded in 1972, producing award-winning films such as *The Unbearable Lightness of Being*, *The English Patient* and *One Flew Over The Cuckoo's Nest*.

Does not accept unsolicited material.

Zero Pictures `MOVIE`

171 Pier Avenue, Suite 317, Santa Monica,
CA 90405
T 310 450 9040
participate@zeropictures.com
www.zeropictures.com

Seeks to offer opportunities to aspiring screenwriters and filmmakers who have the flexibility to work on low budgets.

UK

20th Century Fox Film Co. `MOVIE`

Twentieth Century House, 31–32 Soho Square,
London, W1D 3AP
T 020 7437 7766 F 020 7734 3187
www.fox.co.uk
Contact *Lucy Lumsden*

London office of the American giant.
Does not accept unsolicited material.

Aardman `MOVIE` `TV`

Gas Ferry Road, Bristol, BS1 6UN
T 0117 984 8485 F 0117 984 8486
www.aardman.com

Founded in 1976, Aardman produce multi-award-winning animated feature films and TV programmes. Has won Oscars for *Creature Comforts* and the adventures of *Wallace & Gromit*. Other titles include *Chicken Run* and *Flushed Away*.

Has a policy of not reading unsolicited scripts or story ideas unless they are by established writers and come through an agent. Established writers without an agent should send a CV and a one paragraph description of their idea.

Absolutely Productions Ltd `MOVIE` `TV`

Unit 19, 77 Beak Street, London, W1F 9DB
T 020 7644 5575
www.absolutely.biz

Formed in 1988 by a group of writer/performer/producers including Morwenna Banks, Jack Docherty, Moray Hunter, Pete Baikie, John Sparkes and Gordon Kennedy. Long-established producer for TV, has recently moved into feature films too. First major show was *Absolutely*. More recent productions include *Welcome To Strathmuir* (a new comedy pilot for BBC Scotland) and *Baggage* (BBC Radio 4).

Abstract Images   (TV)

117 Willoughby House, Barbican, London, EC2Y 8BL
☎ 020 7638 5123
productions@abstract-images.co.uk
Managing Director *Howard Ross*

Television production company founded in 1990, producing drama and documentary. Titles include *Bent* and *This is a Man*.

Acacia Productions Ltd   (TV)

80 Weston Park, London, N8 9TB
☎ 020 8341 9392 ℻ 020 8341 4879
em@acaciaproductions.co.uk
www.acaciaproductions.co.uk
MD, Writer and Director *John Edward Douglas Milner*

Founded in 1985, specializing in 'creative documentaries' and news programmes for television.
   Special focus on contemporary and environmental issues and historical documentaries. Titles include *Vietnam After the Fire* and *Spirit of Trees*. Company has won the special jury prize at the Banff International TV and Film Festival and the prize for best film made for TV at the New York International Documentary Festival.
   Only interested in submissions within the company's remit. Does not do drama.

Acrobat Television   (TV)

107 Wellington Road North, Stockport, SK4 2LP
☎ 0161 477 9090 ℻ 0161 477 9191
david.hill@acrobat-tv.co.uk
www.acrobat-tv.co.uk
Contact *David Hill*

All script genres for broadcast and corporate television, including training and promotional scripts, comedy and drama-based material. Output includes *Make a Stand* (Jack Dee, Gina Bellman and John Thompson for Video Arts); *The Customer View* (Roy Barraclough for Air Products); *Serious About Waves* series (Peter Hart for the Royal Yachting Association); *Fat Face Night* series (Extreme).
   No unsolicited manuscripts.

Actaeon Films Ltd   (MOVIE) (TV)

50 Gracefield Gardens, London, SW16 2ST
☎ 020 8769 3339 ℻ 0870 134 7980
info@actaeonfilms.com
www.actaeonfilms.com
Managing Director and Producer *Daniel Cormack*

Production company founded in 2004 by Daniel Cormack to develop and produce feature length theatrical motion pictures. Focus is on films that use elements of established commercial genres in innovative and original ways that challenge and entertain a broad range of audiences. Recent work includes the Tiscali Short Film Award-winning drama *Amelia and Michael* (2007), starring Anthony Head, and the UK Film Council-funded comedy drama *A Fitting Tribute* (2007).
   Runs an open submissions policy and encourages unsolicited screenplays from new writers. Considers original feature length screenplays of any genre and screenplays for single drama or drama series for TV. All screenplays will be read by a professional reader but cannot provide feedback unless there is a strong potential for development and only screenplays submitted with an sae will be returned.

Addictive TV   (TV)

The Old House, 39a North Road, London, N7 9DP
☎ 020 7700 0333
mail@addictive.com
www.addictive.com
Contact *John Sexton*

Launched in 1992. Devises and produces television programming. Focused on music, arts and technology. Recent credits include *Mixmasters* (ITV1), *The Web Review* (ITV1) and *Spaced Out* (Sci-Fi Channel).

After Image   (MOVIE) (TV)

47 Trinity Gardens, London, SW9 8DR
☎ 020 7737 7300
www.afterimage.co.uk

Created by Jane Thorburn and Mark Lucas in 1979. Produces arts-orientated projects for television and video. Best known for art series *After Image* (Channel 4). Now tours *After Image* seasons to arts institutions in the UK as well as in the USA, Italy and France.

All3Media   (MOVIE) (TV)

Berkshire House, 168–173 High Holborn, London, WC1V 7AA
☎ 020 7845 4377 ℻ 020 7845 4399
information@all3media.com
www.all3media.com

Non-Executive Chairman *Sir Robert Phillis*
Chief Executive *Steve Morrison*
Chief Operating Officer *Jules Burns*
Creative Director *David Liddiment*
Finance Director *John Pfeil*

Production and distribution company led by Steve Morrison, David Liddiment, Jules Burns and John Pfeil. Formed following the acquisition of Chrysalis Group's TV division in September 2003. Comprised of a group of production companies from across the UK, The Netherlands, New Zealand and the USA. Members include Company Pictures, Lion Television and South Pacific Pictures. Also includes international distribution subsection to represents third-party producers and broadcasters together with its own production companies.

Amber Films [MOVIE]

5–9 Side, Newcastle-upon-Tyne, NE1 3JE
☎ 0191 232 2000
www.amber-online.com

Film-making arm of a film and photography collective active since 1968. Work is rooted in social documentary and the organization is not-for-profit. Recent output includes *The Bamboozler*, *Shooting Magpies* and *We Did It Together–So Why Do I Feel So Alone?*

Not open to submissions. Work produced from within the collective.

Ambient Light Productions Ltd [MOVIE] [TV]

6 Shipquay Street, Derry, BT48 6DN
☎ 028 7136 3525
info@ambient-light.co.uk
www.ambient-light.co.uk
Managing Director *Tony Doherty*
Director *Eoin Coffey*

Production company founded in 2005 and working in TV, film and corporate. Titles include the documentary *Agnes* and the short film *Peace Cubed*.

Welcomes scripts whether from Bafta-nominated writers or students fresh out of college. Approach by email in the first instance.

AMC Pictures Ltd [MOVIE] [TV]

Little Hemmingford Beaconsfield Road, Farnham Common, Bucks, SL2 3LZ
☎ 01753 646 280 ☏ 01753 648 222
films@amcpictures.co.uk
www.amcpictures.co.uk
Managing Director *Alistair Maclean-Clark*

Production company founded in April 2005. Specializes in film and television drama.

Unsolicited material accepted in writing with a synopsis.

Anglia Factual [TV]

The London Television Centre, Upper Ground, London, SE1 9LT
☎ 020 7620 1620
sue.potter@granadamedia.com
www.angliafactual.com
Contact *Matt Carter*

Semi-independent subsidary of ITV division Granada Media. Established in 2000. Has produced over 1,000 hours of popular documentary for the BBC, ITV, Channel 4, Five, British Sky Broadcasting and Discovery Channel International. Recent output includes *Real Families: My Skin Could Kill Me* (ITV1), *Seeing The Dead* (ITV1) and upcoming *Property Developing Abroad* (Channel 5). Now fast-developing in North America where *Animal Precinct* is one of Discovery's top rated shows.

Anglo-Fortunato Films Ltd [MOVIE] [TV]

170 Popes Lane, London, W5 4NJ
☎ 020 8932 7676 ☏ 020 8932 7491
anglofortunato@aol.com
Contact *Luciano Celentino*

Founded in 1972 for the production of film and television projects. Specializes in action-comedies and psychological thrillers, most notably *Hobo Gallan the Pinch*.

Although keen on encouraging new writing, material is only accepted via agents.

Apocalypso Pictures [MOVIE]

Bretton Lincombe Lane, Boars Hill, Oxfordshire, OX1 5DY
☎ 01865 735332

Company founded by producer Tania Seghatchian and director Pawel Pawlikowski. Best known for BAFTA-winning *My Summer of Love*.

Arcane Pictures [MOVIE]

60–62 Great Titchfield Street, London, W1W 7QG
☎ 020 7636 4996 ☏ 020 7323 0661
info@arcanepictures.com
www.arcanepictures.com

Co-founded by George Duffield and Meg Thomson in 1998. Strives to produce "intelligent, sexy, innovative and commercial films". Interested in documentary, feature and short film projects. Best known for *dot the i* starring Gael Garcia Bernal and *MILK*, an eccentric comedy set on a dairy farm starring Dawn French and Francesca Annis.

Archer Street   MOVIE

Studio 5, 10–11 Archer Street, London, W1D 7A2

T 020 7439 0540 F 020 7437 1182

films@archerstreet.com

Producer of feature films including *And When Did You Last See Your Father*, *Beyond the Sea* and *Girl With a Pearl Earring*.

Arlington Productions Limited   TV

Cippenham Court, Cippenham Lane, Cippenham, Nr. Slough, Berkshire, SL1 5AU

T 01753 516767 F 01753 691785

Produces popular international drama with occasional forays into other areas, for television and the home video market.

Prides itself on its reputation for encouraging new writers but will only look at new material via established agents.

Art & Training Films Ltd   MOVIE TV

PO Box 3459, Stratford upon Avon, CV37 6ZJ

T 01789 294910

andrew.haynes@atf.org.uk

www.atf.org.uk

Contact *Andrew Haynes*

Producer of documentaries, drama, commercials and corporate films and video.

Submissions via agents considered.

The Ashford Entertainment Corporation Ltd   TV

20 The Chase, Coulsdon, Surrey, CR5 2EG

T 020 8668 8188 F 087 0116 4142

info@ashford-entertainment.co.uk

www.ashford-entertainment.co.uk

Managing Director *Frazer Ashford*

Founded in 1996 by award-winning producer Frazer Ashford to produce character-led documentaries for TV. Examples of work include *Serial Killers*, *Great Little Trains* (winner of RTS Best Regional Documentary) and *Streetlife* (winner of bronze award at the Flagstaff Film Festival).

Happy to look at any submissions falling within the remit of character-led documentaries. Send hard copy with an sae.

AV Pictures   MOVIE

Caparo House, 103 Baker Street, London, W1U 6LN

T 020 7317 0140 F 020 7224 5149

info@avpictures.co.uk

www.avpictures.co.uk

Established in 2003 when Victor Film Company was acquired by The Caparo Group. Company's first production was *School for Seduction*, a comedy drama starring Kelly Brook, Emily Woof and Dervla Kirwan.

Avalon   TV

4a Exmoor Street, London, W10 6BD

T 020 7598 7280

mikea@avalonuk.com

www.avalonuk.com/tv

BAFTA-winning television company formed in 1993. Supplies TV programming to all the British terrestrial channels and most leading satellite channels. Best known for light entertainment programmes such as documentary *Frank Skinner on Frank Skinner* (ITV1) or comedy *Fantasy Football* (BBC2). Now broadening output to include drama, factual and formatted shows.

Script, treatment or ideas are welcomed via email.

Baby Cow Productions   MOVIE TV

77 Oxford Street, London, W1D 2ES

T 020 7399 1267 F 020 7399 1262

script@babycow.co.uk

www.babycow.co.uk

Established in 1989 by Steve Coogan and Henry Normal. Has produced for the BBC, ITV and UK digital channels, specializing in cutting edge comedy. Has divisions for animation, radio and films. Productions include *Marion and Geoff*, *Human Remains*, *The Sketch Show*, *24 Hour Party People*, *I'm Alan Partridge*, *Nighty Night*, *The Mighty Boosh*, *I Am Not an Animal* and *Dating Ray Fenwick*.

Dedicated development department always on the lookout for new talent. In first instance, send a synopsis or treatment and a ten page extract, plus a DVD or video if the idea is filmed, and a covering letter. If the idea is of interest, Baby Cow will contact the writer. Include all contact details, including email.

Baker Street Media Finance   MOVIE

96 Baker Street, London, W1U 6TJ

T 020 7487 3677 F 020 7487 5667

enquiries@bakerstreetfinance.tv

www.bakerstreetfinance.tv

Specializes in the co-production, financing and structuring of British feature films, and British-qualifying international co-productions.

When submitting a project, Baker Street will wish to see the latest draft of a script, sales estimates, a projected finance plan, a draft budget and schedule (even if these are preliminary documents) along with a list of any creative attachments, and any available further background material. Unless otherwise requested all initial enquiries regarding script submission should be directed to Anthony Walters or Lucy Main. Unsolicited scripts will not be accepted without sales estimates and finance plan attached. Scripts and accompanying material cannot be returned.

BBC Vision, BBC Audio & Music [TV]

BBC Television Centre, Wood Lane, London, W12 7RJ
T 020 8743 8000 F 020 8749 7520
www.bbc.co.uk
Director-General *Mark Thompson*
Deputy Director-General *Mark Byford*
Director, BBC Vision *Jana Bennett*
Controller, BBC One *Jay Hunt*
Controller, BBC Two *Roly Keating*
Controller, Daytime *Liam Keelan*
Controller, BBC Three *Danny Cohen*
Controller, BBC Four *Janice Hadlow*

The operational structure of the BBC consists of the following groups: BBC Vision Group (which runs three main areas: production, commissioning and services and includes the terrestrial and digital television channels; Drama, Entertainment and Children's; Factual and Learning.); The Audio & Music Group (responsible for TV Music Entertainment; In-house Factual, Specialist Factual and Drama Audio production); and the Journalism Group covers News; Sport; Global News, Nations and Regions. TV and radio commissioning guidance available at bbc.co.uk/commissioning

BBC Writersroom [MOVIE] [TV]

Development Manager, BBC Writersroom, 1st Floor, Grafton House, 379–381 Euston Road, London, NW1 3AU
www.bbc.co.uk/writersroom
Head of Drama Commissioning *Ben Stephenson*
Controller, Fiction *Jane Tranter*
Head of Drama Commissioning and Drama Wales *Julie Gardner*
Continuing Comedy Commissioning *Lucy Lumsden*
Drama Controller *John Yorke*
Head of Series and Serials, BBC Drama Production *Kate Harwood*
Director, BBC Drama Productions *Nicolas Brown*
Executive Producer, East-Enders *Diederick Santer*
Creative Director, New Writing *Kate Rowland*
Head of Comedy *Mark Freeland*
Controller, BBC Children's *Richard Deverell*
Creative Director, CBBC *Anne Gilchrist*
Creative Director, CBeebies *Michael Carrington*
Head of CBBC Drama *Jon East*

Head of CBBC Entertainment *Joe Godwin*
Head of News, Factual & Learning *Reem Nouss*
Head of CBeebies Production & Acquisiti *Kay Benbow*

Champions new writers across all BBC platforms for Drama, Comedy and Children's programmes, running targeted schemes and workshops linked directly to production. It accepts and assesses unsolicited scripts for all departments: film, TV drama, radio drama, TV narrative comedy and radio narrative comedy. The website offers a diary of events, opportunities, competitions, interviews with established writers, submission guidelines and free formatting software. BBC writersroom also has a Manchester base which focuses on new writing in the north of England.

Writers should send hard copies of original, completed scripts to the address above. Before sending in scripts, log on to the website or send an A5 sae. to writersroom at the address above for the latest guidelines on submitting unsolicited work. See also bbc.co.uk/commissioning/structure/public.shtml

BBC New Talent [TV]

www.bbc.co.uk/newtalent

The BBC's search for new talent covers a constantly changing range of outlets that has included radio producers, presenters, young storytellers, filmmakers and comedy writers. Access the website for latest information.

BBC Films [MOVIE]

Grafton House, 379 Euston Road, London, NW1 3AU
T 020 7765 0251 F 020 7765 0278
www.bbc.co.uk/bbcfilms

The feature film-making arm of the BBC, co-producing around eight titles per year. Committed to finding and developing new talent, as well as working with established names. Also produces television dramas. Credits include *Notes on a Scandal*, *The History Boys*, *Mrs Brown*, *Billy Elliot*, *Iris*, *Dirty Pretty Things*, *The Life And Death Of Peter Sellers*, *Becoming Jane*, *Miss Potter*, *Starter for Ten*, *The Other Boleyn Girl*, *The Duchess*, *Brideshead Revisited* and *Revolutionary Road*.

Will not consider unsolicited scripts (unsolicited scripts should be sent to the BBC Writers Room). Scripts sent via production companies and agents should be addressed to the Development Assistant, 3rd Floor, Grafton House, 379 Euston Road, London NW1 3AU. Owing to volume of submissions, cannot guarantee full feedback on projects.

Big Heart Media   (MOVIE) (TV)

Flat 4, 6 Pear Tree Court, London, EC1R 0DW
(T) 020 7608 0352 (F) 020 7250 1138
info@bigheartmedia.com
www.bigheartmedia.com
Contact *Beth Newell*
Contact *Colin Izod*

Producer of drama and documetaries for
television and video.
    Ideas/outlines welcome by email, but not
unsolicted manuscripts. Keen to enourage new
writing.

Blackwatch Productions Limited   (MOVIE) (TV)

2/1 104 Marlboro Avenue, Glasgow, G11 7LE
(T) 0141 339 9996
info@blackwatchtv.com
Company Director *Nicola Black*

Film, television and video producer of drama
and documentary programmes. Output includes
The Paranormal Peter Sellers, *Snorting Coke
with the BBC*, *When Freddie Mercury Met Kenny
Everett*, *Designer Vaginas*, *Bonebreakers*, *Luv
Bytes* and *Can We Carry On, Girls?* for Ch4. Also
coordinates the *Mesh* animation scheme.
    Does not welcome unsolicited manuscripts.

Bona Broadcasting Limited   (MOVIE) (TV)

2nd Floor, 9 Gayfield Square, Edinburgh,
EH1 3NT
(T) 0131 558 1696 (F) 0131 558 1694
enquiries@bonabroadcasting.com
www.bonabroadcasting.com
Contact *Turan Ali*

Producer of award-winning drama and
documentary programmes for BBC Radio, TV
and film projects.
    No unsolicited manuscripts but send a
one-paragraph summary by email in the first
instance. Runs radio and TV drama training
courses in the UK and internationally.

Box   (MOVIE) (TV)

151 Wardour Street, London, W1F 8WE
(T) 020 7297 8040 (F) 020 7297 8041
info@box-tv.co.uk
www.box-tv.co.uk

Founded in 2000 by Gub Neal, formerly Head of
Drama at Channel 4 and Controller of Drama at
Granada Television. Produces film and television
for markets throughout the world. Credits
include *The Last Enemy*, *The Wind in the Willows*,
Sweeney Todd, *No Direction Home-Bob Dylan*,
Case of Evil and *Dice*.
    Unable to accept unsolicited scripts or other
material.

BreakThru Films Ltd   (MOVIE)

3rd Floor, 25 Newman Street, London, W1T 1PN
(T) 020 7580 3688 (F) 020 7580 4445
mail@breakthrufilms.co.uk
www.breakthrufilms.co.uk
Managing Director *Hugh Welchman*

Production company working across formats,
from animation to live action. Credits include the
Oscar-winning *Peter & the Wolf*, *Free Jimmy*, *The
Clap*, *Accident* and *The Most Beautiful Man in the
World*.

The Bureau   (MOVIE)

mail@thebureau.co.uk
www.thebureau.co.uk

Aims to produce films in the crossover arthouse
film sector and to identify and nurture distinctive
talent.

Cactus TV   (TV)

373 Kennington Road, London, SE11 4PS
(T) 020 7091 4900 (F) 020 7091 4901
touch.us@cactustv.co.uk
www.cactustv.co.uk

Founded in 1994 specializing in broad-based
entertainment, features and chat shows. Produces
around 300 hours of broadcast material per year.
Examples of programmes include *Richard & Judy*,
Sir Cliff Richard - The Hits I Missed, *Songs of Bond*
and *The British Soap Awards*.

Calon TV   (MOVIE) (TV)

3 Mount Stuart Square, Butetown, Cardiff,
CF10 5EE
(T) 029 2048 8400 (F) 029 2048 5962
enquiries@calon.tv
www.calon.tv
Head of Development *Andrew Offiler*

Founded in 2005 to produce animation and
live action TV programmes and films for
children, teens and families. Titles include the
Bafta-winning *SuperTed*, an animated version of
Under Milk Wood and *Hilltop Hospital*.
    Happy to view the work of new writers. Send
an email or letter first.

Capitol Films   (MOVIE)

Bridge House, 2nd Floor, 63–65 North Wharf
Road, London, W2 1LA
(T) 020 7298 6200 (F) 020 7298 6201
films@capitolfilms.com
www.capitolfilms.com

Founded in the early 1990s and has since handled
worldwide sales and distribution of over 100
films, the majority of which it has financed or

co-financed. Notable successes include *Gosford Park*, *Jeepers Creepers* and *Lucky Number Slevin*.

Carnival Film & Television   MOVIE TV

47 Marylebone Lane, London, W1U 2NT
T 020 7317 1370 F 020 7317 1380
info@carnivalfilms.co.uk
www.carnivalfilms.co.uk
Head of Development *Clova McCallum*
Managing Director *Gareth Neame*

Founded in 1978, specializing in television drama and feature film. Titles include *Hotel Babylon* (BBC), *Sea of Souls* (BBC), *Empathy* (BBC), *Lifeline* (BBC) and *Whistleblower* (ITV).
    Does not encourage unsolicited material.

Cartwn Cymru   MOVIE TV

12 Queen's Road, Mumbles, Swansea, SA3 4AN
T 07771 640400
production@cartwn-cymru.com
Producer *Naomi Jones*

Animation production company. Output includes: *Toucan 'Tecs* (YTV/S4C); *Funnybones* and *Turandot: Operavox* (both for S4C/BBC); *Testament: The Bible in Animation* (BBC2/S4C); *The Miracle Maker* (S4C/BBC/British Screen/ Icon Entertainment International); *Faeries* (HIT Entertainment plc for CITV); *Otherworld* (animated feature film for S4C Films, British Screen, Arts Council of Wales).

Celador Films   MOVIE

39 Long Acre, London, WC2E 9LG
T 020 7845 6800 F 020 7836 6980
www.celadorfilms.com
Head of Development *Ivana McKinnon*

Production company behind titles such as *Dirty Pretty Things*, *Separate Lies* and *The Descent*. Aims to develop high quality, commercially viable feature films across all genres.
    Will commission projects at all stages of development from verbal pitch to developed screenplay. Submissions should be marked for the attention of Ivana MacKinnon. However, will only accept submissions from established literary and film agents or talent already known to the company or from third party producers seeking possible co-production.

Celador Productions   TV

39 Long Acre, London, WC2E 9LG
T 020 7240 8101 F 020 7845 6976
info@celador.co.uk
www.celadorproductions.com
Head of Entertainment *Ruth Wrigley*

Produces range of TV programmes, from situation comedy to quizzes, game shows, popular factual entertainment and international co-productions. Range includes *24 Carrott Gold*, *All About Me*, *My Blue Heaven*, *Popcorn*, *Schofield's TV Gold*, *The Detectives*, *The Hypnotic World of Paul McKenna*, and *You Are What You Eat*.
    Does not accept unsolicited ideas for comedy or entertainment television formats.

Celtic Films Entertainment Ltd   MOVIE TV

Lodge House, 69 Beaufort Street, London, SW3 5AH
T 020 7351 0909 F 020 7351 4139
info@celticfilms.co.uk
www.celticfilms.co.uk
Managing Director *Stuart Sutherland*
Managing Director *Muir Sutherland*
Development *Steven Russell*

Founded in 1986 producing high-end TV drama and mainstream feature films. Titles include fifteen *Sharpe* TV films, *Girl from Rio* and *Stefan Kiszko: A Life for a Life*.
    Does not invite unsolicited material.

Chameleon Television Ltd   MOVIE TV

Greatminster House, Lister Hill Horsforth, Leeds, LS18 5DL
T 0113 205 0040 F 0113 281 9454
allen@chameleontv.com
Contact *Allen Jewhurst*
Contact *Julia Kirby-Smith*

Film and television drama and documentary producer. Output includes *Edge of the City*; *The Family Who Vanished*; *Killing for Honour*; *College Girls*; *Ken Dodd in the Dock* (all for Ch4); *Diary of a Mother on the Edge*; *Divorces From Hell*; *Shipman*; *Love to Shop*; *The Marchioness* (all for ITV); *Ted & Sylvia – Love, Loss*; *Hamas Bombers* (BBC); *Liverpool Poets* (Ch5).
    Scripts not welcome unless via agents but new writing is encouraged.

Channel 4 Television   TV

124 Horseferry Road, London, SW1P 2TX
T 0845 076 0191
www.channel4.com

Transmits across all of the UK, except for some parts of Wales served by S4C. Public service remit charges the channel with 'the provision of a broad range of high quality and diverse programming'. Does not produce its own programming.

Channel X   TV

3rd Floor, 4 Mile Street, London, N1 7RF
T 020 7566 8160 F 020 7566 8161

info@channelx.co.uk
www.channelx.co.uk

Over 20 years experience of producing comedy and entertainment programmes for all the major UK broadcasters. Specializes in scripted broken and narrative comedy, comedy entertainment formats and gameshows. Examples of recent shows include *Blunder*, *Modern Toss*, *Lucas and Walliams – The Early Years*, *Peter Kay – The Early Years*, *California Dreaming* and *Catterick*.

The Children's Film & Television Foundation Ltd   (TV)

c/o Simon George, Head of Finance, Ealing Studios, Ealing Green, London, W5 5EP
☏ 07887 573479
info@cftf.org
www.cftf.org.uk
Contact *Anna Home*

Involved in the development and co-production of films for children and the family, both for the theatrical market and for TV.

Children's Film Unit   (MOVIE) (TV)

South Way, Leavesden, Herts, WD2 7LZ
☏ 01923 354656
cfilmunit@aol.com
www.childrensfilmunit.com

An educational charity training ten- to sixteen-year-olds in all aspects of film-making. Has also produced a series of well-received feature films for television. Now works exclusively in digital video.

Cleveland Productions   (MOVIE) (TV)

5 Rainbow Court, Oxhey, Watford, Hertfordshire, WD19 4RP
☏ 01923 254000 (F) 01923 254000
michael@gosling.com
Contact *Michael Gosling*

Documentary production company founded in 1977.

Coastal Productions Limited   (TV)

25b Broad Chare Quayside, Newcastle-upon-Tyne, NE1 3DQ
☏ 0191 222 3160 (F) 0191 222 3169
coastalproductions@msn.com
www.coastalproductions.co.uk
Executive Producer *Sandra Jobling*
Contact *Ken Jobling*

Founded in 1996, specializing in television drama. Credits include *Wire in the Blood* (series 1–4), *Rocketman*, *Take Me* and *Grafters*. Company has won 3 RTS awards, a Gold award from the New York Television Festival and was nominated for an Edgar Award.

Always looking for new and innovative writing but accepts approaches by agents only.

Codemasters   (VG)

PO Box 6, Leamington Spa, Warwickshire, CV47 2ZT
☏ 01926 814132 (F) 01926 817595
business.development@codemasters.com
www.codemasters.co.uk

Develops and publishes video games for many major platforms. Output includes *Colin McRae Rally*, *Brian Lara International Cricket*, *Clive Barker's Jericho*, *Maelstrom*, *TOCA Race Driver* and *Operation Flashpoint*. Also maintains operations in the US and throughout Western Europe.

COI   (MOVIE) (TV)

Hercules Road, London, SE1 7DU
☏ 020 7928 2345
www.coi.gov.uk

Government advertising and marketing communications, and public information films.

Collingwood O'Hare Entertainment Ltd   (TV)

10–14 Crown Street, London, W3 8SB
☏ 020 8993 3666 (F) 020 8993 9595
info@crownstreet.co.uk
www.collingwoodohare.com
Head of Development *Helen Stroud*

Founded in 1988, producing award-winning animated series and specials for UK and overseas broadcasters. Examples of recent titles include *The Secret Show* (BBC and Nicktoons USA), *Gordon the Garden Gnome* (BBC), *Yoko! Jakmoko! Toto!* (ITV), *Dennis and Gnasher* (BBC) and *Harry and his Bucketful of Dinosaurs* (C5, NIcktoons).

Does not welcome unsolicited material. Likes to encourage new writing but ability to do so is limited by capacity for development.

Columbia Tristar Motion Picture Group   (MOVIE)

25 Golden Square, London, W1F 9LU
☏ 020 7533 1111
www.sonypictures.co.uk

Owned by Sony Pictures Entertainment. Releases about 25 films per year. Encompasses one production studio (Columbia Pictures) and three film labels (Sony Pictures Classics, Screen Gems and TriStar Pictures). Columbia Pictures develops, produces and distributes main-stream feature films. Founded in 1924. Recent credits include *Memoirs of a Geisha*,

The Da Vinci Code, Casino Royale and *Spiderman 3*. Sony Pictures Classics produces, acquires, finances and distributes independent and art-house films. Screen Gems exists to cover ground between the latter two markets. TriStar Pictures was re-launched in 2004 as a marketing and acquisitions unit with a particular emphasis on genre films. Originally founded in 1982. Output includes *Oliver Twist* and *Running with Scissors*.

The Comedy Unit   (MOVIE) (TV)

6th Floor, 53 Bothwell Street, Glasgow, G2 6TS
℡ 0141 220 6400
scripts@comedyunit.co.uk
www.comedyunit.co.uk
Managing Director *April Chamberlain*
Managing Director *Colin Gilbert*

Producers of comedy entertainment for children's TV, radio, video and film. Output includes: *Still Game*; *Karen Dunbar Show*; *Offside*; *Chewin' the Fat*; *Only An Excuse*; *Yo! Diary*; *Watson's Wind Up*.
    Unsolicited manuscripts welcome by post or email.

Company Pictures   (MOVIE) (TV)

Suffolk House, 1–8 Whitfield Place, London, W1T 5JU
℡ 020 7380 3900 🅕 020 7380 1166
enquiries@companypictures.co.uk
www.companypictures.co.uk
Founder/Executive Producer *George Faber*
Founder/Executive Producer *Charles Pattinson*

One of the UK's largest independent film and television drama production companies. Established in 1998. Won Best Independent Production Company at the 2005 Broadcast Awards and the European Producers of the Year Award at the 2004 Monte Carlo Awards. Production highlights include *Shameless* (Channel 4), *Life and Death of Peter Sellers* (HBO, two Golden Globe Awards, 15 Emmies) and *Elizabeth I* starring Helen Mirren and Jeremy Irons (Channel 4/HBO, three Golden Globe Awards, 9 Emmies).
    Proposals accepted only via agent.

Contentfilm International   (MOVIE)

19 Heddon Street, London, N1B 4BG
℡ 0207 851 6500 🅕 0207 851 6506
london@contentfilm.com
www.contentfilm.com

Launched in 2002. Specializes in high quality, commercial feature films. Past examples include Academy Award-nominated *Transamerica* and *Thank You For Smoking*.

Only accepts unsolicited material via email with the following criteria: film 50% funded, production company in place, significant director and/or cast attached.

Coolabi   (TV)

48 Broadley Terrace, London, NW1 6LG
℡ 020 7258 7080 🅕 020 7258 7090
info@coolabi.com
www.coolabi.com

Originially founded in 1999 as Alibi Communications, producer of prime time television drama and children's television drama. In 2004 merged with Coolebah Limited, a business engaged in licensing and animated children's television production. Became Coolabi in 2005.

Cosgrove Hall Films   (MOVIE) (TV)

8 Albany Road, Chorlton-cum-Hardy, Manchester, M21 0AW
℡ 0161 882 2500 🅕 0161 882 2555
animation@cosgrovehall.com
www.chf.co.uk
Contact *Lee Marriott*

Producer of animated children's television programmes, founded in 1976. Highlights include *Andy Pandy*, *Bill and Ben*, *Count Duckula*, *DangerMouse*, *Fifi and the Flowertots*, *Roald Dahl's BFG*, *The Jungle Book* and *The Wind in the Willows*.

Cougar Films   (MOVIE)

Paramount House, 162–170 Wardour Street, London, W1F 8ZX
℡ 020 7292 6610 🅕 020 7734 7878
admin@cougarfilms.co.uk

Feature film production company whose credits include *Imagine Me and You*.

Cowboy Films   (MOVIE) (TV)

26 Nassau Street, London, W1W 7AQ
℡ 020 7255 6757 🅕 020 7255 1132
charles@cowboyfilms.co.uk
www.cowboyfilms.co.uk

Founded in the 1990s, a feature film and television production company with credits including *The Last King of Scotland* and *Hole*.

The Creative Partnership   (MOVIE) (TV)

13 Bateman Street, London, W1D 3AF
℡ 020 7439 7762
sarah.fforde@thecreativepartnership.co.uk
www.thecreativepartnership.co.uk
Contact *Sarah Fforde*

'Europe's largest "one-stop shop" for advertising and marketing campaigns for the film and television industries.' Clients include most major and independent film companies.

No scripts. 'We train new writers in-house, and find them from submitted CVs. All applicants must have previous commercial writing experience.'

Criterion Games (VG)

Onslow House, Onslow Street, Guildford, Surrey, GU1 4TL
mailbag@criteriongames.com
www.criteriongames.com

Owned by the giant Electronic Arts, specializing in developing leading video games such as the *Burnout* series and *Black*.

Will not read or acknowledge unsolicited approaches.

Cross Day Productions Ltd (MOVIE)

1st Floor, 130–132 Wardour Street, London, W1F 8ZN
T 020 7287 5773 F 020 7287 5774
pippacro@aol.com

Producer of feature films. Credits include the award-winning *Shooting Dogs*.

Crossroads Films (MOVIE) (TV)

2nd Floor, 87 Notting Hill Gate, London, W11 3JZ
T 020 7792 5400 F 020 7792 0592
www.crossroadsfilms.com

Works across a variety of media including film, television, commercials and music video. Film credits include *A Love Song for Bobby Long* and *Snow Angels*.

CS Films (MOVIE)

46 Haycroft Gardens, London, NW10 3BN
T 020 8838 2566 F 020 8838 2566
admin@schoenfeld.co.uk
Contact *Carl Schoenfeld*

In independent producer of feature films and documentaries, founded in 2002. Examples of productions include *The Living and the Dead*, *My Brother Tom* and *A Sarajevo Diary*, which was Bafta- and Prix Europa-nominated.

Is happy to look at unsolicited work. Writers should send a one-page synopsis by email. Their advice to new writers looking to break into the industry is: 'Don't expect much money and keep your day job.'

Cutting Edge Productions Ltd (MOVIE) (TV)

27 Erpingham Road, London, SW15 1BE
T 020 8780 1476 F 020 8780 0102

juliannorridge@btconnect.com
Contact *Julian Norridge*

Corporate and documentary video and television. Output includes a US series on evangelicalism, *Dispatches* on the US tobacco industry plus government videos.

No unsolicited manuscripts; 'we commission all our writing to order, but are open to ideas.'

Dakota Films Ltd (MOVIE)

4A Junction Mews, London, W2 1PN
T 020 7706 9407 F 020 7402 6111
info@dakota-films.demon.co.uk

Film producer whose credits include *Me Without You*, *Fade to Black* and *Head in the Clouds*.

Dan Films (MOVIE)

32 Maple Street, London, W1T 6HB
T 020 7916 4771 F 020 7916 4773
enquiries@danfilms.com
www.danfilms.com

Has worked as main or co-producer on films such as *Creep*, *Severance*, *Summer Things*, *The Great Challenge*, *The Republic of Love* and *Villa des Roses*.

Dazed Film & TV (MOVIE) (TV)

112–116 Old Street, London, EC1V 9BG
T 020 7549 6840 F 020 7336 0966
info.film&tv@dazedgroup.com
www.dazedfilmtv.com
Company Director/Executive Producer *Laura Hastings-Smith*

Aims to create 'intelligent, inspirational and provocative' drama and documentary. Operating since 1997, credits include *The Lives of the Saints*, *Perfect*, *Untold Beauty* and *Gorillaz: Charts of Darkness*.

De Facto Films (MOVIE)

30 Chamberlain Street, Derry, BT48 6LR
T 028 712 60714 F 028 712 60714
defactofilms@eircom.net

Film and video production company whose credits include *Dead Long Enough*.

Dirty Hands Productions (MOVIE)

2nd Floor, 2–4 Noel Street, London, W1F 8GN
T 020 7287 7410
Director *Sir Alan Parker*

Feature film production company founded by British director Sir Alan Parker. Produces Parker films including *The Life of David Gale* (2003), *Angela's Ashes* (1999), *Evita* (1996) and the upcoming *The Ice at the Bottom of the World* (2008).

Diverse Production Limited   TV

6 Gorleston Street, London, W14 8XS
☏ 020 7603 4567 🖷 020 7603 2148
www.diverse.tv

Independent production company specializing in popular prime-time formats, strong documentaries, specialist factual, historical, cultural, religious, arts, music and factual entertainment. Recent Output includes *Ballet Changed My Life: Ballet Hoo!*; *Mission Africa*; *Codex*; *Bear Grylls' Man vs. Wild*; *Tribal Wife*; *Musicality*; *Operatunity*; *Who Wrote the Bible*; *Who You Callin' Nigger?*; *In Search of Tony Blair*; *Britain AD*; *Shock Treatment*; *Beyond Boundaries*; *Escape to the Legion*.

DMS Films Ltd   MOVIE

89 Sevington Road, London, NW4 3RU
☏ 020 8203 5540 🖷 0870 762 5671
danny@dmsfilms.co.uk
www.dmsfilms.co.uk
Producer *Daniel San*

Film producer. Output includes *Understanding Jane*; *Hard Edge*; *Popcorn*.
   Unsolicited screenplays not welcome: phone, fax or email synopsis or outline in first instance.

DNA Films Ltd   MOVIE

15 Greek Street, London, W1D 4DP
☏ 020 7292 8700
info@dnafilms.com
www.dnafilms.com

Production company whose credits include *The Last King of Scotland*, *The History Boys*, *Notes on a Scandal*, *Separate Lies*, *Love Actually* and *28 Days Later*.

Done and Dusted   MOVIE   TV

151 Wardour Street, London, W1F 8WE
☏ 020 7297 8060
sm@doneanddusted.com
www.doneanddusted.com
Executive Producer *Susan Maxwell*

Creating TV and DVD programmes, live shows and events, since 1998. Also has office in New York.

DoubleBand Films   TV

3 Crescent Gardens, Belfast, BT7 1NS
☏ 028 9024 3331 🖷 028 9023 6980
info@doublebandfilms.com
www.doublebandfilms.com
Contact *Michael Hewitt*
Contact *Dermot Lavery*

Specializes in documentaries and drama. Recent productions include *Seven Days that Shook the World* and *War in Mind* (both for Ch4); *Christine's Children* (BBC Northern Ireland; nominated for both the RTS and Celtic Film Festival); *D-Day: Triumph and Tragedy* (BBC NI).
   Does not accept unsolicited scripts.

Drake AV Video Ltd   MOVIE

89 St Fagans Road, Fairwater, Cardiff, CF5 3AE
☏ 029 2056 0333
info@drakeav.com
www.drakeav.com
Contact *Helen Stewart*

Specialists in corporate and educational films since 1973.
   Does not accept unsolicited manuscripts. Sometimes needs writers (writing to a brief) at short notice, so a short message indicating price and availability is useful.

Charles Dunstan Communications Ltd   MOVIE   TV

42 Wolseley Gardens, London, W4 3LS
☏ 020 8994 2328 🖷 020 8994 2328
Contact *Charles Dunstan*

Producer of film, video and TV for documentary and corporate material. Output includes *Renewable Energy* for broadcast worldwide in 'Inside Britain' series; *The Far Reaches* travel series; *The Electric Environment*.
   No unsolicited scripts.

Ecosse Films   MOVIE   TV

Brigade House, 8 Parsons Green, London, SW6 4TN
☏ 020 7371 0290 🖷 020 7736 3436
info@ecossefilms.com
www.ecossefilms.com

Specializes in high-quality drama for film and TV. Titles include *Mrs Brown*, *Monarch of the Glen* and *Becoming Jane*.

Eidos Interactive Ltd   VG

SCi Entertainment Group, Wimbledon Bridge House, 1 Hartfield Road, Wimbledon, SW19 3RU
☏ 020 8636 3000 🖷 020 8636 3001
www.eidosinteractive.co.uk

Founded in 1990 and now part of the SCi Entertainment Group Plc. Eidos incorporates publishing operations in Europe and America, with several development studios including Crystal Dynamics, IO Interactive and Pivotal Games. Credits include *Championship Manager*, *Tomb Raider*, *Who Wants To Be A Millionaire?* and *Commandos*.

Elstree (Production) Company Ltd MOVIE TV

Shepperton Studios, Studios Road, Shepperton, Middlesex, TW17 0QD
T 01932 592680 F 01932 592682
enquiries@elsprod.com
www.elsprod.com
Producer *Greg Smith*
Assistant *Sue Hayworth*

Founded in 1966, producing feature films and films for TV. Titles include *Agnes Brown* and *Animal Farm*.
Does not encourage unsolicited approaches.

Endemol UK Productions TV

Shepherds Building, Central Charecroft Way, London, W14 0EE
T 0870 333 1700 F 0870 333 1800
info@endemoluk.com
www.endemoluk.com

Major independent producer of TV and digital media content, responsible for over 5,000 hours of programming per year. Incorporates several production brands including Brighter Pictures (factual entertainment shows, reality programming, live events and popular documentaries), Cheetah Television (factual), Initial (prime time entertainment, children's and teens, events and arts shows), Zeppotron (comedy) and Showrunner (drama). Credits include *Big Brother*, *Deal Or No Deal*, *Comic Relief Does Fame Academy*, *Restoration Village*, *Soccer Aid*, *Extinct*, *Orange Playlist*, *Rush Hour*, *Lucy: Teen Transsexual* and *34 Stone Teenager: 6 Months On*.

Eon Productions MOVIE

Eon House, 138 Piccadilly, London, W1J 7NR
T 020 7493 7953 F 020 7408 1236

Production company responsible for the *James Bond* series of films.

Eye Film and Television MOVIE TV

Chamberlain House, 2 Dove Street, Norwich, NR2 1DE
T 01603 762 551 F 01603 762 420
production@eyefilmandtv.co.uk
www.eyefilmandtv.co.uk

Independent producers of drama, documentary, corporate, commercial and educational film and television programming. Specializes in development, production and post-production. Past credits include children's drama series *The Secret of Eel Island* (Five), documentary series *Hell to Hotel* (ITV Anglia, UKTV Style) and *A Different Life* (Five, BAFTA-nominated).

Farnham Film Company Ltd MOVIE TV

34 Burnt Hill Road, Lower Bourne, Farnham, GU10 3LZ
T 012 527 10313 F 012 527 25855
info@farnfilm.com
www.farnfilm.com
Contact *Ian Lewis*

Television and film: intelligent full-length film and children's drama.
Unsolicited manuscripts usually welcome but prefers a letter or email to be sent in the first instance. Check website for current requirements.

Fast Films MOVIE TV

Christmas House, 213 Chester Road Castle, Bromwich, Solihull, B36 0ET
T 0121 749 7147/4144
gavinprime@mac.com
Contact *Gavin Prime*

Film and television: comedy, entertainment and animation.
No unsolicited manuscripts.

Feelgood Fiction MOVIE TV

49 Goldhawk Road, London, W12 8QP
T 020 8746 2535 F 020 8740 6177
feelgood@feelgoodfiction.co.uk
www.feelgoodfiction.co.uk
Contact *Laurence Bowen*

Production company launched in 1996. Owned and run by Laurence Bowen and Philip Clarke. Currently developing film and TV projects with, among others, Five, the BBC, ITV, RTE, Paramount Comedy, The Weinstein Company, HBO, UK Film Council and Granada Films. Focuses on drama, comedy, children's TV and feature films. Notable TV credits include co-production *Double Bill* (2002, with Working Title TV), 6 part children's comedy *My Life As A Popat* (2004, BAFTA winner for Best Children's Drama) and *Suburban Shootout* (2006). Feelgood Films, the feature film wing, was launched in 2003 with *Miranda* starring Christina Ricci and Kyle MacLachlan. Has a policy of both nurturing new writers and working with the very best established talent.

Festival Film + TV Ltd MOVIE TV

Festival House, Tranquil Passage, London, SE3 0BJ
T 020 8297 9999 F 020 8297 1155
info@festivalfilm.com
www.festivalfilm.com
Managing Director *Ray Marshall*

Founded in 1992, producing TV dramas and feature films. Has produced fifteen mini series for ITV based on books by Catherine Cookson. Won an International Emmy for *The Black Velvet Gown*.

Always willing to look at professionally-presented material. Adaptations and/or original work. Send treatemnt in writing first. Particularly interested in uplifting, feel-good projects. No horror, supernatural or action thriller.

Film and General Productions Ltd (MOVIE) (TV)

4 Bradbrook House, Studio Place, London, SW1X 8EL

T 020 7235 4495 F 020 7245 9853
cparsons@filmgen.co.uk
Contact *Clive Parsons*

Founded in 1971, specializing in feature films and TV drama.

Films include *Gregory's Girl, Scum, Breaking Glass, Britannia Hospital, True Blue, The Queen's Nose, The Giblet Boys* and *The Greatest Store in the World*.

Open policy on new writing. Send a preliminary email.

Film4 Productions (MOVIE)

T 020 7306 5190 F 020 7306 8368
Film4@channel4.co.uk
www.channel4.com/film/ffproductions
Head of Development *Katherine Butler*

The film arm of Channel 4 Television, dedicated to the production of distinctive and contemporary feature films. Credits include *Dogma, East Is East, Enduring Love, Me, You and Everyone We Know, Sexy Beast, The Future is Unwritten: Joe Strummer, The Motorcycle Diaries, This Is England, Touching the Void, Trainspotting* and *Wish You Were Here*.

Unable to accept unsolicited material.

Firefly Films Limited (MOVIE) (TV)

London W1
T 020 7193 1837 F 0870 974 1165
info@ff-i.com
www.fireflyfilms.info
MD/Producer *Matthew Hobbs*

Output includes *The Good War* and *American Cowslip*.

Does not welcome unsolicited material. Approach via an agent.

The First Film Company Ltd (MOVIE)

3 Bourlet Close, London, W1W 7BQ
T 020 7436 9490 F 020 7637 1290
info@firstfilmcompany.com
Producer *Roger Randall-Cutler*
Producer *Robert Cheek*

Founded 1984. Cinema screenplays.
All submissions should be made through an agent.

Flannel (TV)

21 Berwick Street, London, W1F 0PZ
T 020 7287 9277 F 020 7287 7785
mail@flannel.net
Contact *Kate Haldane*

Producer of drama, documentaries and comedy for television and radio.

No unsolicited manuscripts. 'Keen to encourage new writing, but must come via an agent. Particularly interested in 45–50 minute dramas for radio. Not in a position to produce plays for stage, but very happy to consider adaptations. Welcomes comedy with some track record.'

Flashback Television Ltd (TV)

58 Farringdon Road, London, EC1R 3BP
T 020 7253 8768 F 020 7253 8765
mailbox@flashbacktv.co.uk
www.flashbacktelevision.com
Managing Director *Taylor Downing*
Creative Director *David Edgar*
Director of Production *Tim Bay*

Founded in 1982, award-winning producer of drama, factual entertainment and documentaries for television. Recent credits include *Beau Brummell*.

Does not welcome unsolicited manuscripts.

Focus Films Ltd (MOVIE)

The Rotunda Studios, Rear of 116–118, Finchley Road, London, NW3 5HT
T 020 7435 9004 F 020 7431 3562
focus@focusfilms.co.uk
www.focusfilms.co.uk
Managing Director *David Pupkewitz*
Head of Development *Malcolm Kohll*
Contact *Raimund Berens*

Founded in 1982, an independent feature film development and production company. Output: *The Book of Eve* (Canadian drama); *The Bone Snatcher* (Horror, UK/ Can/SA); *Julia's Ghost* (German co-production); *The 51st State* (feature film); *Secret Society* (comedy drama feature film); *Crimetime* (feature thriller); *Diary of a Sane Man*; *Othello*. Projects in development include *Heaven and Earth; Tainted Desert; The Complete History of the Breast; Triomf*.

No unsolicited scripts.

Mark Forstater Productions Ltd (MOVIE)

11 Keslake Road, London, NW6 6DJ
T 020 8933 5475 F 020 8933 5475
Contact *Mark Forstater*

Active in the selection, development and production of material for film and TV. Output includes *Monty Python and the Holy Grail; The Odd Job; The Grass is Singing; Xtro; Forbidden; Separation; The Fantasist; Shalom Joan Collins; The Silent Touch; Grushko; The Wolves of Willoughby Chase; Between the Devil and the Deep Blue Sea; Doing Rude Things.*

No unsolicited scripts.

Fortissimo Films (UK)   [MOVIE]

34 Tavistock Street, Covent Garden, London, WC2E 7PB
[T] 020 7498 6978 [F] 020 7497 1133
nicole@fortissimo-uk.com
www.fortissimo-uk.com

In recent years the company has become involved in developing and co-producing films and has been involved with titles including *Thomas In Love, Grimm, Springtime In a Small Town, Seven Swords, Invisible Waves* and *Shortbus.*

Does not accept unsolicited scripts.

Forward Films   [MOVIE]

Third Floor, 23 Denmark Street, London, WC2H 8NH
[T] 020 7632 9643 [F] 020 7240 5647
office@forwardfilms.co.uk

Film production company founded in 2004. Does not encourage unsolicited material.

Fragile Films   [MOVIE] [TV]

Ealing Studios, Ealing Green, London, W5 5EP
[T] 020 8567 6655 [F] 020 8758 8658
info@ealingstudios.com
www.fragilefilms.com

Production arm of Ealing Studios. The studio returned to filmmaking after 43 years in 2002 with *The Importance of Being Earnest.* Other recent releases include the animated feature *Valiant* (co-production with Vanguard Films) and *Alien Autopsy* with Ant and Dec. Also expanding into television, most recently with *The Pilot Show* (Channel 4).

Does not accept unsolicited material from un-represented writers. Approaches to the development department should be a one/two page synopsis via email to development@ ealingstudios.com.

FremantleMedia Ltd   [MOVIE] [VG]

1 Stephen Street, London, W1T 1AL
[T] 020 7691 6000
www.fremantlemedia.com
Contact CEO *Tony Cohen*

One of the world's leading providers of television entertainment programmes such as *The Bill, American Idol* and *The Price is Right.*

Fulcrum TV   [TV]

Shepperton Film Studios, Studios Road, Shepperton, Middlesex, TW17 0QD
[T] 01932 593342
www.fulcrumtv.com

Producing for television since 1986. Output includes *Da Vinci's Lost Code, People in Order, State of Texas, The Christmas Truce* and *What Would Jesus Drive?*

Future Films Limited   [MOVIE]

76 Dean Street, London, W1D 3SQ
[T] 020 7009 6600 [F] 020 7009 6602
info@futurefilmgroup.com
www.futurefilmgroup.com

Formed in 2000, providing a full range of services from pre-financing to post-production. Highly experienced as both a co-producer on international productions and as executive producer on other productions.

Noel Gay Television   [TV]

Shepperton Studios, Studios Road, Shepperton, TW17 0QD
[T] 01932 592569 [F] 01932 592172
charles.armitage@virgin.net
CEO *Charles Armitage*

Output includeds *The Fear* (BBC Choice); *Second Chance* (Ch4); *Hububb* Series 1–5 (BBC); *I-Camcorder* (Ch4); *Frank Stubbs Promotes* and *10%ers* Series 2 (both for Carlton/ITV); *Call Up the Stars* (BBC1); *Smeg Outs* (BBC video); *Red Dwarf; Dave Allen* (ITV); *Windrush* (BBC2). The Noel Gay Motion Picture Company has credits including *Virtual Sexuality; Trainspotting* (with Ch4 and Figment Films); *Killer Tongue; Dog Soldiers; Fast Sofa* and *Pasty Faces.* Associate NGTV companies are Grant Naylor Productions and Pepper Productions.

NGTV is willing to accept unsolicited material from writers but 1–2-page treatments only. No scripts.

Glasshouse Entertainment Ltd   [MOVIE]

82 Clerkenwell Road, London, EC1M 5RF
[T] 020 7490 7604
info@glasshousepictures.com
www.glasshousepictures.com

Develops, produces and finances features. Focus on adaptation of strong literary properties. Output includes *Snitch Jacket, Island of the*

Sequined Love Nun and *Little Green Men*. Also has office in Los Angeles.

GMTV ` TV `

London Television Centre Upper Ground, London, SE1 9TT
☏ 020 7827 7000 🖷 020 7827 7001
www.gm.tv
Head of Forward Planning *Terry O'Sullivan*

Launched in 1993, breakfast television programme aimed at housewives with children.

Gold Circle Films ` MOVIE `

10–11 Great Russell Street, London, WC1B 3NH
☏ 020 7631 0173 🖷 020 7631 0189
amartin@goldcirclefilms.com
www.goldcirclefilms.com

Independent film production company founded in 2000 by Norm Waitt. Subsidiary of Gold Circle Entertainment Inc. Recent features include *Because I Said So* (with Diane Keaton) and *My Big Fat Greek Wedding*.
    Unsolicited material is not accepted.

Goldcrest Films International Ltd ` MOVIE ` ` TV `

65–66 Dean Street, London, W1D 4PL
☏ 020 7437 8696 🖷 020 7437 4448
mail@goldcrestfilms.com
www.goldcrestfilms.com
Contact *Laura Murphy*

Since it was established in 1977, Goldcrest Films has become a leading independent film production company winning many prizes at international festivals and 19 Academy Awards and 28 Baftas. Finances, produces and distributes films and television programmes. Output includes *Chariots of Fire, Gandhi, The Killing Fields, A Room With a View, Local Hero, The Mission* and *To End All Wars*. 'We are currently seeking film projects to invest equity in through our Finishing Fund in return for International Sales Rights.'
    Scripts via agents only.

The Good Film Company ` MOVIE `

The Studio, 5–6 Eton Garages, Lambolle Place, London, NW3 4PE
☏ 020 7794 6222 🖷 020 7794 4651
yanina@goodfilms.co.uk
www.goodfilms.co.uk
Executive Producer *Yanina Barry*

Corporate film and video production company founded in 1988. Output includes commercials, music videos, documentaries and photographic stills.
    Unsolicited material not accepted.

Granada Media ` TV `

The London Television Centre Upper Ground, London, SE1 9LT
☏ 020 7620 1620
www.granadamedia.com

Leading commercial television production and distribution company. Part of ITV Productions (the production and resource division) and ITV Worldwide (the distribution, merchandising and international production business) of the UK media company ITV plc. ITV Productions is the largest commercial TV producer in the UK, creating over 3,500 hours of original programming each year. Established for over 50 years. Invests in projects across most genres. ITV Productions is the primary producer for ITV1. Whilst trading as Granada it produces programming for the BBC, Channel 4, five, and satellite and digital channels. Credits include *The Street* (BBC1), *Come Dine With Me, Countdown* (Channel 4) and *Brainiac* (Sky).

Granite Film & Television Productions ` TV `

Easter Davoch Tarland, Aboyne, AB34 4US
☏ 01339 880 175
Contact *Simon Welfare*

Producer of television documentary programmes such as *Nicholas & Alexandra*; *Victoria & Albert* and *Arthur C. Carke's Mysterious Universe*.

Graphite Film and Television ` MOVIE ` ` TV `

72 Newman Street, London, W1T 3EH
☏ 020 7580 5658
contact@graphitefilms.com
Managing Director *Stephen Taylor*

Production company founded in 1994, specializing in film and television drama. Productions include *Get Real*.
    Submissions via agent only.

Green Umbrella Ltd ` TV `

59 Cotham Hill, Cotham, Bristol, BS6 6JR
☏ 0117 906 4336 🖷 0117 923 7003
info@green-umbrella.co.uk
www.green-umbrella.co.uk

Film producers specializing in science and natural history documentaries. Output includes episodes for *The Natural World, Wildlife on One* and original series such as *Living Europe* and *Triumph of Life*.
    Unsolicited treatments relating to natural history and science subjects are welcome.

Greenpoint Films ` MOVIE ` ` TV `

7 Denmark Street, London, WC2H 8LZ
☏ 020 7240 7066 🖷 020 7240 7088

info@greenpointfilms.co.uk
www.greenpointfilms.co.uk

Produces feature films alongside television and radio programming. Feature film credits include *The Only Boy For Me* starring Helen Baxendale and *Hideous Kinky* starring Kate Winslet. Recently aired on ITV was the twelve part comedy series *High Stakes*.

Greenwich Village Productions   `MOVIE`

14 Greenwich Church Street, London, SE10 9BJ
T 020 8853 5100 F 020 8293 3001
gvproductions@fictionfactory.co.uk
www.fictionfactory.co.uk
Contact *John Taylor*

Features, arts and educational projects, webmovies and new-media productions. Recent output includes *Athletes* and *Secret Roads*.

Manuscripts via agents only or from writers with a professional track record in the chosen medium.

Gruber Films Ltd   `MOVIE`

2 Sheraton Street, London, W1F 8BH
T 08703 669 313
richard.holmes@gruberfilms.com

Feature film production company, whose credits include *The Great Pretender*.

Hammer Productions   `MOVIE` `TV`

info@hammerfilms.com
www.hammerfilms.com

Legendary production house for horror films, operating from 1934 until the 1980s. In 2007 the company and its back catalogue was bought by Cyrte Investments, headed by John De Mol, with the intention of developing and producing new titles. To this end an initial fund of US$50 million was made available. The new company's first title was *Beyond the Rave*, a contemporary vampire story filmed in association with MySpaceTV.

Hammerwood Film Productions   `MOVIE` `TV`

info@filmangel.co.uk
www.filmangel.co.uk

Film, video and TV drama. Output includes *Iceni* (film; co-production with Pan-European Film Productions and Boudicca Film Productions Ltd); *Boudicca – A Celtic Tragedy* (TV series). In preproduction: *The Black Egg* (witchcraft in 17th century England); *The Ghosthunter*; *Iceni* (documentary of the rebellion of AD61); *No Case to Answer* (legal series).

Study the website before submitting material.

Harbour Pictures   `MOVIE` `TV`

6 Providence Villas Studio, Brackenbury Road, London, W6 0BA
T 020 8749 4100 F 020 8740 1937
info@harbourpictures.com
www.harbourpictures.com
Producer/Chief Executive *Nicholas Barton*
Producer/Director of Development *Suzanne Mackie*
Finance Director *Amanda Calder*
Assistant to Producers *Debbie Lowndes*
Research Assistant *Lucy Geering*
General Dogsbody *Milo Lowndes*

Originally set up as the film production arm of ABTV Ltd (now Harbour Pictures Holdings Ltd). Began feature film development and production in 1999 with *Calendar Girls* (co-produced with Buena Vista), the seventh highest-grossing British film ever in the UK: over £20 million taken at the UK box office and approximately $100 million internationally. Follow-up *Kinky Boots* (co-produced with Miramax) released in 2006. Currently two other films in development for Miramax plus others for BBC Films, The UK Film Council, Pathé and Hanway Films. Previous television drama and documentary credits for ABTV/Harbour Pictures include: *Byzantium – The Lost Empire* (Discovery/Channel 4), *The Vanishing Man* (ITV), *Bye Bye Baby* (Channel 4 film by Jack Rosenthal, winner of 1992 Prix Europa and 1993 Writer's Guild Award).

Harcourt Films   `MOVIE` `TV`

58 Camden Square, London, NW1 9XE
T 020 7266 5880
jmarre@harcourtfilms.com
www.harcourtfilms.com
Director/Writer/Producer *Jeremy Marre*

Production company run by Jeremy Marre concentrating on documentary features for television. Strong links to the international market. Recent credits include *Soul Brittannia* (BBC), *Reigning in Hell* (Channel 4, Court TV in USA), *What's Going On* (BBC, PBS) and *The Real Phil Spector* (Channel 4). To date Marre has won one Grammy and two Emmies.

Hartswood Films Ltd   `MOVIE` `TV`

Twickenham Studios, The Barons, St Margarets Twickenham, Middlesex, TW1 2AW
T 020 8607 8736 F 020 8607 8744
films.tv@hartswoodfilms.co.uk
www.hartswoodfilms.co.uk
Contact *Elaine Cameron*

Established in the 1980s, an award-winning independent producer of drama, light

entertainment and documentary films.
Production credits include *Jekyll, After Thomas, Wonderful You, The English Wife, A Woman's Guide to Adultery, Men Behaving Badly, Coupling, Carrie & Barry, Is It Legal?* and *The Savages*.
Documentaries include *The Welsh Great Escape, In Love With Elizabeth* and *Going to Chelsea*.

Hat Trick Productions Ltd [TV]

10 Livonia Street, London, W1F 8AF
T 020 7434 2451 F 020 7287 9791
info@hattrick.com
www.hattrick.com
Managing Director *Jimmy Mulville*

Set up in 1986, producing TV programmes in the fields of comedy, drama and entertainment.
Among its large collection of award-winning shows are *Clive Anderson Talks Back, Drop the Dead Donkey, Father Ted, Game On, Harry Enfield's Television Programme, Have I Got News For You, Room 101, The Kumars at No. 42, Trevor's World of Sport, Underworld* and *Whose Line Is It Anyway?*
   Will not consider unsolicited scripts or treatments. Material should be sent via a recognized agent only.

Headgear Films Limited [MOVIE]

4th Floor, Douglas House, 3 Richmond Buildings (off Dean Street), London, W1D 3HE
T 020 7734 3566 F 020 3230 1059
maria@headgearfilms.com
www.headgearfilms.com

Founded in 2002, the company has quickly gained a reputation for developing, producing and distributing feature films for the international market. Examples of titles include *Rabbit on the Moon, Chunky Monkey* and *Fingers X'D*.

Healthcare Productions Limited [TV]

The Great Barn, Godmersham Park, Canterbury, Kent, CT4 7DT
T 01227 738279
penny@healthcareproductions.co.uk
www.healthcareproductions.co.uk
Contact *Penny Webb*

Established in 1989, producing health-based dramas and documentaries for TV and the DVD market.
Has won numerous awards from the BMA.
   Happy to look at new material, which should be submitted by email.

Heritage Theatre Ltd [MOVIE] [TV]

Unit 1, 8 Clanricarde Gardens, London, W2 4NA
T 020 7243 2750 F 020 7792 8584
rm@heritagetheatre.com

www.heritagetheatre.com
Contact *Robert Marshall*

Founded in 2000 to produce and distribute distinguished theatre productions via DVD or broadcast TV. Titles include *Primo* and *The Rivals*.
   Does not invite unsolicited submissions.

Heyday [MOVIE]

5 Denmark Street, London, WC2H 8LP
T 020 7836 6333 F 020 7836 6444
office@heydayfilms.demon.co.uk

Founded in 1997, the company's first film was *Ravenous* and it has subsequently worked on the *Harry Potter* series of movies.

Holmes Associates [MOVIE] [TV]

The Studio, 37 Redington Road, London, NW3 7QY
T 020 7813 4333
holmesassociates@blueyonder.co.uk
Contact *Andrew Holmes*

Prolific originator, producer and packager of documentary, drama and music television and films. Output has included *Ashes and Sand* (Film 4); *Chunky Monkey* (J&V Films); *Prometheus* (Ch4 'Film 4'); *The Shadow of Hiroshima* (Ch4 'Witness'); *The House of Bernarda Alba* (Ch4/WNET/Amaya); *Piece of Cake* (LWT); *The Cormorant* (BBC/Screen 2); *John Gielgud Looks Back; Rock Steady; Well Being; Signals; Ideal Home?* (all Ch4); *Seven Canticles of St Francis* (BBC2).
   Submissions only accepted by email in synopsis form.

Holy Cow Films [MOVIE]

222 Dalling Road, London, W6 0ER
T 020 8735 9161 F 020 8748 1976

Feature film production company whose credits include *Backwoods*.

Hourglass Productions [TV]

27 Princes Road, Wimbledon, London, SW19 8RA
T 020 8540 8786
productions@hourglass.co.uk
www.hourglass.co.uk
Company Partner *Martin Chilcott*

Founded in 1984, specializing in science and medical documentaries for TV and DVD. Recent titles include *Science from Nature*. Has been BAFTA-nominated and awarded a gold medal at the New York Film Festival.
   Interested in leading-edge scientific proposals.
Approach in writing.

Hurricane Films   MOVIE TV

19 Hope Street, Liverpool, L1 9BQ
℡ 0151 707 9700 ℻ 0151 707 9149
info@hurricanefilms.co.uk
www.hurricanefilms.net

Founded in 2000, working across short films, single dramas, documentaries and feature films. Output includes *Dockers: Writing the Wrongs*, *Nightclub: Tales from the Grafton*, *The Last Straw*, *Under The Mud* and *Warship*.

Iambic Productions   TV

89 Whiteladies Road, Clifton, Bristol, BS8 2NT
℡ 0117 923 7222 ℻ 0117 923 8343
admin@iambic.tv
www.iambicproductions.com

Independent production company specializing in arts, music and entertainment programmes since 1988. A new division, IP Entertainment, works in drama and factual entertainment. Credits include *Kiss Me Kate*, *The Merchant Of Venice*, *The Mystery of Charles Dickens*, *Romeo Et Juliette* and *The Truth about Boy Bands*.

Icon Films   MOVIE TV

1–2 Fitzroy Terrace, Bristol, BS6 6TF
info@iconfilms.co.uk
www.iconfilms.co.uk
Contact *Harry Marshall*

Film and TV documentaries. Output includes *Nick Baker's Weird Creatures* (five/Animal Planet/Granada International); *Tom Harrisson – The Barefoot Anthropologist* (BBC).
   Specializes in factual documentaries.
   Open-minded to new documentary proposals.

The Illuminated Film Company   MOVIE TV

2 Glenthorne Mewsoff, Glenthorne Road, London, W6 0LJ
℡ 020 8748 3030 ℻ 020 8748 3725
info@illuminatedfilms.com
www.illuminatedfilms.com
Producer *Iain Harvey*

Set up in 1993. Producers of *The Very Hungry Caterpillar and Other Stories*, *T.R.A.N.S.I.T.*, *War Game*, *Christmas Carol - The Movie* and *Little Princess*.
   Will look at unsolicited material but contact by email first.

Imari Entertainment Ltd   TV

PO Box 158, Beaconsfield, HP9 1AY
℡ 01494 677147 ℻ 01494 677147
info@imarientertainment.com
www.imarientertainment.com
Contact *Jonathan Fowke*

TV and video producer, covering all areas of drama, documentary and corporate productions.

Impact Pictures   MOVIE

3 Percy Street, London, W1T 1DE
℡ 020 7636 7716 ℻ 020 7636 7814
production@impactpix.com

Established in 1989. Film production company whose credits include *Resident Evil*, *Resident Evil: Apocalypse* and *The Dark*.

Indigo Television Ltd   TV

76 Marylebone High Street, London, W1U 5JU
℡ 020 7486 4443
info@indigotelevision.co.uk
www.indigotelevision.co.uk

Specializing in prime time event television and documentaries.

International Media   MOVIE TV

6 Kingly Street, London, W1B 5PF
℡ 020 7292 4790
www.internationalmedia.de

Global, independent film company based in Munich, London and Los Angeles. Develops, finances, produces and distributes film and television projects. Intermedia Films, the company's traditional business, concentrates on high-quality motion pictures. Intermedia Cinema serves the high-budget action and horror market, while Intermedia TV provides programming for the international market. Output includes Oliver Stone's *Alexander* (starring Colin Farrell, Angelina Jolie and Sir Anthony Hopkins), *Terminator 3* (starring Arnold Schwarzenegger), *Life of David Gale* (starring Kate Winslet) and *The Wedding Planner* (starring Jennifer Lopez).

Intrepido   MOVIE

℡ 020 8879 7100 ℻ 020 8879 7320
mail@intrepido.co.uk
www.intrepido.co.uk

Independent film production company, specializing in distinctive shorts and feature films for the domestic and international market. Credits include *Breaking Out*, *El Hoppo!*, *Floating*, *He Said*, *Heavy Metal Drummer*, *Love Letter*, *One Minute Past Midnight* and *Take Me Back*.

Ipso Facto Films   MOVIE

11–13 Broad Court, London, WC2B 5PY
℡ 020 7240 6166 ℻ 020 7240 6160
info@ipsofactofilms.com
www.ipsofactofilms.com

Founded in 1993 and has produced over 10 features and over 50 shorts and documentaries. Credits include *Irina Palm, School For Seduction, Short Order, Bye Bye Blackbird, Headrush* and *Gone For A Dance*.

Isis Productions   MOVIE   TV

387b King Street, Hammersmith, London, W6 9NJ
T 020 8748 3042 F 020 8748 7634
hello@isis-productions.com
www.isisproductions.co.uk
Director *Nick de Grunwald*
Director *Jamie Rugge-Price*

Formed in 1991, Isis Productions focuses on the production of music and documentary programmes. Output includes *Imagine – Yusuf Islam* (BBC); *Pet Shop Boys – A Life in Pop; Rufus Wainwright; Brian Ferry – The Dylan Sessions; Ray Davies – The World from My Window; James Brown – Soul Survivor* (Ch4); *Bernie Taupin* (ITV 'South Bank Show'); *Iron Maiden, Judas Priest* (Five 'Rock Classics'); Films on *Deep Purple, Metallica, Def Leppard, Lou Reed, Elton John, Elvis Presley, Sex Pistols* (ITV 'Classic Albums 3'); *Simply Red, Nirvana, Cream, Pink Floyd, Motorhead* ('Classic Albums 4'); *England's Other Elizabeth – Elizabeth Taylor* (BBC 'Omnibus').

Isolde Films   MOVIE   TV

28 Twyford Avenue, London, W3 9QB
T 020 8896 2860
isolde@btinternet.com
www.tonypalmer.org
Contact *Michela Antonello*

Film and TV documentaries. Output includes *Wagner; Margot; Menuhin; Maria Callas; Testimony; In From the Cold; Pushkin; England, My England* (by John Osborne).
  Unsolicited material is read, but send a written outline first.

ITV Productions   TV

The London Television Centre Upper Ground, South Bank, London, SE1 9LT
T 020 7620 1620 F 020 7261 3041
www.itv.com
Controller of ITV Productions (Drama), London *Kate Bartlett*
Director *John Whiston*

ITV Productions is the largest commercial TV production company in the UK. Produces original programmes, co-productions and TV movies for ITV channels and other broadcasters, both in the UK and abroad. Output includes *Hornblower; Poirot; Miss Marple; Touching Evil; Where the Heart Is; The Last Detective; Jericho*.

JAM Pictures and Jane Walmsley Productions   MOVIE   TV

8 Hanover Street, London, W1S 1YE
T 020 7290 2676 F 020 7256 6818
producers@jampix.com
Contact *Jane Walmsley*
Contact *Michael Braham*

JAM Pictures was founded in 1996 to produce drama for film, TV and stage. Projects include: *Hillary's Choice* (TV film, A&E Network); *Son of Pocahontas* (TV film, ABC); *Rudy: the Rudy Giuliani Story* (TV film, USA Network); *One More Kiss* (feature, directed by Vadim Jean); *Bad Blood* (UK theatre tour). Jane Walmsley Productions, formed in 1985 by TV producer, writer and broadcaster, Jane Walmsley, has completed award-winning documentaries and features such as *Hot House People* (Ch4).
  No unsolicited manuscripts. 'Letters can be sent to us, asking if we wish to see manuscripts; we are very interested in quality material, from published or produced writers only, please.'

Juniper Communications Ltd   TV

52 Lant Street, London, SE1 1RB
T 020 7407 9292 F 020 7407 3940
juniper@junipertv.co.uk
www.junipertv.co.uk
Company Director *Belkis Bhegani*

Founded in the early 1980s, a TV production company which has been nominated for Baftas and won several RTS awards. Also does some radio work.
  Specializes in documentaries, dram-docs, studio-based discussions, political programmes, history, science, arts and popular culture. Productions include *The Trouble with Atheism, Aberfan: The Untold Story, Dispatches: What Muslims Want, Iraq - The Reckoning, The Edwardians* and *London Architecture*.
  Contact by phone email or post in the first instance.

Justice Entertainment   TV

PO Box 4377, London, W1A 7SX
info@timwestwood.com
www.timwestwood.com

Television and radio production company.

Kaos Films   MOVIE

info@kaosfilms.co.uk
www.kaosfilms.co.uk

Set up to manage the annual British Short Screenplay Competition, in association with the National Film and Television School, and produce the winning scripts. Subsequently the company began feature film production as part of an on-going initiative is to find hi-concept, commercial screenplays (independent of the Short Screenplay Competition).

Unsolicited material accepted subject to guidelines detailed on the website. Email to hi-concept@kaosfilms.co.uk. All genres considered. Competition details found under "Awards".

Kelpie Films   MOVIE

The Mansion House, 1 Ardgowan Square, Greenock, PA16 8NG
℡ 0800 840 2815 Ⓕ 0871 522 7635
www.kelpiefilms.com

Producing award-winning animation, documentary and drama. Credits include *And So Goodbye*, *Cannonman* and *The Mermaid*.

Keo Films.com Ltd   TV

101 St John Street, London, EC1M 4AS
℡ 020 7490 3580 Ⓕ 020 7490 8419
keo@keofilms.com
www.keofilms.com
Contact *Katherine Perry*

Television documentaries and factual entertainment. Output includes *Atlantic Britain* and *Surviving Extremes* (both for Ch4/NatGeo Europe); *Beyond River Cottage*; *Where's Your F***ing Manners?*; *Road Trip*; *Tales From River Cottage*; *How To Be a Man*; *Running for God*; *Heavy*; *My Body My Business*; *A Dangerous Obsession*; *Sperm Bandits* (all for Ch4); *10 Years Younger* (Discovery Health).

No unsolicited manuscripts.

Kingfisher Television Productions   TV

Martindale House, The Green Ruddington, Nottingham, NG11 6HH
℡ 0115 945 6581
info@kingfishertv.co.uk
www.kingfishertv.co.uk
Contact *Tony Francis*

Broadcast television production company, specializing in factual, countryside and wildlife series.

Kismet Film Company   MOVIE

25A Old Compton Street, London, W1D 5JW
℡ 020 7734 0099 Ⓕ 020 7734 1222
kismetfilms@dial.pipex.com

Produces feature films. Previous titles include *Born Romantic*, *Hypnotic* and *The River King*.

Kudos Film & Television   MOVIE TV

12–14 Amwell Street, London, EC1R 1UQ
℡ 020 7812 3270 Ⓕ 020 7812 3271
info@kudosfilmandtv.com
www.kudosfilmandtv.com

Established in 1992, producing TV programmes and feature films. Production credits include the TV series *Spooks*, *Life On Mars*, *Ashes to Ashes* and *Hustle* and films such as *Eastern Promises*, *Pure*, *Comfortably Numb*, *Among Giants* and *Meeting People is Easy*.

Docs not accept unsolicited scripts.

Lagan Pictures Ltd   MOVIE TV

21 Tullaghbrow, Tullaghgarley, Ballymena, BT42 2LY
℡ 028 2563 9479/077 9852 8797 Ⓕ 028 2563 9479
laganpictures@btinternet.com
Producer/Director *Stephen Butcher*

Film, video and TV: drama, documentary and corporate. Output includes *A Force Under Fire* (Ulster TV). In development: *Into the Bright Light of Day* (drama-doc); *The £10 Float* (feature film); *The Centre* (drama series).

'We are always interested in hearing from writers originating from or based in Northern Ireland or anyone with, preferably unstereotypical, projects relevant to Northern Ireland. We do not have the resources to deal with unsolicited manuscripts, so please write with a brief treatment/synopsis in the first instance.'

Landseer Productions Ltd   TV

140 Royal College Street, London, NW1 0TA
℡ 020 7485 7333 Ⓕ 1 866 469 9445
ken@landseerfilms.com
www.landseerfilms.com
Director *Ken Howard*

Specializes in documenatries, childrens' televion and music programmes. Recent productions include *Living the Dream* for the South Bank Show. Has won many international awards including Baftas, Emmys, Royal Television Society awards and the New York Festival Gold Medal.

Approach with ideas (not manuscripts) via email.

Lilyville Screen Entertainment Ltd   TV

7 Lilyville Road, London, SW6 5DP
℡ 020 7471 8989
tony.cash@btclick.com
MD *Tony Cash*

Drama and documentaries for TV. Output includes *Poetry in Motion* (series for Ch4); *South Bank Show: Ben Elton and Vanessa Redgrave*; *Musique Enquête* (drama-based French language series, Ch4); *Sex and Religion* (ITV); *Landscape and Memory* (arts documentary series for the BBC); Jonathan Miller's production of the *St Matthew Passion* for the BBC; major documentary on the BeeGees for the *South Bank Show*.

Scripts with an obvious application to TV may be considered. Interested in new writing for documentary programmes.

Lime Pictures Limited   `TV`

Campus Manor, Childwall Abbey Road, Childwall, Liverpool, L16 OJP
℡ 0151 722 9122 ℻ 0151 722 1969
www.limepictures.com

Formerly Mersey Television before changing its name in 2006. Specializes in drama productions for television. Examples of output include *Bonkers*, *Hollyoaks* and *The Outsiders*.

Lionsgate Films   `MOVIE`

Ariel House, 74a Charlotte Street, London, W1T 4QJ
℡ 020 7299 8800
www.lionsgatefilms.co.uk

Independent producer and distributor of motion pictures, television programming, home entertainment, family entertainment and video-on-demand content. Prides itself on producing 'original, daring, quality entertainment' for an international market. Past credits include *Bend It Like Beckham* and *Good Night and Good Luck*.

London Scientific Films Ltd   `MOVIE`

Dassels House, Dassels, Braughing, Ware, SG11 2RW
℡ 01763 289905
lsf@londonscientificfilms.co.uk
www.londonscientificfilms.co.uk
Contact *Mike Cockburn*

Film and video documentary and corporate programming.
No unsolicited manuscripts.

Lucida Productions   `TV`

5 Alleyn Crescent, London, SE21 8BN
℡ 020 8761 4344
pj.lucida@tiscali.co.uk
Contact *Paul Joyce*

Television and cinema: arts, adventure, current affairs, documentary, drama and music. Output has included *Motion and Emotion: The Films of Wim Wenders*; *Dirk Bogarde – By Myself*; *Sam Peckinpah – Man of Iron*; *Kris Kristofferson – Pilgrim*; *Wild One: Marlon Brando*; *Stanley Kubrick: 'The Invisible Man'*; *2001: the Making of a Myth* (Ch4); *Mantrap – Straw Dogs, the final cut* (with Dustin Hoffman). Restoration of the Director's Cut of *The Devils* plus the

documentary *Hell on Earth* with Ken Russell and Vanessa Redgrave. Currently in development for documentary projects.

Lumina   `MOVIE` `TV`

3rd Floor, 1A Adpar Street, London, W2 1DE
℡ 020 7535 6714 ℻ 020 7563 7283
ivana@lumina-films.com
www.lumina-films.com

Involved in financing and producing features, series and documentaries aimed at young adults. Titles include *Straightheads*, the series *City of Men* and *Favela Rising*.

Malone Gill Productions Ltd   `TV`

27 Campden Hill Road, London, W8 7DX
℡ 020 7937 0557 ℻ 020 7460 3750
malonegill@aol.com
Contact *Georgina Denison*

Mainly documentary but also some drama. Output includes *The Face of Russia* (PBS); *Vermeer* ('South Bank Show'); *Highlanders* (ITV); *Storm Chasers*; *Nature Perfected* and *The Feast of Christmas* (all for Ch4); *The Buried Mirror: Reflections on Spain and the New World* by Carlos Fuentes (BBC2/Discovery Channel).

Approach by letter with proposal in the first instance.

Marchmont Films   `MOVIE`

24 Three Cups Yard, Sandland Street, London, WC1R 4PZ
admin@marchmontfilms.com
www.marchmontfilms.com
Development Executive *Daniel Hayes*
Development Executive *Beverley Hills*

Founded in 2002, making films across genres including drama, comedy and thriller. Titles include *Out in the Cold* and *The Green Wave*.

Recieved 2,000 script submissions in 2006. Developing slate of low budget features in syndication with producers worldwide. Welcomes new writing. Visit website for details of current schemes and policies.

Material Entertainment   `MOVIE`

3rd Floor, 101–102 Jermyn Street, London, SW1Y 6EE
℡ 020 7808 3999 ℻ 020 7839 3514
info@material-ent.com
President *Robert Jones*

Joint feature film production venture between New Line Cinema in the US and Entertainment Film Distributors in the UK. Launched in 2005. Ultimately hopes to produce up to four films a year. First film was *Run Fat Boy Run* directed by David Schwimmer. Also a children's fantasy project, *Dream*.

Maverick Television (TV)

Progress Works, Heath Mill Lane, Birmingham, B9 4AL

☎ 0121 771 1812 🖷 0121 771 1550
mail@mavericktv.co.uk
www.mavericktv.co.uk
Head of Production *Allan Tott*

Established in 1994, Maverick has a strong reputation for popular factual programming as well as drama. It is now one of network television's most prolific independent suppliers. Output includes *10 Years Younger*; *Who'll Age Worst?*; *Bollywood Star*; *Fat Chance*; *Born Too Soon*; *VeeTV*; *Trade Secrets*; *Embarrassing Illnesses*; *10 Things You Didn't Know About ...*; *How To Live Longer*; *The Property Chain*; *Male, 33, Seeks Puberty*; *Extreme Engineering*; *Picture This: Accidental Hero*; *Up Your Street*; *The Property Chain*; *Motherless Daughters*; *Highland Bollywood: Black Bag*; *Health Alert: My Teenage Menopause*; *Long Haul*; *Learning to Love the Grey*.

Maya Vision International Ltd (MOVIE) (TV)

6 Kinghorn Street, London, EC1A 7HW
☎ 020 7796 4842 🖷 020 7796 4580
info@mayavisionint.com
www.mayavisionint.com
Contact *Tamsin Ranger*

Film and TV: drama and documentary. Output includes *Saddam's Killing Fields* (for 'Viewpoint', Central TV); *3 Steps to Heaven* and *A Bit of Scarlet* (feature films for BFI/Ch4); *A Place in the Sun* and *North of Vortex* (dramas for Ch4/Arts Council); *The Real History Show* (Ch4); *In Search of Myths and Heroes*; *In Search of Shakespeare*; *In the Footsteps of Alexander the Great*; *Conquistadors* (BBC documentaries); *Hitler's Search for the Holy Grail*; *Once Upon a Time in Iran* (Ch4 documentaries).
  Absolutely no unsolicited material; commissions only.

MBP TV (MOVIE)

Saucelands Barn, Coolham, Horsham, RH13 8QG
☎ 01403 741620 🖷 01403 741647
info@mbptv.com
www.mbptv.com
Contact *Phil Jennings*

Maker of film and video specializing in programmes covering equestrianism and the countryside.
  No unsolicited scripts, but always looking for new writers who are fully acquainted with the subject.

Melendez Films (MOVIE) (TV)

Julia House, 44 Newman Street, London, W1T 1QD
☎ 020 7323 5273 🖷 020 7323 5373
Contact *Steven Melendez*

Independent production company specializing in 2D animation. Also involved in production and film design for clients in England, Spain, Sweden, India and the US, plus website design and 3D animation on the web. Clients include book publishers, TV companies and advertisers. Winner of international awards for films, particularly of classic books, stories and comic characters.
  Will look at unsolicited projects in outline or synopsis form only. Enclose sae.

Mendoza Film Productions (MOVIE)

3–5 Barrett Street, London, W1U 1AY
☎ 020 7935 4674 🖷 020 7935 4417
office@mendozafilms.com
www.mendozafilms.com
Contact *Wynn Wheldon*
Contact *Debby Mendoza*

Commercials, title sequences (e.g. Alan Bleasdale's G.B.H.); party political broadcasts. Currently in pre-production on a feature-length comedy film. Involved with the Screenwriters' Workshop.
  Unsolicited manuscripts welcome but 'comedies only, please'. Material will not be returned without sae.

Mentorn (TV)

43 Whitfield Street, London, W1T 4HA
☎ 020 7258 6700 🖷 020 7258 6888
www.mentorn.co.uk

Independent television producers founded in 1985, with offices in London, Oxford and Glasgow. Produces across a range of genres including drama, current affairs, factual and entertainment. Titles include *The Government Inspector*, *The Hamburg Cell*, *A Very Social Secretary* and the documentaries *The Boy Who Gave Birth to His Twin* and *Terror in Moscow*.

Merchant Ivory (MOVIE)

46 Lexington Street, London, W1F 0LP
☎ 020 7437 1200 🖷 020 7734 1579
contact@merchantivory.com
www.merchantivory.com

Production company founded in 1961 by James Ivory and Ismail Merchant. Strong reputation for producing high-quality features, short films and documentaries. The company's output includes *A Room With a View, Howards End,*

The Remains of the Day, Jefferson in Paris, Surviving Picasso, The Mystic Masseur, Le Divorce, Heights and *The White Countess*.

Met Film Productions [MOVIE] [TV]

Ealing Studios, Ealing Green, London, W5 5EP
T 020 8280 9125 F 020 8280 9111
www.metfilmproduction.co.uk
Managing Director *Jonny Persey*

Founded in 2007 by the team formerly at APT Films. Develops and produces commercially-driven feature films for the national and international markets. Credits at APT included *Deep Water, Wondrous Oblivion* and *Solomon & Gaenor*. Met titles include *French Film, Heavy Load, Little Ashes* and *Sweet Swan of Avon*.

Midsummer Films Productions [MOVIE]

33 St Lawrence Terrace, London, W10 5SR
T 020 8932 8870 F 020 8932 8871
sales@midsummerfilms.com
www.midsummerfilms.com
Head of Development *Terhi Kylliainen*

Film production company founded in 1997.
Specializes in action, horror and destinctive niche films. Titles include *An American Haunting*.
Email with a CV and synopsis in the first instance.

Minds Eye Films [MOVIE]

2 Hazledene Road, London, W4 3JB
www.mindseyefilms.com

Film production company, founded by Johnny Kevorkian and Neil Murphy. Seeking to develop a diverse slate of feature films, including horrors, sci-fi and thrillers. Production credits include *The Disappeared, Fractured, Seizures* and *The Wake*.

Mirage [MOVIE]

Old Chapel Studios, 19 Fleet Street, London, NW3 2QR
T 020 7284 5588 F 020 7284 5599
tbricknell@mirageenterprises.co.uk

Film production company whose credits include *Birthday Girl, Heaven, Cold Mountain* and *Breaking and Entering*.

Mirage Films [MOVIE] [TV]

5 Wardor Mews, London, W1F 8AL
T 020 7734 3627 F 020 7734 3735
production@miragefilms.net
www.miragefilms.net

Film and video production company focusing on documentary alongside TV commercial and corporate production. Experienced in foreign adaptations.

Missing In Action Films Ltd [MOVIE]

c/o Flat 1, 2 Whitfield Street, London, W1T 2RB
T 020 7323 2895
info@miafilms.co.uk
www.miafilms.co.uk

A short and feature film production company, founded by Mia Bays in 2003. Credits include *Six Shooter, Ex Memoria* and *Scott Walker–30 Century Man*.

Mission Pictures [MOVIE]

23 Golden Square, London, W1F 9JP
T 020 7734 6303 F 020 7734 6202
info@missionpictures.net

Film production company whose past credits include *Thunderpants, Gladiatress, Millions* and *Piccadilly Jim*.

Mob Film Company [MOVIE] [TV]

10–11 Great Russell Street, London, WC1B 3NH
T 020 7580 8142 F 020 7255 1721
mail@mobfilm.com
www.mobfilm.com

Founded in 1998 to produce commercials, documentary, television drama and film. Offices in London and Leeds. Examples of output includes: the films *One More Kiss, The Virgin of Liverpool* and *Scenes of a Sexual Nature*; the TV drama *Hogfather*; and the factual programmes *Working The Thames* and *Skin Deep*.
Scripts (with synopsis) should be emailed or posted to Andrew Boswell (andrew@mobfilm.com).

Moonstone Films [TV]

London
T 020 8144 9940 F 087 0005 6839
tonystark@moonstonefilms.co.uk
www.moonstonefilms.co.uk
Executive Producer *Tony Stark*

Founded in 1989, producing documentaries for television. Has made programmes for the BBC and Channel 4. Output includes *Arafat Investigated, Arafat's Authority, Under Pressure, The Mind of a Martyr* and *Collaborators*.
Happy to look at scripts/ideas. Approach by email.

Mosaic Films [TV]

4th Floor, Shacklewell Studios, 28 Shacklewell Lane, London, E8 2EZ
T 020 7923 2994 F 020 7923 2994
info@mosaicfilms.com
www.mosaicfilms.com

Specializes in high-end documentaries for broadcasters including the BBC, Channel 4 and ITV. Particular expertise in international co-productions. Output includes *A Year in Tibet, Blood Matters, Made in India, Mish Kids, The Thames Through Time* and *To Courtney, with Love.*

Mulholland Pictures   (MOVIE)

19 Barnsbury Park, London, N1 1HQ
(T) 020 7607 7120 (F) 020 7700 4428
dejong@mulhollandpictures.com
www.mulhollandpictures.com
General Manager *Ate de Jong*

Founded in 1999, a Dutch company with a London office. Productions include *Left Luggage, The Discovery of Heaven* and *Fogbound.* Also co-produced *Enigma.*
   Will look at one-page synopses but does not encourage unsolicited material.

Neal Street Productions   (MOVIE)

26–28 Neal Street, London, WC2H 9QQ
(T) 020 7240 8890 (F) 020 7240 7099
post@nealstreetproductions.com
www.nealstreetproductions.com

Independent film and theatre production company established in 2003 by Sam Mendes, Pippa Harris and Caro Newling. Past productions include *Jarhead.*
   Does not accept unsolicited material and will immediately return to sender, unread.

Neon   (TV)

Studio Two, 19 Marine Crescent, Kinning Park, Glasgow, G51 1HD
(T) 0141 429 6366 (F) 0141 429 6377
stephy@go2neon.com
www.go2neon.com
Contact *Stephanie Pordage*

Television and radio: drama and documentary producers. Output includes *Brand New Country; Asian Overground; Peeking Past the Gates of Skibo.* Supports and encourages new writing 'at every opportunity'.
   Welcomes unsolicited material but telephone in the first instance.

Number 9 Films   (MOVIE)

Linton House, 24 Wells Street, London, W1T 3PH
(T) 020 7323 4060 (F) 020 7323 0456
info@number9films.co.uk
Contact *Stephen Woolley*
Contact *Elizabeth Karlsen*

Leading feature film producer. Output includes *Breakfast on Pluto; Stoned; Mrs Harris.* Forthcoming productions: *And When Did You*

Last See Your Father?; How to Lose Friends and Alienate People; Edith and the Lonely Doll.
   No unsolicited material.

October Films Ltd   (MOVIE) (TV)

Lymehouse Studios, 30/31 Lyme Street, London, NW1 0EE
(T) 020 7284 6868 (F) 020 7284 6869
info@octoberfilms.co.uk
www.octoberfilms.co.uk

Independent film and TV producer operating since 1989. Offices in London and Manchester. Produces for a range of domestic and international broadcasters. Output includes documentary, factual entertainment, popular current affairs, docu-drama, history and science. Titles include *Godless in America, The Fundamentalists, Srebrenica: Never Again?, The Boy Who Lived Before, The Fight for Ground Zero, The Last Slave* and *Meeting The Taleban.*

Omnivision Ltd   (TV)

Pinewood Studios, Iver Heath, Buckinghamshire, SL0 0NH
(T) 01753 656 329 (F) 01753 631 145
info@omnivision.co.uk
www.omnivision.co.uk
Managing Director *Christopher Morris*
Senior Editor *Nick Long*

A TV production and post-production company founded in 2000. Produces TV documentaries for the UK and abroad, along with news coverage and tv series. Also produces corporate DVDs for the health industry.
   Welcomes approaches by letter or email. Offers quick decisions on material.

Orlando TV Productions   (TV)

Up-the-Steps, Little Tew, Chipping Norton, OX7 4JB
(T) 01608 683218 (F) 01608 683364
info@orlandomedia.co.uk
www.orlandomedia.co.uk
Contact *Mike Tomlinson*

Producer of TV documentaries and digital multimedia content, with science, health and information technology subjects as a specialization.
   Approaches by established writers/journalists to discuss proposals for collaboration are welcome.

Orpheus Productions   (TV)

6 Amyand Park Gardens, Twickenham, TW1 3HS
(T) 020 8892 3172 (F) 020 8892 4821
richard-taylor@blueyonder.co.uk
Contact *Richard Taylor*

Television documentaries and corporate work. Output has included programmes for the BBC, ITV and Ch4 as well as documentaries for the United Nations, the Shell Film Unit and Video Arts.

Unsolicited scripts are welcomed with caution. 'Our preference is for the more classically structured documentary that, while being hard-hitting, explores the subtleties and the paradox of an issue – and is not presented by unqualified celebrities.'

Outcast Production (MOVIE)

92 Buckhold Road, London, SW18 4AP
andythewise@aol.com
Contact *Andreas Wisniewski*

Low-budget feature films.

No unsolicited manuscripts; send synopsis or treatment only. 'We are actively searching for and encouraging new writing.'

Oxford Film and Television (MOVIE) (TV)

6 Erskine Road, London, NW3 3AJ
℡ 020 7483 3637 Ⓕ 020 7483 3567
email@oftv.co.uk
www.oftv.co.uk
Creative Director *Nicolas Kent*
Head of Drama *Mark Bentley*
Head of Production *Annie Lee*
Head of Documentaries *Patrick Forbes*

Independent film and television production company. Produced over 100 hours of drama, docu-drama and documentaries for the BBC, Channel Four, ITV, PBS, Showtime, Lifetime, A&E, HBO, Bravo, ABC Australia, Discovery and The Learning Channel. Produced six feature films including the Oscar-winning *Restoration* and the Oscar and BAFTA-nominated *Hilary and Jackie*. To date has won two Oscars, four BAFTAs, two Indies, an RTS award, the Silver Hugo at the Chicago Film Festival and a Cable ACE.

Paladin Invision (TV)

8 Barb Mews, London, W6 7PA
℡ 020 7348 1956 Ⓕ 020 7371 2160
pitv@pitv.com
www.pitv.com
Founder/Producer *William Clan*
Founder/Producer *Clive Syddall*

Co-venture between two influential British documentary-makers: William Cran and Clive Syddall. Combined 500 hours of television programming produced. Focuses on factual projects including history, current affairs, religion, music & arts documentaries. Sold both at home and internationally. Output includes *The Age of Aids* (WGBH/PBS/Channel

4/Granada International), *Do You Speak American?* (WNET-BBC/CBCPBS), *Extreme Oil* (WNET-PBS/BBC/CBC/C4 International) and landmark six-part series *Commanding Heights: The Battle for the World Economy*. Awards received from BAFTA, Royal Television Society, the Chicago, San Francisco, and New York Television Festivals, the Peabody Award for Broadcasting and four American Emmy awards.

Paper Moon Productions (TV)

Wychwood House, Burchetts Green Lane, Littlewick Green, Nr. Maidenhead, SL6 3QW
℡ 01628 829819 Ⓕ 01628 829819
david@paper-moon.co.uk
Contact *David Haggas*

Broadcast documentaries and corporate communications. Recent output includes *Bilbo & Beyond*, an affectionate glimpse into the life and work of the dedicated philologist and fantasy writer J.R.R. Tolkien.

Parallax East Ltd (MOVIE)

Victoria Chambers, St Runwald Street, Colchester, CO1 1HF
℡ 01206 574909 Ⓕ 01206 577669
assistant@parallaxindependent.co.uk
Contact *Sally Hibbin*

Founded in 2005, producing films including *Almost Adult* and *Yasmin*.

Very receptive to new writing but doesn't encourage unsolicited approaches.

Passion Pictures (MOVIE)

3rd Floor, 33–34 Rathbone Place, London, W1T 1JN
℡ 020 7323 9933 Ⓕ 020 7323 9030
info@passion-pictures.com
www.passion-pictures.com
Managing Director *Andrew Ruhemann*

Documentary and drama includes: *One Day in September* (Academy Award-winner for Best Feature Documentary, 2000); also commercials and music videos.

Unsolicited manuscripts welcome.

Pathé Pictures (MOVIE)

Kent House, 14–17 Market Place, Great Titchfield Street, London, W1W 8AR
℡ 020 7323 5151 Ⓕ 020 7631 3568
press.office@pathe-uk.com
www.pathe.co.uk

Feature film production company. 4–6 titles per year. Examples of recent credits include *Enduring Love*, *Bride and Prejudice - A Bollywood Musical*, *The Magic Roundabout*, *Mrs Henderson Presents*, *The Queen* and *Big Nothing*.

Happy to consider any material that has representation from an agent or production company.

Pelicula Films   (TV)

59 Holland Street, Glasgow, G2 4NJ
T 0141 287 9522
Contact *Mike Alexander*

Television producer. Maker of drama documentaries and music programmes for the BBC and Ch4. Output includes *As an Eilean* (From the Island), *The Trans-Atlantic Sessions 1 & 2, Nanci Griffith, Other Voices 2* and *Follow the Moonstone*.

Photoplay Productions Ltd   (MOVIE) (TV)

21 Princess Road, London, NW1 8JR
T 020 7722 2500 F 020 7722 6662
info@photoplay.co.uk
www.photoplay.co.uk
Contact *Patrick Stanbury*

Documentaries for film, television and video plus restoration of silent films and their theatrical presentation. Output includes *The Cat and the Canary*; *Orphans of the Storm*; *Cecil B. DeMille: American Epic* and the 'Channel 4 Silents' series of silent film restoration, including *The Wedding March* and *The Iron Mask*. Recently completed *Garbo* and *I'm King Kong!*

No unsolicited manuscripts; 'we tend to create and write all our own programmes.'

Picture Palace Films   (MOVIE) (TV)

13 Egbert Street, London, NW1 8LJ
T 020 7586 8763 F 020 7586 9048
info@picturepalace.com
www.picturepalace.com
Producer and Chief Executive *Malcolm Craddock*
Development Assistant *Cathy Simpson*

Founded in 1971, producing films and TV dramas. Works include *Acid House*, the *Sharpe* films (for ITV), *Frances Tuesday*, *A Life for a Life* (for ITV) and *Rebel Heart* (for BBC1).

Welcomes new writing in principle (though admits to being 'often disappointed') via agents.

Planet24 Pictures Ltd   (TV)

Octagon House, Fir Road, Stockport, SK7 2NP
T 0870 765 8780
planet24picture@aol.com
www.planet24pictures.co.uk
Contact *Mercedes de Dunewíc*

Producer of television and video documentaries plus community/infomercials. Output includes *Jack the Ripper: The Conspiracies*; *Truth to Tell* ; *Pitmen and Politics: The County Built on Coal*.

Manuscripts, plot outlines and screenplays considered but email for information first. 'Welcome the input of new writers with radical points of view.'

Plantagenet Films Limited   (MOVIE) (TV)

Ard-Daraich Studio, B Ardgour, Nr Fort William, PH33 7AB
T 01855 841384 F 01855 841384
plantagenetfilms@aol.com
Contact *Norrie Maclaren*

Film and television: documentary and drama programming such as *Dig* (gardening series for Ch4), various *Dispatches* for Ch4 and *Omnibus* for BBC.

Keen to encourage and promote new writing; unsolicited manuscripts welcome.

Portobello Pictures Ltd   (MOVIE) (TV)

12 Addison Avenue, Holland Park, London, W11 4QR
T 020 7605 1396 F 020 7605 1391
mail@portobellopictures.com
www.portobellopictures.com
Producer *Eric Abraham*
Associate Producer *Kate McCullagh*
Head of Development *Chris Ciancimino*

Founded in 1985, producing all kinds of drama for film, TV and theatre. Productions include Roald Dahl's *Danny the Champion of the World, Kolya* (which won Best Foreign Language Oscar in 1997), *Dark Blue World* and *Dalziel and Pascoe*.

Interested in new writing and actively scouts new writing platforms at festivals, on television, in the theatre and via agents. Does not invite unsolicited material though.

Powerstone Entertainment Ltd   (MOVIE) (TV)

info@powerstonefilms.com
www.powerstonefilms.com

Develops and produces feature films and drama, aiming for universal stories with international appeal. Particular interest in films for children, youth and families but also caters for older audiences.

Pozzitive Television   (TV)

5th Floor, Paramount House, 162–170 Wardour Street, London, W1F 8AB
T 020 7734 3258 F 020 7437 3130
pozzitive@pozzitive.co.uk
www.pozzitive.co.uk
Contact *David Tyler*

Producer of comedy and entertainment for television and radio. Output includes *Dinner Ladies; Coogan's Run; The 99p Challenge;*

The Comic Side of 7 Days; Armando Iannucci's Charm Offensive.

Unsolicited manuscripts of TV and radio comedy welcome. 'No screenplays or stage plays or novels, please. Send hard copy of full sample script. We read everything submitted this way. Sorry, we don't return scripts unless you send an sae.'

Prescience Film Finance Ltd   MOVIE

Canon House, 27 London End, Beaconsfield, Bucks, HP9 2HN
T 01494 670737 F 01494 670740
info@presciencefilmfinance.co.uk
www.presciencefilmfinance.co.uk
Director *Paul Brett*

Specialist film company founded in 2004. Works with leading UK and international producers, distributors and sales agents. Has working relationships with a range of international studios (Hollywood majrs and mini-majors), as well as leading film and television producers.

Recent titles include *Tideland* (2005), *The Ferryman* (2007) and *How About You* (2007).

Welcomes unsolicited approaches.

Princess Productions   TV

3rd Floor, Whiteley's Centre, 151 Queensway, London, W2 4YN
T 020 7985 1985 F 020 7985 1986
formatsales@princesstv.com
www.princesstv.com

TV production company established in 1996, producing entertainment programmes across a range of genres. Titles include *The Wright Stuff*, *The Friday Night Project*, *Lily Allen & Friends*, *Mobile Act Unsigned* and *Something for the Weekend*. Now part of the Shine Group.

Priority Pictures   MOVIE

9 Grafton Mews, London, W1T 5HZ
T 0207 380 3983 F 0207 380 3981
www.prioritypictures.co.uk

Founded by Marion Pilowsky and Colin Leventhal in 2004. Develops, finances and produces feature films for a worldwide audience. Credits include *The All Together*, *Nina's Heavenly Delights*, *Stingray* and *Teacher Boy*.

Qwerty Films   MOVIE

42–44 Beak Street, London, W1F 9RH
T 020 7440 5920 F 020 7440 5959
info@qwertyfilms.com

Feature film producer. Production credits include *Alien Autopsy*, *Aspects of Love*, *Severance* and *Stage Beauty*.

Sarah Radclyffe Productions   MOVIE

5th Floor, 83–84 Berwick Street, London, W1V 3PL
T 020 7437 3128 F 020 7437 3129
srpltd@globalnet.co.uk

Film production company established by the co-founder of Working Title. Credits include *Love's Brother*, *Free Jimmy* and *Tara Road*.

Ragdoll Ltd   TV

Timothy's Bridge Road, Stratford-upon-Avon, Warks, CV37 9NQ
T 01789 404100 F 01789 404136
info@ragdoll.co.uk
www.ragdoll.co.uk

Founded by Anne Wood. Notable titles include *Teletubbies*, *Brum*, *Boohbah*, *Rosie & Jim*, *BadJelly the Witch*, *Pob* and *Open a Door* (an international exchange of five minute films). *In The Night Garden* and *Tronji* are broadcast on the BBC.

RDF Media Group   TV

The Gloucester Building, Kensington Village, London, W14 8RF
T 020 7013 4000 F 020 7013 4001
www.rdfmedia.com

Has several programming production operations. RDF Television is the largest production company within the group, with offices in London and Bristol. Works across a range of genres. Credits include *Wife Swap*, *Faking It*, *Ladette to Lady*, *Scrapheap Challenge*, *Shipwrecked* and *How To Be A Property Developer*. The Foundation, based at The Maidstone Studios, specializes in pre-school and family entertainment. Output includes *Mister Maker*, *Big Barn Farm* and *Escape from Scorpion Island*. IWC Media, with offices in Glasgow and London, produces factual, factual entertainment and drama programmes. Titles includes *Location, Location, Location*, *Fallen Angel*, *Mission Implausible*, *A Room with a View*, *Mountain* and *Crimes That Shook The World*. Presentable is based in Cardiff, producing entertainment, music, comedy, arts, history, features and documentary programmes. Particular reputation for poker programmes. Touchpaper Television is a drama producer, with credits including *The Queen's Sister*, *Rocket Man*, *The Best Man* and *City of Vice*. The Comedy Unit, based in Glasgow, produces radio and television. Output includes *Still Game*, *The Karen Dunbar Show* and *Empty*.

Recorded Picture Company   MOVIE

24 Hanway Street, London, W1T 1UH
T 020 7636 2251 F 020 7636 2261
rpc@recordedpicture.com
www.recordedpicture.com

Film production company established in 1971. Credits include *Merry Christmas, Mr. Lawrence, The Last Emperor* (winner of nine Oscars), *Crash, Sexy Beast, Tideland* and *Fast Food Nation*.

Red Planet `TV`

13 Doolittle Hill, Ampthill Froghall Road, Bedford, MK45 2ND
T 01525 408970 F 01525 408971
info@redplanetpictures.co.uk
www.redplanetpictures.co.uk

Independent production company of writer Tony Jordan. Output includes *Holby Blue, Echo Beach* and *Moving Wallpaper*.

Red Production Company `TV`

c/o Granada TV, Quay Street, Manchester, M60 9EA
T 0161 827 2530 F 0161 827 2518
info@redlimited.co.uk
www.redproductioncompany.com
Chief Executive *Nicola Shindler*
Managing Director *Andrew Critchley*

Founded in 1998, aims to bring original and intersting voices to the screen in award-winning productions.
   Titles include *New Street Law, Clocking Off, The Second Coming, Queer as Folk, The Mark of Cain* and *Casanova*.
   Welcomes new writing and reads all submissions. Keen to work with fresh, new talent. Approach with ideas via post only (include sae).

Redweather `TV`

Easton Business Centre, Felix Road, Bristol, BS5 0HE
T 0117 941 5854 F 0117 941 5851
enquiries@redweather.co.uk
www.redweather.co.uk

Broadcast documentaries on arts and disability, corporate video and CD-ROM.

Renaissance Films `MOVIE`

34/35 Berwick Street, London, W1F 8RP
T 020 7287 5190 F 020 7287 5191
info@renaissance-films.com

Film production company whose credits include *Henry V, Peter's Friends, Much Ado About Nothing, The Luzhin Defence, The Reckoning, Inbreeds, We Don't Live Here Anymore* and *Candy*.

Revolution Films `MOVIE`

9a Dallington Street, London, EC1V 0BQ
T 020 7566 0700 F 020 7566 0701
email@revolution-films.com
www.revolution-films.com
Director *Michael Winterbottom*

Founded by producer Andrew Eaton and director Michael Winterbottom in 1994. Focuses mostly on feature-length dramatic projects. Recent output includes *The Cock and Bull Story, The Road to Guantanamo* and *A Mighty Heart*.
   No unsolicited material.

Richmond Films & Television `MOVIE` `TV`

PO Box 33154, London, NW3 4AZ
T 020 7722 6464
mail@richmondfilms.com
Head of Development *John Calkins*

Founded in 1986 and has won Bafta and RTS awards. Credits include *Press Gang, Wavelength* and *in2minds*.
   Does not welcome unsolicited material. Will look at if sent by an agent. Seeks short treatments with all info about submissions of project and reaction to it.

Rocket Pictures `MOVIE` `TV`

1 Blythe Road, London, W14 0HE
T 020 7603 9530 F 020 7348 4830
luke@rocketpictures.co.uk

Film and TV production company established by Elton John in 1996. Credits include *Women Talking Dirty, It's a Boy Girl Thing* and *Him and Us*.

Rocliffe `MOVIE`

PO Box 37344, London, N1 8YB
scripts@rocliffe.com
www.rocliffe.com

A production company that has produced the short films *Chicken Soup, No Deposit, No Return* and *The Welcome Committee* (commissioned by the European Commission, CICEB and The British Council to celebrate the enlargement of Europe). Currently developing its feature film slate and keen to build relationships with writers and directors.

RS Productions `MOVIE` `TV`

191 Trewhitt Road, Newcastle-upon-Tyne, NE6 5DY
T 07710064632
enquiries@rsproductions.co.uk
www.rsproductions.co.uk
Contact *Mark Lavendar*

Feature films and television: drama series/serials and singles. TV documentaries and series. Working with established and new talent.

Ruby Films MOVIE

26 Lloyd Baker Street, London,
WC1X 9AW
T 020 7833 9990 F 020 7837 5862
info@rubyfilms.co.uk
Contact *Faye Ward*

Established in 1999 by Alison Owen and Neris
Thomas. Production credits include *Sylvia, Love
+ Hate, Love and Other Disasters, Brick Lane* and
The Other Boleyn Girl.

Sands Films MOVIE TV

119 Rotherhithe Street, London, SE16 4NF
T 020 7231 2209 F 020 7231 2119
info@sandsfilms.co.uk
www.sandsfilms.co.uk
Contact *Christine Edzard*
Contact *Olivier Stockman*

Film and TV drama. Output includes *Little
Dorrit; The Fool; As You Like It; A Dangerous
Man; The Long Day Closes; A Passage to India;
Topsy Turvy; Nicholas Nickleby; The Gangs of
New York; The Children's Midsummer Night's
Dream.*
 No unsolicited scripts.

Scala Productions MOVIE

2nd Floor, 37 Foley Street, London, W1W 7TN
T 020 7637 5720 F 020 7637 5734
scalaprods@aol.com
Head of Production *Ian Prior*
Chairman *Nik Powell*

Founded in 1992, producing feature films. Set up
by ex-Palace Production's Nik Powell and Stephen
Woolley. Credits include the Oscar-nominated
and BAFTA-winning *Little Voice*, Shane Meadows
*24 7: Twenty Four
Seven*, Fred Schepisi's *Last Orders, Backbeat,
Leo* and *Ladies in Lavendar* (which made over
US$30m).
 A keen supporter of new writing through its
relationships with leading talent agents, links to
established and emerging academic institutions
and its patronage of theatre nationwide. Does not
invite unsolicited submissions.

Scope Productions Ltd MOVIE

180 West Regent Street, Glasgow, G2 4RW
T 0141 221 4312
laurakingwell@scopeproductions.co.uk
www.scopeproductions.co.uk
Contact *Laura Kingwell*

Corporate film and video and multimedia
communications for clients across all
sectors.

Screen First Ltd MOVIE TV

The Studios, Funnells Farm, Down Street, Nutley,
East Sussex, TN22 3LG
T 01825 712034
paul.madden@virgin.net
Producer/Director *Paul Madden*

Founded in 1985 to make films and television
programmes.
 Focus on animation, arts, children's
programmes, documentary and drama. Examples
of output include *Ivor the Invisible.*
 Does not welcome unsolicited approaches.

Screen Ventures Ltd MOVIE TV

49 Goodge Street, London, W1T 1TE
T 020 7580 7448 F 020 7631 1265
info@screenventures.com
www.screenventures.com
Contact *Christopher Mould*
Contact *Michael Evans*

Film and TV sales and production:
documentary, music videos and drama.
Output includes *Life and Limb* (documentary,
Discovery Health Channel); *Pavement
Aristocrats* (SABC); *Woodstock Diary; Vanessa
Redgrave* and *Genet* (both for LWT 'South
Bank Show'); *Mojo Working; Burma: Dying
for Democracy* (Ch4); *Dani Dares* (Ch4
series on strong women); *Pagad* (Ch4 news
report).

Screenhouse Productions Ltd TV

Chapel Allerton House, 114 Harrogate Road,
Leeds, LS7 4NY
T 0113 266 8881 F 0113 266 8882
paul.bader@screenhouse.co.uk
www.screenhouse.co.uk
Contact *Paul Bader*
Contact *Barbara Govan*

Specializes in science TV, documentary, stunts
and events, including outside broadcasts. Output
includes *Stardate* (BBC2 astronomy series),
Zapped (Discovery/US/Canada), *The Man Who
Invented the Aeroplane* (UKTV/BBC North);
Science Shack, Local Heroes (BBC2, presented by
Adam Hart-Davis).
 'More likely to consider written up proposals.'

Screenprojex.com – a division of Screen Production Associates Ltd MOVIE

13 Manette Street, London, W1D 4AW
T 020 7287 1170 F 020 7287 1123
info@screenprojex.com
screenprojex.com
Contact *Doug Abbott*
Contact *John Jaquiss*

Feature films: *The Fourth Man*; *The Case*; *Black Badge*; *The Truth Game*; *Club Le Monde*; *Midnight Warriors*; *Strong Language*; *Holding On*; *Chunky Monkey*; *Sixty-three Closure*.

No unsolicited manuscripts. Send preliminary letter outlining project and CV.

September Films Ltd   [MOVIE] [TV]

Glen House, 22 Glenthorne Road, London, W6 0NG
☏ 020 8563 9393 🖷 020 8741 7214
september@septemberfilms.com
www.septemberfilms.com
Chairman *David Green*
Director of Production *Elaine Day*

Founded in 1985, working in film and television.

Specializes in popular factual entertainment, reality and entertainment formats, documentaries, feature films and TV movies. Offices in London and LA. Recent titles include *Beauty and the Geek*, *Bridezillas*, *Haunted Homes* and *Breathtaking*. Other highlights include *Ozzy Osbourne Uncut* (winner of Montreux Bronze Rose Award), the award-winning feature *House of America* and the Oscar-nominated *Solomon and Gaenor*.

Does not usually accept unsolicited manuscripts. Has own in-house development team.

Serendipity Picture Company   [TV]

Media Cabin, 11 Lyndhurst Road, Westbury on Trym, BS9 3QY
☏ 0117 908 2711
tony@serendipitypictures.com
Director *Tony Yeadon*

Founded in 1984. Specializes in television documentaries with focus on travel, food and history. Was awarded first runner up prize by the One World Broadcasting Trust for TV Documentaries at BAFTA.

Only interested in documentary ideas. Email in first instance.

Shine Entertainment   [TV]

Newcombe House, 43–45 Notting Hill Gate, Notting Hill, London, W11 3LQ
☏ 020 7985 7000 🖷 020 7985 7001
info@shinelimited.com
www.shinelimited.com

Founded by Elisabeth Murdoch in 2001. Makes a range of programmes for the UK, US and international markets including drama, entertainment, factual, features and format entertainment programming. Examples of titles include *Sinchronicity*, *Sugar Rush*, *The Story of Light Entertainment* and *1970s: That Was The Decade That Was*.

Does not accept unsolicited manuscripts.

Sianco Cyf   [TV]

Pen-y-graig Llanfaglan, Caernarfon, LL4 5RF
☏ 01286 676100 🖷 01286 677616
sian@sianco.tv
Contact *Siân Teifi*

Children's, youth and education programmes, children's drama, people-based documentaries for adults.

Does not accept any unsolicited scripts.

Silent Sound Films   [MOVIE] [TV]

Cambridge Court, Cambridge Road, Frinton-on-Sea, CO13 9HN
☏ 01255 676381
thj@silentsoundfilms.co.uk
www.silentsoundfilms.co.uk
Managing Director *Timothy Foster*

Production company founded in 1998 with background in fine art and art-house films. Also involved in television arts documentaries, with co-production contacts in France, the Netherlands and Hungary.

No manuscripts. Prefer to be approached via email with 50 word soundbite. Very rarely takes up unsolicited material.

Sixteen Films   [MOVIE] [TV]

2nd Floor, 187 Wardour Street, London, W1F 8ZB
☏ 020 7734 0168 🖷 020 7439 4196
eimhear@sixteenfilms.co.uk
www.sixteenfilms.co.uk
Director *Ken Loach*
Producer *Rebecca O'Brien*
Screenwriter *Paul Laverty*
Assistant to Ken Loach and Paul Laverty *Ann Cattrall*
Assistant to Rebecca O'Brien *Eimher McMahon*

Established by Ken Loach and Rebecca O'Brien in spring 2002 following the dissolution of Parallas Pictures. Concentrates on film and television drama. Recent production *The Wind That Shakes The Barley* won Best Film at the 2006 Cannes Film Festival.

Not currently seeking new projects. However, unsolicited submissions (treatment or screenplay) are accepted via post FAO Alistair Griggs. Responses take a minimum of six weeks.

Skyline Productions   [MOVIE] [TV]

10 Scotland Street, Edinburgh, EH3 6PS
☏ 0131 557 4580 🖷 0131 556 4377
leslie@skyline.uk.com
www.skyline.uk.com
Producer/Writer *Leslie Hills*

Produces film and television drama and documentary.

Slate Films   MOVIE TV

20 Great Chapel Street, London,
W1F 8FW
T 020 7734 1217 F 020 7287 9622
info@slatefilms.com
www.slatefilms.com
Executive/Producer *Andrea Calderwood*
Head of Development *Vicky Patterson*

International film and television production
company established by Andrea Calderwood
in 2000. Projects cover a range of budgets and
markets, from mainstream cinema comedies to
edgy first feature films, BBC1 series to Channel 4
drama serials. Notable credits include *Once Upon A
Time In The Midlands* starring Robert Carlyle (Film
Four) and *Last King of Scotland* (co-production
feature film directed by Kevin MacDonald).

Sly Fox Films   MOVIE TV

The Far Barn, Foxhole Lane, Cranbrook,
TN18 5NJ
T 01580 752839
info@slyfoxfilms.com
www.slyfoxfilms.com
Producer *Linda James*

Founded in 1982 by Stephen Bayly and Linda
James, who have produced over 50 hours of prime
time drama, six TV movies and seven feature films.
Recent titles include *Slatehead* (a rock climbing
drama) and *The Icarus Girl* (a supernatural drama).
  Does not encourage unsolicited material. The
principals predominately generate projects from
their own ideas or by optioning books to adapt.

SMG Productions & Ginger Productions   TV

Pacific Quay, Glasgow, G51 1PQ
T 0141 300 3000
website@smgproductions.tv
www.smgproductions.tv
Managing Director *Elizabeth Partyka*
Head of Drama *Eric Coulter*

SMG Productions, which incorporates Ginger
Productions, makes programmes for the national
television networks, including ITV, Ch4 and
Sky. Specializes in drama, factual entertainment
and children's programming. Output includes
Taggart; Rebus; Our Daughter Holly; Club Reps.
(Ginger Productions at 3 Waterhouse Square,
138–142 Holborn, London EC1N 2NY. T 020 7882
1020 W ginger.tv E production@ginger.com)

So Television Ltd   TV

18 Hatfields, London, SE1 8GN
T 020 7960 2000 F 020 7960 2095
info@sotelevision.co.uk
www.sotelevision.co.uk

Set up in 2000 by Graham Norton and Graham
Stuart, producing entertainment programmes
for TV. Titles include *The Graham Norton Show*,
School's Out and *Bring Back Dallas*.

Somethin' Else   TV

20–26 Brunswick Place, London, N1 6DZ
T 020 7250 5500 F 020 7250 0937
info@somethinelse.com
www.somethinelse.com
Contact *Jez Nelson*

Producer of television, video and radio
documentaries, DVD and interactive content.
  Ideas for TV shows welcome; send letter in the
first instance.

Specific Films   MOVIE

25 Rathbone Street, London, W1T 1NQ
T 020 7580 7476 F 020 7636 6886
info@specificfilms.com
Managing Director *Michael Hamlyn*

Film production company whose titles include
Priscilla, Queen of the Desert.

Spellbound Productions Ltd   MOVIE TV

90 Cowdenbeath Path, Islington, London,
N1 0LG
T 020 7713 8066 F 020 7713 8067
phspellbound@hotmail.com
Contact *Paul Harris*

Specializes in feature films for cinema and drama
for television. Keen to support and encourage
new writing.
  Material will only be considered if in correct
screenplay format and accompanied by sae.

Spice Factory   MOVIE

14 Regent Hill, Brighton, BN1 3ED
T 01273 739182 F 01273 749122
info@spicefactory.co.uk
www.spicefactory.co.uk
Development Co-ordinator *Shirine Best*
Head of Development *Lucy Shuttleworth*

Established in 1994, has now produced,
co-produced or co-financed more than forty
films (with over US$300m. of production
spend).
  Recent titles include *Perfect Creature* and
Merchant of Venice.
  Does not welcome unsolicited approaches.

'Spoken' Image Ltd   VG MOVIE

8 Hewitt Street, Manchester, M15 4GB
T 0161 236 7522 F 0161 236 0020
info@spoken-image.com

www.spoken-image.com
Contact *Geoff Allman*

Film, video and TV production for documentary and corporate material. Specializes in high-quality brochures and reports, CD-ROMs, exhibitions, conferences, film and video production for broadcast, industry and commerce.

Stagescreen Productions   MOVIE TV

12 Upper Saint Martins Lane, London, WC2H 9JY
℡ 020 7497 2510 🄵 020 7497 2208
info@stagescreenproductions.com
www.stagescreenproductions.com
CEO *Jeffrey Taylor*
Development Executive *John Segal*

Founded in 1989 and has worked on a number of prestigious films and television programmes. Particular expertise in arranging international co-productions. Output includes *What's Cooking* (a comedy-drama that opened at Sundance), the adventure film *Young Alexander* and *Jekyll*, a six-hour series for the BBC.
    Does not accept unsolicited material.

Tony Staveacre Productions   TV

Channel View, Blagdon, BS40 7TP
℡ 01761 462161 🄵 01761 462161
newstaving@btinternet.com
Contact *Tony Staveacre*

Producer of dramas and documentaries as well as music, arts and comedy progammes. Recent Output includes *Tango Maestro* (BBC4) and *The Old Boys Band* (BBC1).
    No unsolicited manuscripts.

Stirling Film & TV Productions Limited   TV

137 University Street, Belfast, BT7 1HP
℡ 028 9033 3848 🄵 028 9043 8644
anne@stirlingtelevision.co.uk
www.stirlingtelevision.co.uk
Contact *Anne Stirling*

Producer of broadcast and corporate programming– documentary, sport, entertainment and lifestyle programmes.

Straight Forward Film & Television   MOVIE TV

Building 2, Lesley Office Park, 393 Hollywood Road, Belfast, BT4 2LS
℡ 028 9065 1010 🄵 028 9065 1012
enquiries@straightforwardltd.co.uk
www.straightforwardltd.co.uk
Contact *John Nicholson*
Contact *Ian Kennedy*

Northern Ireland-based production company specializing in documentary, feature and lifestyle series for both regional and network transmission. Output includes *We Shall Overcome* (winner of Best Documentary at 1999 Celtic Television Festival for BBC); *Conquering the Normans* (Ch4 Learning – history of Normans in Ireland); *Gift of the Gab* (Ch4 Learning – contemporary Irish writing); *Sportsweek* (BBC Radio Ulster); *On Eagle's Wing* (full stage musical/TV material; story of the Scots/Irish in America; *Fire School; Mission Employable; Sweet Child of Mine; School Challenge*, 3rd series; *World Indoor Bowls* (all for BBC NI); *Awash With Colour* (series, BBC Daytime).

Sunset + Vine Productions Ltd   TV

Elsinore House, 77 Fulham Palace Road, London, W6 8JA
℡ 020 7478 7400 🄵 020 7478 7412
enquiries@sunsetvine.co.uk
www.sunsetvine.co.uk

Established in 1983, a leading independent supplier of sports programmes. Produces over 2,000 hours per year.

Table Top Productions   MOVIE TV

1 The Orchard, London, W4 1JZ
℡ 020 8742 0507 🄵 020 8742 0507
top@tabletopproductions.com
www.tabletopproductions.com
Production Manager *Ben Berry*

Film and television production company, founded in 1989 by producer/director/writer Alvin Rakoff and his late wife, the actress Jacqueline Hill. Output includes *Paradise Postponed, The Adventures of Don Quixote, A Voyage Round My Father, The First Olympics 1896, Dirty Tricks, A Dance to the Music of Time* and *Too Marvellous for Words.*
    Does not accept unsolicited manuscripts.

talkbackTHAMES   TV

20–21 Newman Street, London, W1T 1PG
℡ 020 7861 8000 🄵 020 7861 8001
reception@talkbackthames.tv
www.talkbackthames.tv
Chief Executive Officer *Lorraine Heggessey*
Chief Operating Officer *Sara Geater*

talkackTHAMES Productions is a FremantleMedia company. Works across comedy, comedy drama , drama, entertainment, documentary and lifestyle programmes. Output includes *The Apprentice*; *Green Wing*; *The Bill*; *The X Factor*; *Never Mind the Buzzcocks*; *Grand Designs*; *Unteachables*; *Property Ladder.*

Talking Heads Productions VG MOVIE TV

2–4 Noel Street, London, W1F 8SB
T 020 7292 7575 F 020 7292 7576
johnsachs@talkingheadsproductions.com
www.talkingheadsproductions.com
Contact *John Sachs*

Film and television production company founded in 1993. Recently co-produced *Merchant of Venice*, starring Al Pacino.
 Unsolicited material accepted via email.

Tandem TV & Film Ltd TV

Charleston House, 13 High Street, Hemel Hempstead, Herts, HP1 3AA
T 01442 261576 F 01442 219250
info@tandemtv.com
www.tandemtv.com
Director *Terry Page*
Creative Director *Barbara Page*
Production Controller *Jevan Green*

Over twenty-five years experience of producing TV and corporate films. Examples of output include *One for the Road*, a story about an alcoholic produced for broadcast TV.
 Does not invite unsolicited material.

Taylor Made Broadcasts Ltd TV

3B Cromwell Park, Chipping Norton, OX7 5SR
T 01608 646444
post@tmtv.co.uk
Contact *Trevor Taylor*

Producer of documentaries for radio and television.
 No unsolicited manuscripts.

Telemagination MOVIE TV

Royalty House, 72–74 Dean Street, London, W1D 3SG
T 020 7434 1551 F 020 7434 3344
mail@tmation.co.uk
www.telemagination.co.uk
Managing Director *Beth Parker*

Children's animation production company best known for *Animals of Farthing Wood*, *Little Ghosts* and *The Cramp Twins*.
 Submission guidelines available upon request.

Tern Television Productions Ltd TV

73 Crown Street, Aberdeen, AB11 6EX
T 01224 211123 F 01224 211199
aberdeen@terntv.com
www.terntv.com
Contact *David Strachan*
Contact *Gwyneth Hardy*
Contact *Harry Bell*

Broadcast, television and corporate video productions. Specializes in factual entertainment. Currently developing drama. Also at: 4th Floor, 114 Union Street, Glasgow G1 3QQ; T 0141 204 1717; E glasgow @ terntv.com. And: 1st Floor, Cotton Court, 38–42 Waring Street, Belfast BT1 2ED; T 02890 241433; E belfast @ terntv.com.

Testimony Films · TV

12 Great George Street, Bristol, BS1 5RH
T 0117 925 8589
steve.humphries@testimonyfilms.com
www.testimonyfilms.com
Contact *Steve Humphries*

TV documentary producer. Specializes in social history exploring Britain's past using living memory. Output includes *Hooked: History of Addictions; Married Love* (both Ch4 series); *A Secret World of Sex* (BBC series); *The 50s & 60s in Living Colour; Some Liked It Hot* (both ITV series).
 Welcomes ideas from those working on life stories and oral history.

Thin Man Films MOVIE

9 Greek Street, London, W1D 4DQ
T 020 7734 7372 F 020 7287 5228
info@thinman.co.uk

Producer of feature films including *All or Nothing* and *Vera Drake*.

Tiger Aspect Productions TV

7 Soho Street, London, W1D 3DQ
T 020 7434 6700 F 020 7434 1798
general@tigeraspect.co.uk
www.tigeraspect.co.uk
Head of Entertainment *Clive Tulloh*
Head of Comedy *Sophie Clarke-Jervoise*

An independent television producer working across a range of genres including comedy, drama, entertainment, factual, animation, wildlife (Tigress) and feature films (Tiger Aspect Pictures). Examples of titles include *Mr. Bean, Lenny Henry in Pieces, Harry Enfield and Chums, Murphy's Law, Rescue Me, The Vicar of Dibley, Fat Friends, Teachers, Murder, Omagh, Gimme Gimme Gimme, A Place in France, Country House* and *Streetmate*.

Tigerlily Films Limited MOVIE TV

Studio 17, The Whitecahpel Centre, Myrdle Street, London, E1 1HL
T 020 7247 1107 F 020 7247 2008
info@tigerlilyfilms.com
www.tigerlilyfilms.com

Founded in 2000, a film and television production company working across a range of genres including documentary, drama, feature films and children's programming. Titles include the children's drama *Patrick's Planet*, the feature film *Transit*, the documentaries *37 Uses for a Dead Sheep*, *Arte* and *Alexis Arquette: She's My Brother*. Does not accept any unsolicited material.

Touch Productions Ltd   `TV`

18 Queen Square, Bath, BA1 2HN
T 01225 484666 F 01225 483620
erica@touchproductions.co.uk
www.touchproductions.co.uk
Contact *Erica Wolfe-Murray*
Contact *Malcolm Brinkworth*

Over the last 20 years, Touch has made a wide range of programmes including award-winning investigations, popular documentaries, medical and science films, revelatory history productions as well as observational, social, religious and arts programmes. Current commissions include *The Human Footprint* (a Ch4 documentary special) and various series and documentaries for the BBC, National Geographic, TLC and Animal Planet. Other projects include *Transplanting Memories?*; *The Boy Who Couldn't Stop Running*; *Parish in the Sun*; *Revival* and *Angela's Dying Wish*.

Transatlantic Films Production and Distribution Company   `TV`

Studio 1, 3 Brackenbury Road, London, W6 0BE
T 020 8735 0505 F 020 8735 0605
revel@transatlanticfilms.com
www.transatlanticfilms.com
Executive Producer *Revel Guest*

Producer of TV documentaries. Output includes *Belzoni* (Ch4 Schools); *Science of Sleep and Dreams*; *Science of Love* and *Extreme Body Parts* (all for Discovery Health); *Legends of the Living Dead* (Discovery Travel/S4C International); *2025* (Discovery Digital); *How Animals Tell the Time* (Discovery); *Trailblazers* (Travel Channel). No unsolicited scripts. Interested in new writers to write 'the book of the series', e.g. for Greek Fire and History's Turning Points, but not usually drama script writers.

Travel Channel   `TV`

64 Newman Street, London, W1T 3EF
T 020 7636 5401 F 020 7636 6424
katie.isworth@travelchannel.co.uk
www.travelchannel.co.uk
Marketing Executive *Katie Isworth*
Head of Sales and Marketing *Andrea Schumacher*

Founded in 1994, specializing in travel programmes.

Happy to receive unsolicited approaches by email.

TV Choice Ltd   `TV`

PO Box 597, Bromley, BR2 0YB
T 020 8464 7402 F 020 8464 7845
tvchoiceuk@aol.com
www.tvchoice.uk.com
Contact *Norman Thomas*

Produces a range of educational videos for schools and colleges on subjects such as history, geography, business studies and economics.
   No unsolicited manuscripts; send proposals only.

Twenty Twenty Television   `TV`

20 Kentish Town Road, London, NW1 9NX
T 020 7284 2020 F 020 7284 1810
pch@twentytwenty.tv
www.twentytwenty.tv
Managing Director *Peter Casely-Hayford*
Co-Founder and Consultant Executive Producer *Claudia Milne*
Creative Director/Executive Producer *Jamie Isaacs*
Head of Factual Programming/Executive Producer *Tim Carter*
Senior Producer *Sam Whittaker*

Large independent television production company. Concentrates on documentaries, lifestyle programmes, popular drama and living history series. Established in 1982. Broadcast internationally by networks including the BBC, CBBC, ITV, Channels 4, Five, ABC, The Discovery Channel, Turner Original Productions, Sundance Channel and CNN. Recent successes include *The Choir* (BBC2), *Bad Lads Army* (ITV1) and *How to Divorce without Screwing up you Children* (Channel 4). Both *Brat Camp* and *That'll Teach Em* have now aired a third season.
   Drama enquiries and scripts should be emailed to jamieisaacs@twentytwenty.tv.

Twofour Broadcast   `TV`

5th Floor, 6–7 St Cross Street, London, EC1N 8UA
T 020 7438 1800 F 020 7438 1850
enq@twofour.co.uk
www.twofour.co.uk
Managing Director *Melanie Leach*
Creative Director *Stuart Murphy*
Director of Programmes *Joe Houlihan*

Television production company founded in 1987. Specializes in factual entertainment, drama and comedy.
   Unsolicited material or ideas accepted via email.

Tyburn Film Productions Limited [MOVIE]

Cippenham Court, Cippenham Lane, Cippenham, Nr. Slough, Berkshire, SL1 5AU
T 01753 516767 F 01753 691785

A subsidiary of Arlington Productions Limited, specializing in theatrical films.

The UK Film and TV Production Company PLC [MOVIE]

3 Colville Place, London, W1T 2BH
T 020 7255 1650
Head of Development *Henrietta Fudakowski*

In 2006 *Tsotsi*, UKFTV's first wholly-produced film won the Oscar for best foreign language film. Other titles include *Bugs! in 3D*.

Telephone ahead of sending a submission to see if script might be suitable for the company.

Vera Productions [TV]

66–68 Margaret Street, London, W1W 8SR
T 020 7436 6116 F 020 7436 6117
Contact *Phoebe Wallace*

Produces television comedy such as *Bremner, Bird and Fortune*.

Vertigo Films [MOVIE]

The Big Room Studios, 77 Fortess Road, London, NW5 1AG
T 020 7428 7555 F 020 7485 9713
scripts@vertigofilms.com
www.vertigofilms.com

A development, finance, production and distribution media company founded in July 2002. Major titles include *The Football Factory*, *It's All Gone Pete Tong*, *A Good Woman*, *Dirty Sanchez*, *Outlaw* and *Dogging–A Love Story*.

Waddell Media [TV]

Strand Studios, 5/7 Shore Road, Holywood, County Down, BT18 9HX
T 028 9042 7646 F 028 9042 7922
info@waddellmedia.com
www.waddellmedia.com

A leading producer of factual, lifestyle and entertainment programming for the UK, Irish and international markets. Founded in 1988, producing around 50 hours of television each year for most of the leading UK broadcasters. Also has major commissions in the US for Discovery, National Geographic and A&E.

Wall to Wall [TV]

8–9 Spring Place, London, NW5 3ER
T 020 7485 7424 F 020 7267 5292
mail@walltowall.co.uk
www.walltowall.co.uk
Chief Executive *Alex Graham*

Factual and drama programming. Output includes *Who Do You Think You Are?* ; *New Tricks*; *A Rather English Marriage*; *Glasgow Kiss*; *Sex, Chips & Rock 'n' Roll*; *The 1940s House*; *Body Story*; *Neanderthal*; *The Mafia*; *Not Forgotten*; *H.G. Wells*.

Walsh Bros Ltd [MOVIE] [TV] [VG]

4 Trafalgar Grove, Greenwich, London, SE10 9TB
T 020 8858 6870
info@walshbros.co.uk
www.walshbros.co.uk
Producer/Director *John Walsh*
Producer *David Walsh*
Producer/Head of Development *Maura Walsh*

Founded in 1996, working in film and TV. BAFTA-nominated producers of television, film, drama and documentaries. Output includes the film *Monarch*, the BAFTA-winning *Don't Make Me Angry* (for Channel 4), Grierson-nominated *Headhunting the Homeless*, *Trex* (factual series on teenagers at work in China, Mexico, Vancouver and Alaska), *Trex2* (follow-up series covering Romania, India, Iceland and Louisiana), *Boyz & Girlz* (dairy farm documentary series), *Cowboyz & Cowgirlz* (US sequal to hit series of Brit teens working on a ranch in Montana). Also arts documentaries: *The Comedy Store*, *Ray Harryhausen* (the work of Hollywood special effects legend). Drama: *The Sleeper*, *The Sceptic and the Psychic*, *A State of Mind*.

Will look at new writers but contact by email first. Initial enquiries to development@walshbros.co.uk.

WarpFilms [MOVIE] [TV]

The Workstation, Paternoster Row, Sheffield, S1 2BX
T 0114 213 0333
productions@warpfilms.com
www.warpfilms.com

WarpFilms was founded in 2000 and makes features, short films, music videos and TV programmes. Titles include *Dead Man's Shoes* (Bafta-nominated), *My Wrongs* (Bafta-winning) and *This is England* (named Best Film at the BIFAs).

Is open to new writing submitted via agents or personal recommendation.

Warp X [MOVIE]

The Workstation, Paternoster Row, Sheffield, S1 2BX
F 0114 221 0484

info@warpx.co.uk
www.warpx.co.uk

Warp X is a pioneering digital film studio based in Sheffield with offices in London and Nottingham. It is the sister company of Warp Films. Warp X has been financed by Film4, UK Film Council, EM Media and Screen Yorkshire to make six low budget features between 2006 and 2008 to be distributed by Optimum Releasing. Budgets between £400,000 and £1m.

Interested in 'genre films with a twist'. Does not accept unsolicited scripts.

Working Title Films Ltd   MOVIE   TV

Oxford House, 76 Oxford Street, London, W1D 1BS
Ⓣ 020 7307 3000 Ⓕ 020 7307 3001
www.workingtitlefilms.com
Head of Development (Films) *Natascha Wharton*
Television *Simon Wright*
Development Executive (Films) *Amelia Granger*
Development Executive (Films) *Rachael Prior*

Feature film producer. Output includes *Definitely Maybe*; *Hot Fuzz*; *Atonement*; *Elizabeth*; *The Golden Age*; *Mr Bean's Holiday*; *United 93*; *Smokin' Aces*; *Sixty Six*; *Catch a Fire*; *Gone*; *Nanny McPhee*; *Pride & Prejudice*; *The Interpreter*; *Wimbledon*; *Bridget Jones 2: Edge of Reason*; *Shaun of the Dead*; *Love Actually*; *Thunderbirds*; *Ned Kelly*; *Johnny English*; *Bridget Jones's Diary*; *Captain Corelli's Mandolin*; *Ali G Indahouse*; *Billy Elliot*; *Notting Hill*; *Elizabeth*; *Fargo*; *Dead Man Walking*; *French Kiss*; *Four Weddings and a Funeral*; *The Hudsucker Proxy*; *The Tall Guy*; *Wish You Were Here*; *My Beautiful Laundrette*. Television (drama, family/children's entertainment, comedy) Output *The Robber Bride*; *Perfect Strangers*; *The Other Woman*; *Dr Jekyll & Mr Hyde*; *Ready When You Are Mr McGill*; *Come Together*; *Lucky Jim*; *Randall & Hopkirk (deceased) I &II*; *The Last of the Blonde Bombshells*; *Doomwatch*; *Tales of the City*; *Lano and Woodley I &II*; *The Baldy Man I &II*; *The Borrowers I &II*; *News Hounds*.

No unsolicited manuscripts at present.

World Productions   TV

16 Dufours Place, London, W1F 7SP
Ⓣ 020 7734 3536
jemma@world-productions.com
www.world-productions.com
Executive Producer *Tony Harnett*
Office Manager *Helen Saunders*

Established in 1990. Has produced over 250 hours of network drama: renewable one hour series, mini-series and single films. Over 40 hours programming produced in 2006. Exists as a company of producers working across all genres. Creators of *Lilies* (BBC1), *Goldplated* (Channel 4), *Rough Diamond* (Channel 4), *This Life* and *Ballykissangel*.

Submissions via agent only.

Wortman UK / Polestar Pictures   MOVIE   TV

48 Chiswick Staithe, London, W4 3TP
Ⓣ 020 8994 8886
nevillewortman@beeb.net
Senior Producer, Development *Neville Wortman*
Producer/Director *Matthew Wortman*

Founded in 1989, producing feature films and mainstream television.

Specializes in drama and documentary, with examples of output including *Eruption Mount St Helena*, *Munich Air Crash*, *Murder in Paradise*, *Julius Caesar*, *Nuclear Spy Race* and *Hitler Bomb Plot*. Programmes have been shown on BBC, Channel 4 and Discovery, among others.

Open to new writing but only through established agents. send a synopsis and 2–6 pages of dialogue.

Zephyr Films Limited   MOVIE

33 Percy Street, London, W1T 2DF
Ⓣ 020 7255 3555 Ⓕ 020 7255 3777
info@zephyrfilms.co.uk
www.zephyrfilms.co.uk

Independent production company specializing in raising finance and providing production services. Production credits include *Daniel and the Superdogs*, *Mathilde*, *Asylum*, *Penelope*, *The Last Legion*, *Hannibal Rising*, *Virgin Territory* and *White Rose*.

Canada

100 Percent Film & Television Inc.   MOVIE   TV

119 Spadina Avenue, PO Box 21, Toronto, Ontario, M5T 2T2
Ⓣ 416 304 5225 Ⓕ 416 304 1222
info@onehundredpercent.ca
www.onehunredpercent.ca

Emmy Award-winning film and TV production company. Credits include *The Newsroom* and *At The Hotel*.

Affinity Productions   TV

2740B Queensview Drive, Ottawa, Ontario, K2B 2A2
Ⓣ 613 820 4660 Ⓕ 613 820 5020
info@affinityproductions.tv
www.affinityproductions.tv

Television production company producing for domestic and international broadcasters.

Afterdark Productions Inc. [MOVIE] [TV]

Suite 303, 1668 Barrington Street, Halifax, Nova Scotia, B3J 2A2
T 902 423 3497 F 902 484 6880
info@afterdarktv.com
www.afterdarktv.com

Film and television production house specializing in the creation, development and production of popular appeal films, television and documentaries. Titles include *A Bug and a Bag of Weed*, *The Life and Times of Super 8* and *Afterdark TV*.
    Does not accept unsolicited material unless accompanied by a submission release form.

Alberta Filmworks [MOVIE] [TV]

1310–11th Street, SW., Calgary, Alberta, T2R 1G6
T 403 777 9900 F 403 777 9914
mail@albertafilmworks.com
www.albertafilmworks.com

An independent film and television production founded in the late 1980s. Production credits include *Brokeback Mountain*, *The Christmas Blessing* and *The Ron Clark Story*.

Amberwood Entertainment Corp. [TV]

987 Wellington Street, 2nd Floor, Ottawa, Ontario, K1Y 2Y1
T 613 238 4567 F 613 233 3857
info@amberwoodanimation.com
www.amberwoodanimation.com
Sales & Acquisitions *Jonathan Wiseman*

Founded in 1997, specializing in, but not limited to, animation. Produces across genres and formats, including features and series. Titles include *The Secret World of Benjamin Bear*, *The Snow Queen*, *Zeroman* and *Hoze Houndz*.

Anaïd Productions Inc. [MOVIE] [TV]

208 3132 Parsons Road, Edmonton, Alberta, T6N 1L6
T 780 413 9285 F 780 465 0580
anaid@anaid.com
www.anaid.com

Founded in 1993, an award-winning film and television production company dedicated to the development and production of quality dramatic and factual programming. Includes documentaries and lifestyle series, reality and game shows, children's and adult drama, mini series and feature/television movies. Examples of output include *Mentors*, *The Tourist*, *Taking It Off*, *The Family Restaurant* and *X-Weighted*.
    Submissions guidelines and proposal submission agreement available from website.

Will not accept proposals for 'one-off' or single, one-hour documentaries.

Anagram Pictures [MOVIE] [TV]

3rd Floor, 291 East 2nd Avenue, Vancouver, British Columbia, V5T 1B8
T 604 720 9021 F 604 730 9042
inform@anagrampictures.ca
www.anagrampictures.ca
Writer/Director/Producer *Andrew Currie*
Producer *Mary Anne Waterhouse*
President/Producer/Writer *Blake Corbet*
Writer/Director/Producer *Trent Carlson*

Feature-film and tv movie production company. Launched in 1997. Focuses on drama and comedy.
    No unsolicited material.

Angel Entertainment Corporation [MOVIE] [TV]

2710 Millar Avenue, Saskatoon, Saskatchewan, S7K 4J4
T 306 244 8691 F 306 933 3183
bob@angelentertainment.ca
www.angelentertainment.ca
Partner/Producer *Bob Crowe*
Partner/Producer *Wally Start*

An international motion picture company producing market-driven feature films, documentaries and television programming. Examples of output include *Shadow Puppets* (a sci-fi horror-thriller), *Rabbit Fall* (a dramatic series) and the documentary series, *Spirit Creations*.

Apartment 11 Productions [TV]

828 St. Catherine Street East, Suite 200, Montreal, Quebec, H2L 2E3
T 514 282 0776 F 514 282 0796
www.apartment11.tv

Children's television producer, with titles including *Prank Patrol* and *Mystery Hunters*.

Aquila Pictures [MOVIE] [TV]

130 Bloor Street West, Suite 600, Toronto, Ontario, M5S 1N5
info@aquilapictures.com
www.aquilapictures.com

Primarily concerned with the development and production of feature films and television programmes. Recent productions include *The House*, *Gangster Exchange* and *Tomkat: Animated*.

Avanti Pictures Corporation [MOVIE] [TV]

810–289 Alexander Street, Vancouver, British Columbia, V6A 4H6
T 604 609 0339

info@avantipics.com
www.avantipics.com

Founded in 1996 as a base for like-minded producers and other creative people to produce high quality dramas and documentaries for television and festivals.

Happy to receive ideas, programme suggestions and literary materials. Study company's policy and agreement on website. Will review and consider submissions only if this policy and agreement is accepted.

Avrio Filmworks Inc.   (MOVIE)

5865 Marine Drive, West Vancouver, British Columbia, V7W 2S1
T 702 768 1918 F 702 877 4703
cmason@avriofilmworks.com
www.avriofilmworks.com

Founded by Michael Derbas to make and distribute feature films and participate in worldwide co-productions. Production credits include *Sub-Zero* and *Premonition*.

Bardel Entertainment Inc.   (MOVIE) (TV) (VG)

548 Beatty Street, Vancouver, British Columbia, V6B 2L3
T 604 669 5589 F 604 669 9079
info@bardel.ca
www.bardelentertainment.com

Producing animated content for over twenty years, employing up to three hundred people. Has worked on over a dozen feature films and produced approximately three hundred half-hours of television programming, as well as working on the production of seventeen gaming, interactive or multimedia properties. Projects include *The Christmas Orange*, the *Silverwing* series and trilogies, and (in association with MegaBlocks) the *Dragons* TV movies. Currently in production with Leaping Lizards Productions and Avrill Stark Entertainment on *Zeke's Pad*.

A release form must be signed and compelted before any submissions are accepted. Email to request a release form before sending anything in. Owing to volume of submissions, turnaround time is usually 6–8 weeks for a response.

Barna-Alper Productions Inc.   (TV)

366 Adelaide Street West, Suite 700, Toronto, Ontario, M5V 1R9
T 416 979 0676 F 416 979 7476
info@bap.ca
www.bap.ca

Founded in 1980, producing dramas and documentaries. Examples of output include CBC's *DaVinci's Inquest, Shania: A Life In Eight*

Albums, the mini-series *Whiskey Echo* and *Turning Points of History*.

Big Motion Pictures Limited   (TV)

5 Pleasant Street, 2nd Floor, PO Box 202, Chester, Nova Scotia, B0J 1J0
T 902 275 1350 F 902 275 1353
contact@bigmotionpictures.com
www.bigmotionpictures.com

Independent television company founded in 1999. Their first production was *Task Force*. Other titles include the TV movie *Sleep Murder* and the series *Guy & a Girl* and *Snakes & Ladders*.

Big Soul Productions Inc.   (TV)

401 Richmond Street West, Suite B102, Toronto, Ontario, M5V 3A8
T 416 598 7762 F 416 598 5392
comments@bigsoul.net
www.bigsoul.net

Established in 1999, a leader in Aboriginal content programming. Has produced three series of the award-winning *The Seventh Generation*.

BioWare Corp.   (VG)

200, 4445 Calgary Trail, Edmonton, Alberta, T6H 5R7
T 780 430 0164 F 780 439 6374
contact@bioware.com
www.bioware.com

An electronic entertainment company specializing in creating computer and console video games, founded in 1995. Aims to 'deliver the best story-driven games in the world'. Recently part of a US$300 million merger with Pandemic Studios. Examples of output include *Jade Empire, Star Wars: Knights Of The Old Republic, Neverwinter Nights Series* and *Shattered Steel*.

All unsolicited submissions will be destroyed.

Black Walk   (MOVIE) (TV)

99 Sudbury Street, Unit 201, Toronto, Ontario, M6J 3S7
T 416 533 5864 F 416 533 2016
admin@blackwalk.com
www.blackwalk.com

Founded in the 1990s as a music video production company. Since 2002 has focused on movie and television production, with four films completed.

Breakthrough Films & Television Inc.   (TV)

122 Sherbourne Street, Toronto, Ontario, M5A 2R4
T 416 766 6588 F 416 769 1436
business@breakthroughfilms.com
www.breakthroughfilms.com

Over twenty years experience of producing television programmes including children's, documentary, factual, drama and lifestyle series.

Brightlight Pictures Inc. [MOVIE] [TV]

The Bridge Studios, 2400 Boundary Road, Burnaby, British Columbia, V5M 3Z3
T 604 628 3000 F 604 628 3001
info@brightlightpictures.com
www.brightlightpictures.com

Founded in 2001 to develop, finance and produce independent feature films and television projects for domestic and international markets. Production credits include *White Noise*, *Dungeon Siege*, *Edison* and *Wicker Man*.
Does not accept unsolicited materials.

Canadian Accents Inc. [MOVIE] [TV]

26 Yarmouth Road, Toronto, Ontario, M6G 1W6
T 416 653 0702 F 416 658 3176
info@canadianaccents.ca
www.canadianaccents.ca

Founded in 2003, a film and television production company focussing on comedy. Titles include *Women Fully Clothed: All Dresses Up & Places to Go* and *Is it Art?*

Canamedia Film Productions Inc. [TV]

381 Richmond Street East, Suite 200, Toronto, Ontario, M5A 1P6
T 416 483 7446 F 416 483 7529
canamed@canamedia.com
www.canamedia.com

Founded in 1978. Works across range of genres including TV movies, drama series, children's drama, documentaries, travel and lifestyle.

Capri Films Inc. [MOVIE] [TV]

259 Lakeshore Boulevard East, 2nd Floor, Toronto, Ontario, M5A 3T7
T 416 535 1870 F 416 535 3414
info@caprifilms.com
www.caprifilms.com

Creates, develops and produces feature film and television projects. Produces three to four film and television projects per year. Examples of output include *Tideland* and the mini-series *Karol*.

Cheap and Dirty Productions [MOVIE] [TV] [VG]

1874 Grant Street, Vancouver, British Columbia, V5L 2Y8
T 604 872 7006 F 604 872 7086
cheapanddirty@telus.net
www.cheapanddirty.ca

Company founded in 2001 to produce film, television, features, documentaries and video games. Also available to assist in development. Notable projects include *Fetching Cody* and *A Safer Sex Trade*. Interested in all genres. Material accepted via agent only.

Ciné Qua Non Média [TV]

445 Rue St-Pierre, Bureau 402, Montréal, Québec, H2Y 2M8
T 514 271 4000 F 514 271 4331
info@cqnmedia.com
Producer and President *Michel Ouellette*
Co-ordinator *Stéphanie Murphy*

Founded in 1982. Focuses on films made for television. Noted for blending genres such as documentary, performing arts and fiction. Recent releases include *Velasquez's Little Museum* and the feature documentary *Mary Shelley*.
Welcomes unsolicited material via email. Must include letter of presentation, one-liner and synopsis.

Crawley Films Ltd [MOVIE] [TV]

PO Box 398, Merrickville, Ontario, K0G 1N0
bstevens@titlecraft.com
President and CEO *Bill Stevens Jr*
V. P. Distribution *Caroline Stevens*
V. P. Production *Will Stevens III*

One of Canada's oldest production companies, founded in 1939. Has won an Academy Award and over 100 awards from around the world. Produces drama, documentary and animation, concentrating on family values.
Welcomes unsolicited material in the form of a one-page concept outline.

Crescent Entertainment Ltd [MOVIE] [TV]

343 Railway Street, Suite 304, Vancouver, British Columbia, V6A 1A4
T 604 357 3606 F 604 357 3605
crescent@crescent.ca
www.crescent.ca

Established in 1990, an award-winning television and film production company. Specializes in developing and producing original feature films, television and documentaries. Television projects include *Terminal City*, *Happy Land* and *Dead Zone*. Films include *Mirage*, *Moving Malcolm*, *Naked Frailties* and *Showdown at Williams Creek*.
Email synopses/concepts and all relevant details for consideration.

CTVglobemedia Inc. [TV]

9 Channel Nine Court, Scarborough, Ontario, M1S 4B5
T 416 332 5000

ctvglobemediacommunications@ctvglobemedia.com
www.ctvglobemedia.com

One of Canada's leading multimedia companies, with production and broadcasting interests in radio and television. Took over Chum Ltd, and its associated radio and TV networks, in 2007.

Darius Films Inc. MOVIE TV

326 Carlaw Avenue, Suite 105, Toronto, Ontario, M4M 3N8
T 416 922 0007 F 416 406 0034
info@dariusfilms.com
www.dariusfilms.com

Founded in 1996, a film and television production company. Titles include *Weirdsville*, *A Lobster's Tale*, *The Limb Salesman*, *Motel* and *Caprice*.

Decode Entertainment Inc. TV VG

512 King Street East, Suite 104, Toronto, Ontario, M5A 1M1
T 416 363 8034 F 416 363 8919
decode@decode-ent.com
www.decode.tv
Development Manager *Nancy Yeaman*

Children's and family production company founded in 1997. Interested in live action and animated television series projects. Also extends to computer games. Output includes *Chop Socky Chooks*, *Naturally Sadie* and *Franny's Feet*.

No unsolicited material via post or email. Welcomes phone calls to pitch potential projects. Will respond accordingly.

Devine Entertainment MOVIE TV

2 Berkeley Street, Suite 504, Toronto, Ontario, M5A 2W3
T 416 364 2282 F 416 364 1440
info@devine-ent.com
www.devine-ent.com

Develops and produces children's and family films for global television and cable markets and the international home video markets. Has won numerous international awards including five Emmy Awards and five Gemini Awards. Focus is on historical and timeless stories. Examples of output include *Beethoven Lives Upstairs*, *The Artists' Specials* (consiting of one hour episodes each focusing on a famous artist) and *Bailey's Billion$*.

Dream Street Pictures Inc. MOVIE TV

75 Archibald Street, Moncton, New Brunswick, E1C 5J2
T 506 854 1057 F 506 382 4144
tammy@dreamstreetpictures.com
www.dreamstreetpictures.com

Film and TV production company whose credits include *Planet Luxury*, *Open Heart*, *Neutral Ground* and *Record Man: the Life & Times of Sam Sniderman*.

Epitome Pictures Inc. MOVIE TV

220 Bartley Drive, Toronto, Ontario, M4A 1G1
T 416 752 7627 F 416 752 7837
info@epitomepictures.com
www.epitomepictures.com

Over twenty-five years experience of television production, including documentaries, drama and soap operas. Output includes *Instant Star* and *Degrassi: The Next Generation*.

Fast Productions Ltd MOVIE TV

8039 Osler Street, Vancouver, British Columbia, V6P 4E3
T 778 786 1628 F 778 786 1585
fast-productions@shaw.ca
Producer *Cheryl-Lee Fast*

Formed in 2003. Develops, produces and distributes documentaries and feature length films. Film output focuses on drama, comedy, thriller and romantic comedy. Also develops television projects and reality series. Past work includes *The Zero Sum*, *Cries in the Dark*, *Papal Chase* and *Swimming Lessons*.

No interested in first time writers. Prefers to be approached via email.

Flamingo Neck Pictures Inc. MOVIE TV

108 Victoria Park Avenue, Toronto, Ontario, M4E 3R9
T 416 693 5952 F 416 693 9132
scarab23@sympatico.ca
www.creepy.tv
Executive Producer *Brian O'Dea*
Producer *Bill Burke*

Founded in 2001, television and film production company specializing in drama and docu-dramas. Recent credits include *Creepy Canada* and *High: Confessions of a Pot Smuggler*.

Welcomes unsolicited manuscripts by email - 'if it's good, that's the only issue'.

Force Four Productions TV

#202 - 221 E. 10th Avenue, Vancouver, British Columbia, V5T 4V1
T 604 669 4424 F 604 669 4535
tv@forcefour.com
www.forcefour.com

Television production company set up in 1983, with an output embracing drama, lifestyle, documentary and reality series.

Fremantle Corporation (TV)

23 Lesmill Road, Suite 201, Toronto, Ontario, M3B 3P6
T 416 443 9204 F 416 443 8685
www.fremantlecorp.com

The Fremantle Corporation, founded in 1952, is a distribution company that co-producers programmes. Has an extensive library of programming including daytime drama *All My Children*, the comedy *Girls Behaving Badly* and *Candid Camera*.

Front Street Pictures (MOVIE) (TV)

202–2339 Columbia Street, Vancouver, British Columbia, V5Y 3Y3
T 604 257 4720 F 604 257 4739
www.frontstreetpictures.com

Motion picture production company founded in 1999, specializing in independent films for worldwide distribution. Recent productions include the films *The Deal* and *We Don't Live Here Anymore*, and the made-for-television movies *Her Sister's Keeper*, *The Secret of Hidden Lake* and *7 Things to Do Before I'm 30*.

Actively seeking new creative talent for both television and film. To submit logling and synopsis, download and sign the submissions form on the website. This form needs to be signed and included with the submission or it will be rejected. Cannot guarantee a response or give specific feedback on any rejected material but will make contact if interested in the work. Does not keep or return material but disposes of it so that no-one has access to it anymore.

Galafilm Inc. (MOVIE) (TV)

5643 Clark Street, Suite 300, Montreal, Quebec, H2T 2V5
T 514 273 4252 F 514 273 8689
info@galafilm.com
www.galafilm.com

Founded in 1990, an independent film, television and new media production company. Works across genres including documentary series and one-offs, children's programming, television drama and feature film.

Ginty Films (MOVIE) (TV)

483 Euclid Avenue, Toronto, Ontario, M6G 2TI
T 416 992 5438 F 416 924 3229
rwginty@aol.com
www.robertginty.com

Film and television production company formed in 1982. Specializes in drama, such as *The Young Churchill* for ITV.

Submissions only via agent.

Haddock Entertainment Inc. (TV)

#810–207 West Hastings Street, Vancouver, British Columbia, V6B 1H7
T 604 681 1516 F 604 684 3530
office@haddockentertainment.com
www.haddockentertainment.com

Specializes in popular prime time drama. Founded by Chris Haddock, the creative force behind the award-winning *Da Vinci's Inquest*, *The Handler*, *Da Vinci's City Hall* and *Intelligence*.

Does not accept or consider materials, creative ideas or suggestions of any nature other than those specifically commissioned or requested by Haddock Entertainment or its affiliated companies.

Hamilton-Mehta Productions (MOVIE)

221 Lyon Street North, Suite 2603, Ottawa, Ontario, K1R 7X5
T 613 782 2220 F 613 782 3131
dch888@rogers.com
Contact *David Hamilton*

Producers of film dramas such as *Water*.
Does not welcome unsolicited approaches.

Handel Productions Inc. (MOVIE) (TV)

424 Rue Guy, Suite 110, Montreal, Québec, H3J 1S6
T 514 487 1881 F 514 487 7796
info@handelproductions.ca
www.handelproductions.ca

Leading producer of documentaries and factual programmes, established in 1985. Production credits include *How William Shatner Changed the World*, *Faith and Fortune: The Reichmann Story*, *Nowhere Else to Live* and *A Passage from Burnt Islands*.

Horizon Motion Pictures Inc. (MOVIE)

1271 Howe Street, Suite 111, Vancouver, British Columbia, V6Z1R3
T 604 632 1707 F 604 632 1711
film@horizonmotionpictures.com
www.horizonmotionpictures.com

Formed in 1989 as a feature film development and production company, debuting with *By the Sword*. It has subsequently developed a distribution wing for third-party productions.

ImagiNation Film & Television Productions Inc. (MOVIE) (TV)

3rd Floor, 10318–82nd Avenue (The Griffith Block), Edmonton, Alberta, T6E 1Z8
T 780 439 8755 F 780 430 1871
imagi@telusplanet.net
www.imaginationfilm.tv

Develops and produces dramatic long-form television, theatrical movies and socio-political documentaries. Titles include *A Kid's View* (*I, II* and *"of the US"*), *Great Canadian Ghost Stories* and *100 Days in the Jungle*.

imX Communications Inc. MOVIE TV

1556 Queen Street, Halifax, Nova Scotia, B3J 2H8
T 902 422 4000 F 902 422 4427
imx@imx.ca
www.imxcommunications.com

Television and film production company established by Christopher Zimmer. Past productions include *Margaret's Museum, Love and Death on Long Island, The Real Howard Spitz, The Weight of Water, The River King* and *Partition*.

Infinity Filmed Entertainment Group Ltd MOVIE TV

1412 West 7th Avenue, Suite 100, Vancouver, British Columbia, V6H 1C1
T 604 681 5650 F 604 681 5664
www.infinityfilms.ca

Covers the spectrum from features and documentaries to television comedy, drama and variety. Titles include the film *Almost Heaven*, the documentaries *Comedic Genius, Love & Duty* and *Race of the Century*, and the series *Improv Comedy, The Jim Byrnes Show, Dawn Wells* and *Reel Adventures*.

Inner City Films Inc. MOVIE TV

842 King Street West, Toronto, Ontario, M5V 1P1
T 416 368 3339 F 416 368 5234
info@innercityfilms.ca
www.innercityfilms.ca

Founded in 1987, a television and film production company with offices in Los Angeles, London and Johannesburg. Has produced more than 100 hours of prime-time television programming across a range of genres, from drama series to lifestyle programming.

Insight Production Company Ltd. TV

489 Kings Street West, Suite 401, Toronto, Ontario, M5V 1K4
T 416 596 8118 F 416 596 8270
jbrunton@insighttv.com
www.insighttv.com
President and CEO, Executive Producer *John Brunton*
Executive Vice President, Executive Producer *Barbara Bowlby*

Large multi-faceted television production company founded in 1978. Involved in all genres

of programming. Recent releases include *Falcon Beach* (drama), *Hatching, Matching and Dispatching* (comedy), *Canadian Idol* (variety), *Comedy Gold* (documentary) and *MuchMusic VJ Search* (reality).

Unsolicited material accepted when accompanied by a signed policy agreement. Further details found on the website.

International Documentary Television Corporation MOVIE TV

73 Alexander Street, Vancouver, British Columbia, V6A 1B2
T 604 664 0450 F 604 664 0451
hello@doctv.com
www.doctv.com
Chairman *Bob Duncan*

Television production company specializing in full-length network documentary programming. Does biographical portraits, issue-driven documentary and topics of interest to the producers which can be programmed by network television. Currently developing a situation comedy series and a MOW, *The Search for D. B. Cooper*.

Will view all submissions.

International Keystone Entertainment MOVIE

300–2339 Columbia Street, Vancouver, British Columbia, V5Y 3Y3
T 604 873 9739 F 604 873 5919
films@keypics.com
www.keypics.com

Producers of feature films for worldwide markets. Keystone Family Pictures wing makes films for 5–11 year olds and their parents while Keystone Pictures focuses on drama, action and thriller titles.

Keatley Entertainment Ltd MOVIE TV

718–510 West Hastings Street, Vancouver, British Columbia, V6B 1L8
T 604 291 9789 F 604 291 9759
andrew@keatleyentertainment.com
www.keatleyentertainment.com

Specializes in the development and production of feature films and television series for domestic and international markets. Is the producer of the award-winning Cold Squad television series.

Knight Enterprises Inc. TV

307–99 Fifth Avenue, Ottawa, Ontario, K1S 5P5
T 613 730 1728 F 613 730 0182
info@knight-tv.com
www.knight-tv.com
Senior Producer *Kathy Doherty*
Head of Business Affairs *Sheldon Speers*

Television and new media company specializing in lifestyle programming. Launched in 1997. Recent releases include *License to Grill*, *This Food That Wine* and *Junk Brothers*.

Unsolicited submissions accepted by letter. Must be accompanied by a signed Proposal Submission Agreement (obtained by emailing info@knight-tv.com).

Lumanity Productions Inc. MOVIE TV

313 Brock Avenue, Main Floor, Toronto, Ontario, M6K 2M6
T 416 531 9691 F 416 946 1854
rbudreau@lumanityproductions.com
www.lumanityproductions.com

Established in 2002 by Robert Budreau to produce quality films for theatrical and television release. Has worked on films, television dramas and long-form documentaries.

Pleased to receive ideas, programme suggestions and literary materials but study the submissions policy and agreement on the website first.

Make Believe Media Inc. MOVIE TV

Suite 201, 145 Keefer Street, Vancouver, British Columbia, V6A 1X3
T 604 874 9498 F 604 874 9484
info@makebelievemedia.com
www.makebelievemedia.com

Established by Lynn Booth in 1999 to make documentaries and non-fiction programming. Production credits include *The Perfect Husband: The Life and Times of Tim Horton*, *The Whistleblower*, *Mandrake the Magician*, *Portraits from the Frontier* and *Paris Stories*.

Media Headquarters Film & Television Inc. TV

760 Bathurst Street, Studio 2, Toronto, Ontario, M5S 2R6
T 416 537 8384 F 416 537 8602
contact@mediahqs.net
www.mediahqs.net

Produces documentaries and prime-time specials for domestic and international markets. Credits inlcude *Le Mozart Noir: Reviving A Legend*, *Souvenir of Canada* and *Modern Love*.

Milagro Films Inc. MOVIE TV

4446 St-Laurent Boulevard, Suite 806, Montréal, Québec, H2W 1Z5
T 514 939 9969 F 514 985 2563
info@milagrofilms.ca
www.milagrofilms.ca
Head of Development *Flavie Langois*

Established in 1997. Concentrates on television drama and feature films, notably *Jericho Mansions*.

Unsolicited submissions via email. Must be accompanied by a signed release (available from the website).

Minds Eye Entertainment Ltd. MOVIE TV

480 Henderson Drive, Regina, Saskatchewan, S4N 6E3
T 306 359 7618 F 306 359 3466
mindseye@mindseyepictures.com
www.mindseyepictures.com
President and CEO *Kevin DeWalt*

Television and film production company founded in 1986. Concentrates on live action drama, children's and youth drama and lifestyle programming. Past releases include *The Englishman's Boy* (TV miniseries), *Just Cause* (22 x 1hr drama) and *Falling Angels* (feature film). Has won over 50 national and international awards, including both Genie and Gemini awards.

Unsolicited synopses accepted only when accompanied by signed submission releases. More information available on the website.

Morag Productions/Passage Films MOVIE TV

Box 52, St. John's, Newfoundland, A1C 5H5
T 709 739 0447 F 709 739 0467
info@morag.ca
www.morag.ca

Film and television production company, specializing in documentary and drama, including the Gemini Award-winning mini-series *Random Passage*.

Muse Entertainment Enterprises MOVIE TV

3451 Rue St-Jacques, Montreal, Quebec, H4C 1H1
T 514 866 6873 F 514 876 3911
jprupas@muse.ca
www.muse.ca
VP (Development & Distribution) *Jesse Prupas*
Director (Communications & Documentary Division) *Betty Palik*
President (Production, Quebec) *Irene Litinsky*

Independent film and television producer established in 1998. Has produced or co-produced over 85 TV movies and mini-series, 9 television series and 17 feature films. Works across genres including drama, family programming, event television, documentaries and comedy. Credits include *Human Trafficking* and *Killer Wave*.

Contact only via an agent and with a release form. Study website to see if your material might be suitable. Company receives so much material that a response is not guaranteed.

MVP Entertainment Inc. [MOVIE] [TV]

Suite 400–1450, Creekside Drive, Vancouver,
British Columbia, V6J 5B3
Ⓣ 604 731 9194 Ⓕ 604 731 7174
info@mvpfilm.com
www.mvpfilm.com

Produces for television and cinema. Titles include
The Wild Guys.

Nelvana Limited [MOVIE] [TV]

42 Pardee Avenue, Toronto, Ontario,
M6K 1X8
Ⓣ 416 535 0935 Ⓕ 416 530 2832
webmaster@nelvana.com
www.nelvana.com

Major producer of animated films and
programmes for the family market.

Neophyte Productions Inc. [MOVIE]

581 Markham Street, Toronto, Ontario,
M6G 2L7
Ⓣ 416 944 9892 Ⓕ 416 944 9916
info@neophyteproductions.com

Specializes in film and video production,
including short and feature length films. Its first
film was *The Onion* by Sheila Heti.

Network Entertainment Inc. [MOVIE] [TV]

116 West 6th Avenue, Vancouver, British
Columbia, V5T 1M6
Ⓣ 604 739 8825 Ⓕ 604 739 8835
info@networkentertainment.ca
www.networkentertainment.ca

Specialists in sports and entertainment
broadcasts, established in 1999. Works across
genres including documentary and film.

Nightingale Company [MOVIE] [TV]

588 Richmond Street West, Toronto, Ontario,
M5V 1Y9
Ⓣ 416 628 1355 Ⓕ 416 628 1505
sam@dnightingale.com
www.dnightingale.com

Film and television production company founded
in 2000, producing across genres including
features, documentaries and children's series.

Norstar Filmed Entertainment Inc. [MOVIE] [TV]

148 Yorkville Avenue, 2nd Floor, Toronto,
Ontario, M5R 1C2
Ⓣ 416 961 6278 Ⓕ 416 961 5608
www.norstarfilms.com

Founded in 1997, producing and distributing
feature films for domestic and international

markets. Production credits include the
supernatural thriller *The Marsh*, *Almost Heaven*
and the Gemini Award-winning television series,
The Eleventh Hour.

Always interested in new projects and potential
co-productions.

Omni Film Productions Ltd [TV]

204–111 Water Street, Vancouver, British
Columbia, V6B 1A7
Ⓣ 604 681 6543 Ⓕ 604 688 1425
info@omnifilm.com
www.omnifilm.com
President *Michael Chechik*
Vice President *Brian Hamilton*
Vice President *Gabriela Schonbach*
Vice President of Finance *Andrea Droege*

One of Canada's largest independent television
producers founded, in 1979. Exists as a television
production, post-production and distribution
branch alongside sister companies Water Street
Pictures, Water Street Releasing and Omni Post.
Programmes are screened in 175 countries and
translated into over 20 languages. Specializes in
drama, comedy, documentary, and factual and
lifestyle series. Recent titles include *Make Some
Noise* (CBC), *Dragon Boys* (CBC), *Robson Arms*
(CTV, The Comedy Network) and *Greenpeace:
Making a Stand* (Global).

Unsolicited pitches accepted via short emailed
introduction. If interested, the company will
request a synopsis, outline, treatment and script.
Preference given to established writers or projects
with broadcaster interest.

ORCA Productions [MOVIE] [TV]

3425 West 2nd Avenue, Vancouver, British
Columbia, V6R 1J3
Ⓣ 604 732 9387 Ⓕ 604 732 3587
www.orcaproductions.com

Over twenty years experience of producing
award-winning feature films, television dramas
and documentaries for the international market.
Recent productions include *Would Be Kings*.

Original Pictures Inc. [MOVIE] [TV]

602–191 Lombard Avenue, Winnipeg, Manitoba,
R3B 0X1
Ⓣ 204 940 3310 Ⓕ 204 943 5502
jessicak@originalpicturesinc.com
www.originalpicturesinc.com
Executive Assistant *Jessica Krawec*

Established in 2000. Produces drama projects
for film and television. Past examples include
Falcon Beach and *A Bear Named Winnie*. Claims
7 Geminis and 29 Gemini nominations to date.
Focuses mostly on international projects with

potential for an American sale and/or substantial European sales or funding.

Original Pictures does not accept unsolicited scripts. If you wish to make a submission send a synopsis with a signed release form, which can be downloaded from the website.

Panacea Entertainment (MOVIE) (TV)

2nd Floor, 9876A 33 Avenue, Edmonton, Alberta, T6N 1C6
(T) 780 490 1220 (F) 780 490 5255
www.panaceaentertainment.com

Produces for television and screen. Film credits include *Best Served Cold* and *Generation X*. TV credits include *A Total Write-Off*, *Catching the Chameleon* and *Children & War*.

Interested in seeing projects at an advanced stage of development i.e. draft screenplay or teleplay. Before sending a script, approach by email providing a short description of the project and a couple of introductory lines about the author. If interested in idea, the company will ask for a hard copy of the script and a completed and signed submission form.

Peace Arch Motion Pictures Inc. (MOVIE) (TV)

1867 Yonge Street, Suite 650, Toronto, Ontario, M4S 1Y5
(T) 416 783 8383 (F) 416 783 8384
mail@peacearch.com
www.peacearch.com

Production company specializing in feature film and television projects. Credits include *Absolon*, *Crime Spree* and the *Nature Unleashed* series.

Peace Point Entertainment Group (TV)

78 Berkeley Street, Toronto, Ontario, M5A 2W7
(T) 416 365 7734 (F) 416 365 7739
info@peacepoint.tv
www.peacepoint.tv

Mid-sized television production company with credits including *Ed's Up!*, *Barnstormers* and *Devil's Perch*.

Point Grey Pictures Inc. (MOVIE) (TV)

#4–1214 West 7th Avenue, Vancouver, British Columbia, V6H 1B6
(T) 604 221 4426 (F) 604 742 9957
info@pointgreypictures.com
www.pointgreypictures.com

Produces documentaries for television and cinema release. Headed by Oscar-winning John Zaritsky. Output includes *Radiation Roulette*, *College Days, College Nights*, *No Kidding* and *Men Don't Cry*.

Principia Productions Ltd (MOVIE) (TV)

2531 Bellevue Avenue W., Vancouver, British Columbia, V7V 1E3
(T) 604 834 8084 (F) 614 921 9134
princip@shaw.ca

Producers of dramas and documentaries for television and the cinema since 2000. Credits include the movie *A Simple Curve*.

Prospero Pictures (MOVIE) (TV)

1200 Bay Street, Suite 400, Toronto, Ontario, M5R 2A5
(T) 416 926 0853 (F) 416 920 8373
martin.katz@prosperopictures.com
www.prosperopictures.com

A one-stop world-wide productions services and financing boutique. Offers financing of Canadian film and television productions and treaty co-productions.

Protocol Entertainment Inc. (MOVIE) (TV)

80 Spadina Avenue, Suite 405, Toronto, Ontario, M5V 2J4
(T) 416 966 2711 (F) 416 599 6100
www.protocolent.com

Founded in 1993 to develop, finance, produce and market films, series and mini-series for the North American market. Production credits include *Train 48*, *The Saddle Club*, *Police Academy: The Series* and *Goosebumps*.

Queen Street Entertainment (MOVIE) (TV)

23 Lesmill Road, Suite 203, Toronto, Ontario, M3B 3P6
(T) 416 691 6655 (F) 416 691 8419
www.queenstreetentertainment.com

Produces live action feature films and television productions for the whole family under the Knightscove Family Films banner. Titles include *Blizzard*, *Kart Racer* and *Virginia's Run*.

Before reviewing or considering any material, the person submitting must download the release form from the website and mail it together with a hard copy of script or other written creative material, marked for the attention of 'The Great Canadian Movie'. Cannot review materials submitted by email or fax.

Raven West Films Ltd (MOVIE) (TV)

701–207 West Hastings Street, Vancouver, British Columbia, V6B 1H7
(T) 604 681 7121 (F) 604 681 7173
info@ravenwestfilms.com
www.ravenwestfilms.com
President/Producer/Director *Carl Bessai*
Associate Producer *Rajvinder Uppal*

Drama and documentary filmmaking company founded in 1999. Past successes include *Johnny* (Jury Prize winner at 1999 Toronto International Film Festival), *Lola* (Berlin, Sundance and Toronto film festivals, 2001), *Emile* starring Ian McKellen (Toronto, 2003), and *Unnatural and Accidental* (Toronto, 2006).

Encourages new writing. Welcomes unsolicited material via email.

Reel Girls Media Inc. (TV)

2nd Floor, 9860A 33 Avenue, Edmonton, Alberta, T6N 1C6
T 780 488 0440 F 780 452 4980
info@reelgirlsmedia.com
www.reelgirlsmedia.com

Produces factual and non-factual programmes for television and new media. Examples of titles include *WildFiles.TV*, *Booked.TV* and *Stories from the Seventh Fire*.

Remstar Productions (TV)

85 St-Paul Street West, Montreal, Quebec, H2Y 3V4
T 514 847 1136
www.remstarcorp.com

Produces theatrical and television programming for domestic and international markets. Credits include *No Good Deed*, *Dangerous Liason*, *The Five of Us*, *An American Haunting* and *Ma Fille, Mon Ange*.

Rhombus Media Inc. (MOVIE) (TV)

99 Spadina Avenue, Suite 600, Toronto, Ontario, M5V 3P8
T 416 971 7856 F 416 971 9647
info@rhombusmedia.com
www.rhombusmedia.com

Over twenty-five years experience of producing feature films, documentaries and performing arts programmes. Has won an Oscar for *The Red Violin* and was nominated for *Making Overtures*. Other credits include the *Yo-Yo Ma - Inspired By Bach* series, *The Firebird*, *Slings & Arrows* and *Childstar*.

Rink Rat Productions, Inc. (MOVIE) (TV)

683 Water Street, 2nd Floor, St. John's, Newfoundland and Labrador, A1E 1B5
T 709 739 9055 F 709 739 9065
msexton@nfld.com
www.rinkratproductions.com
Producer *Mary Sexton*

Formed in 1993. Produces comedy, drama, arts and documentary projects for film and television. Previously gained a Gemini Award for *Tommy...*

A Family Portrait and two nominations for *Ron Hynes: The Irish Tour*. Other past productions include *To Think Like A Composer* and reality show *Canadian Idol: Audition Tour* (Seasons 1–3). Also runs Dark Flowers Productions (drama and documentary) and 2M Innovative.

Accepts unsolicited mailed proposals. No email. Previous credentials are an asset.

S & S Productions Inc. (TV)

Dufferin Liberty Centre, 219 Dufferin Street, Suite 100A, Toronto, Ontario, M6K 3J1
T 416 260 0538 F 416 260 1628
www.ssp.ca

Television producer for Canadian and international audiences, in business since 1980. Flagship title is the comedy series, *The Red Green Show*. Other credits include *History Bites*, *An American in Canada*, the animated series *Sons of Butcher* and the feature film *Red Green's Duct Tape Forever*.

Send a one-page summary of show proposal or script, copy of the proposal or script, contact name, address, phone number and email address, and a signed and dated copy of a release form. A proposal that is not accompanied with this form will not be read.

Savi Media Inc. (MOVIE) (TV)

68 Water Street, Suite 403, Vancouver, British Columbia, V6B 1A4
T 604 662 8438
mail@savi-media.com
www.savi-media.com

Founded in 2000, developing and producing feature films and television programmes. Film credits include *Crusade*, *Flower and Garnet* and *Society Rules*. Television productions include *Baby Game*, *Chinatown Diaries*, *Little India* and *My Tango With Porn*.

Actively seeking to expand portfolio, especially in television series and television longform.

Screen Door (MOVIE) (TV)

18 Gloucester Lane, 3rd Floor, Toronto, Ontario, M4Y 1L5
T 416 535 7402 F 416 535 1839
www.screendoor.org

Independent production company founded in 1999 and specializing in award-winning dramatic films, television series and mini-series.

Serendipity Point Films (MOVIE)

9 Price Street, Toronto, Ontario, M4W 1Z1
T 416 960 0300 F 416 960 8656
www.serendipitypoint.com

Boutique motion picture production company formed in 1998. Examples of output include *Fugitive Pieces*, *Where the Truth Lies*, *Being Julia* and *Ararat*.

Cannot consider or accept and materials, ideas or proposals other than those specifically requested or commissioned by the company.

Shaftesbury Films `MOVIE` `TV`

163 Queen Street East, Suite 100, Toronto, Ontario, M5A 1S1
☏ 416 363 1411 🖷 416 363 1428
mailbox@shaftesbury.org
www.shaftesbury.org

Produces feature films and movies and series for television, aimed at prime time and children's audiences. Examples of output include the films *Camilla*, *Conquest* and *Painted Angels*, and the television productions *11 Cameras*, *The Jane Show* and *ReGenesis*.

Shavick Entertainment `MOVIE` `TV`

112 West 6th Avenue, Vancouver, British Columbia, V5Y 1K6
☏ 604 874 4300 🖷 604 874 4305
info@shavickentertainment.com
www.shavickentertainment.com

Film and television production company. Examples of output include *Young Blades* and *Third Man Out*.

Shoes Full of Feet `MOVIE`

246 Dunview Avenue, Toronto, Ontario, M2N 4J2
☏ 416 512 0084
info@shoesfulloffeet.com
www.shoesfulloffeet.com
President *Kris Booth*
CEO *Bryce Mitchell*
COO *Raj Panikkar*

Feature film production company founded in 1999. Focuses on narrative projects.
No unsolicited material.

Showdog Productions `TV`

100 Cameron Street, PO Box 788, Moncton, New Brunswick, E1C 8N6
☏ 506 857 0297 🖷 506 959 1319
bob@showdogproductions.com
www.showdogproductions.com
President *Bob Sissons*
Producer *Jack Coreau*
Production Manager *Tony Sissons*

Animation company founded in 2002 specializing in children's entertainment programmes and educational DVDs. Most recently produced the 2D animated children's TV series *Rugg Buggs*.

Material accepted via email or post. All submissions reviewed and responses sent within four weeks.

Sienna Films `MOVIE` `TV`

260 Spadina Avenue, Suite 504, Toronto, Ontario, M5T 2E4
☏ 416 703 1126
siennainfo@siennafilms.com
www.siennafilms.com

Independent film and television production company.
Not currently accepting unsolicited material.

Soapbox Productions Inc. `TV`

Baker House, 650 Keith Road West, North Vancouver, British Columbia, V7M 1M6
☏ 604 983 2555 🖷 604 983 2558
info@soapboxproductions.ca
www.soapboxproductions.ca

Formed in 1990, output includes drama, comedy, documentary, music, children's and reality series. Credits include *Northwood*, *Cosmic Highway*, *Double Exposure*, *O.Com: Cybersex Addiction* and *Shakin' All Over*.

Spire Entertainment Inc. `MOVIE` `TV`

25 Isabella Street, Toronto, Ontario, M4Y 1M7
☏ 416 964 3247 🖷 416 964 0964
info@spirefilms.com
www.spirefilms.com
President and CEO, Executive Producer *Jasbir "Jazz" Mann*

Film and television production company established in 2003. Interested mainly in drama, comedy, short film, feature film and sitcom projects. Recent titles include *Code* and *Sohni Sapna*.

Welcomes unsolicited material via email and post. Any stage from treatment to script acceptable.

Storyline Entertainment Inc. `TV`

115 Palmerston Avenue, Toronto, Ontario, M6J 2J2
☏ 416 603 8333 🖷 416 603 6318
lisa@storylineentertainment.com
www.storylineentertainment.ca
Contact *Lisa Valencia-Svensson*

Founded in 2000, producing documentaries and non-fiction programming for television. Credits include *A Whale of a Tale*, *Hitler's Canadians*, *The Secret of the Snake Goddess* and *Tiger Spirit*.

Strada Films Inc. `MOVIE`

788 King Street West, 2nd Floor, Toronto, Ontario, M45V 1N6

☏ 416 642 2005 🖷 416 642 1970
info@stradafilms.ca
www.stradafilms.ca

Feature film production company.
   Unable to accept unsolicited screenplays or project proposals.

Sudden Storm Productions Inc. MOVIE TV

1 Deer Park Crescent, Suite 703, Toronto, Ontario, M4V 3C4
☏ 416 927 9342
information@suddenstorm.ca
www.suddenstorm.ca

Formed in 2001, a feature film and television production, service and finance company. Has provided financial and consulting services to projects such as the television series *The Newsroom* and *Whistler*. Continues to develop features and movies-of-the-week in addition to its other services.

Sullivan Entertainment Inc. MOVIE TV

110 Davenport Road, Toronto, Ontario, M5R 3R3
☏ 416 921 7177 🖷 416 921 7538
inquire@sullivan-ent.com
www.sullivan-ent.com

Over 20 years experience as a developer, producer and distributor of series, mini-series and movies for television and international release. Examples of output include *Anne of Green Gables*, *Road to Avonlea* and *Butterbox Babies*.

Summer Pictures Inc. MOVIE

91 Beresford Drive, Richmond Hill, Ontario, L4B 4J5
☏ 905 883 5561 🖷 905 787 1240
summer@summerpictures.biz
www.summerpictures.biz

Specializes in feature film-making. Also offers in-house post-production.

Temple Street Productions MOVIE TV

119 Spadina Avenue, Suite 705, Toronto, Ontario, M5V 2L1
☏ 416 591 0065 🖷 416 591 0075
info@templestreetproductions.com
www.templestreetproductions.com

Film and television production company with credits including *Queer as Folk V*, *Darcy's Wild Life II* and *Blueprint for Disaster II*.

Tricon Films & Television TV

372 Richmond Street West, Suite 200, Toronto, Ontario, M5V 1X6
☏ 416 341 9926 🖷 416 341 0173
info@triconfilms.com
www.triconfilms.com

Founded in 2000, developing and producing television prgrammes for international markets. Specializing in lifestyle and feature length documentaries.

Triptych Media Inc. MOVIE TV

788 King Street West, Toronto, Ontario, M5V 1N6
☏ 416 703 8866 🖷 416 703 8867
info@triptychmedia.ca
www.triptychmedia.ca

Produces film and television dramas. Particularly know for adaptations of literary works by Canadian writers including Carol Shields, Barbara Gowdy, David Adams Richards, Matt Cohen and Michel Marc Bouchard. Examples of output include *The Republic of Love, Falling Angels, The Hanging Garden, Lucky Girl* and *The Bookfair Murders*.
   Email a one page synopsis and cover letter detailing project history, plus relevant attachments and details of key personnel.

True West Films MOVIE

2050 Scotia Street, Suite 201, Vancouver, British Columbia, V5T 4T1
☏ 604 879 4590
www.truewestfilms.com
President/Producer *Elizabeth Yake*
Director of Development *Gosia Kamela*

Film production company founded in April 2002 specializing in drama, comedy and POV documentary. Past work includes *It's All Gone Peter Tong* and *Everything's Gone Green*.
   Unsolicited material accepted in one-page pitch format. Scripts not read unless requested.

Universal Studios Canada Inc. MOVIE

4–2450 Victoria Park Avenue, Willowdale, Ontario, M2J 4A2
☏ 416 491 3000 🖷 416 491 2857
www.universalstudioscanada.com

Canadian arm of one of the world's great film producing and distributing companies.
   Policy not to accept or consider creative materials, ideas, or suggestions other than those specifically requested. Do not send any original creative materials such as screenplays, stories, original artwork, etc. Anything sent may be used by Universal or its affiliates for any purpose, including, but not limited to, reproduction, disclosure, transmission, publication, broadcast and posting.

White Iron Pictures Inc. MOVIE TV

533, 1201–5th Street, SW, Calgary, Alberta, T2R 0Y6
☏ 403 298 4700 🖷 403 233 0528

info@whiteiron.tv
www.whiteiron.tv

Established in 1990, working with many North American networks as well as independent filmmakers. Works across genres, media and disciplines. Produces television programming and documentaries.

Wizz Films Inc. (MOVIE)

418 Sherbrook Street East, Suite 300, Montréal, Québec, H2L 1J6
T 514 932 4191 F 514 932 7277
info@wizzfilms.com
www.wizzfilms.com
Contact *Danny Bergeron*

Feature film production company focused on family animation and live action projects. Founded in 2002.
    Welcomes new scripts and co-production projects via email.

XGenStudios Inc. (VG)

Suite 385, 11215 Jasper Avenue, Edmonton, Alberta, T5K 0L5
T 888 808 XGEN
Games@XGenStudios.com
www.XGenStudios.com
PR Director *Dan Greig*

Founded in 2003, games production company specializing in free web browser games.
    Welcomes unsolicited submissions. New writing ideas should include mock screen shots and an indication of how the game will work on a technical level.

ZIJI Film & Television Productions Ltd (MOVIE) (TV)

The Roy Building, Suite 422, 1657 Barrington Street, Halifax, Nova Scotia, B3J 2A1
T 902 425 5001 F 902 429 0077
info@ziji.ca
www.ziji.ca

Producers of award-winning documentaries, feature films and television programmes. Titles include *Words of My Perfect Teacher*, *Warrior Songs* and *Regarding: Cohen*.

Ireland

2000 AD Productions (MOVIE)

Ardmore Studios, Herbert Road, Bray, County Wicklow
T 087 297 9131 F 01 276 9546
Adrian@2000adproductions.com
www.2000adproductions.com
Producer *Adrian Devane*

Established in 2000 by freelance producer Adrian Devane. Currently working on several features with Irish writers.
    Welcomes unsolicited manuscripts. Prefers to be approached with a synopsis, then treatment, then script.

Abú Media (MOVIE) (TV)

Teach Indreabháin, Inverin, County Galway
T 091 505100 F 091 505135
info@abumedia.com
www.abumedia.com

Established in 2000, an award-winning television and film production company producing a range of media content, including drama, documentary and light entertainment for English and Irish speaking markets. Also specializes in co-productions and co-financing international projects. Production credits include the TV dramas *Teenage Cics* and *Cut & Dry*, the documentary series *Coiscéim* and the short film, *Lasair*.

Accomplice Television (TV)

The Barracks, 76 Irishtown Road, Dublin 4
T 01 660 3235 F 01 660 3238
office@accomplice-tv.com
www.accomplice-tv.com

Founded in 2001, producing drama, documentary and comedy for TV. Production credits include *Dan & Becs* and *Bachelors Walk 3*.

Adare Productions (MOVIE) (TV)

Adare House, 35A Patrick Street, Dun Laoghaire, County Dublin
T 01 284 3877 F 01 284 1814
adare@eircom.net
Managing Director *Brian Graham*

Producing high-end television programmes since 1993. Specializing in the creation of entertainment formats, the company has it's own in-house studio directors and on-line editors as well as a creative team of producers and researchers.
    Does not accept unsolicited manuscripts.

Animo Television (TV)

4 Windmill Lane, Dublin 2
T 01 671 3004 F 01 679 7046
info@animo.ie
www.animo.ie
Director *Grainne Barron*

Established in 2003 to produce high-end factual and entertainment programming for the Irish and international market. Has since stablished itself in its home market with a leading celebrity entertainment series, seven primetime documentary series and a number of one-off television event specials and mini animation series. Programmes include *Three 60, Far Away Up Close, Anonymous* and *Just for Laughs.*

Athena Media TV

Digital Depot, Digital Hub, Thomas Street, Dublin 8
T 01 488 5850
info@athenamedia.ie
www.athenamedia.ie
Managing Director *Helen Shaw*

Formed in 2003.
Specializes in television, primarily factual documentaries but other genres including comedy and sport. Recent titles include *Headbanging to Beethoven, Last Chance Catholics* and *The Smallest Bundles* (Doc Shorts).

Barley Films MOVIE

2 Rogan's Court, Patrick Street, Dun Laoghaire, Dublin
T 01 214 5940
info@barleyfilms.net
www.barleyfilms.net

Founded in 2002, producing animated films. Output includes *Boys Night Out* and *Agricultural Report*, both nominated for awards by the International Animated Film Society. Its first feature length project, *Little Caribou*, is in production.

Bootstrap Films Limited MOVIE

Grange House, 60 Beaumont Avenue, Churchtown, Dublin 14
T 01 298 7466 F 01 298 7472
info@bootstrapfilms.com
www.bootstrapfilms.com

Established in 2004 as an independent production company. Run by John Phelan who has raised over €30 million for international projects shooting in Ireland. Raises funds via the tax incentive scheme for filming in Ireland and through other international sources.

Boulder Media MOVIE TV

14 Hawkins Street, Dublin 2
T 01 677 9775
pete@bouldermedia.tv
www.bouldermedia.tv

Major animation studio established in 2000. Produces work for clients including Cartoon Network, Disney and Nickelodeon. Also produces its own projects, as well as short films.

Brown Bag Films TV

Block F, 1st Floor, Smithfield Market, Smithfield, Dublin 7
T 01 872 1608
studio@brownbagfilms.com
www.brownbagfilms.com
Producer *Cathal Gaffney*
Head of Development *Gerald Murphy*
Creative Director *Darragh O Connell*

Established in 1994, producing primarily pre-school animation (such as *Wobbly Land*) and television commercials. Received an Academy Award nomination in 2002.
Does not accept unsolicited manuscripts but interested in developing relationships with experienced children's writers for television.

CABOOM TV

10 Stephens Green, Dublin 2
T 01 672 7077 F 01 672 7043
damo@caboom.ie
www.caboom.ie
Creative Director *Damian Farrell*

Produces entertainment, documentary, short film, commercials and animation programmes. Also has an office in LA.
Welcomes outlines or treatments only.

Campbell Ryan Productions Ltd MOVIE TV

M305 Media Cube, Kill Avenue, Dun Laoghaire, County Dublin
T 01 440 4205 F 01 443 0639
info@campbellryanproductions.com
www.crfilms.net
Manager/Director *Triona Campbell*
Co-ordinator *Trevor Curran*

Established in 2000, its debut feature film, *The Crooked Mile*, won the Tribeca First Look award.
Specializes in television and film dramas.
Welcomes unsolicited manuscripts and prefers to be emailed with a synopsis.

Cartoon Saloon MOVIE TV

The Maltings, Tilbury Place, James Street, Kilkenny
T 056 776 4481 F 056 772 0089
info@cartoonsaloon.ie
www.cartoonsaloon.ie
Studio Manager *Kairen Waloch*

Award-winning animation company producing for TV and cinema. Output includes the television series *Skunk Fu!* and the feature film *The Secret of Kells.*

Claddagh Films `MOVIE` `TV`

Somerset Studios, Aughinish (Kinvara PO),
County Clare
℡ 065 7078454 🖷 065 7078242
www.claddagh.ie

A small, independent film and television
production company based in Galway. Previous
films include *An Autumn Affair, Packy's Cousin,
A Place in my Heart, A Talk in the Dark, The Lift*
and *The Biscuit Eaters.*

Coco Television `TV`

49–50 Berystede, Leeson Park, Dublin 6
℡ 01 497 0817 🖷 01 497 0796
info@cocotelevision.ie
www.cocotelevision.ie

Formed in 1986, a leading cross-genre television
production company serving domestic and
international markets. Specialists in lifestyle,
reality, documentary and event TV. Examples
of output include *Treasure Island, Cabin Fever,
CrimeCall, Desperate Houses, The Great Escape,
House Hunters, A Gap in the Mountain, Any
Given Sunday* and *Made in America.*

Crannog Films `MOVIE` `TV` `VG`

33 Clarendon Street, Dublin 2
℡ 01 707 1612 🖷 01 707 1614
crannogfilms@indigo.ie
Director *Conor Harrington*
Director *Sophie Loughnan*
New Development *Medb Riordan*

Produces for both TV and film, as well as for
the computer games market with a co-venture
partner.
 Approach through an agent or via referral.

Digital Animation Media Limited `MOVIE` `TV` `VG`

Digital Depot, Roe Lane, Thomas Street, Dublin 8
℡ 01 489 3644
info@digitalanimationmedia.com
www.digitalanimationmedia.com

Produces animated films and series for TV, and
is involved in computer game development.
TerraGlyph is the wing that develops, produces
and finances animated features films, television
series and other niche market animation titles. Its
output includes 26 episodes of *The Island of Inis
Cool, The Wilde Stories* trilogy and the movies,
Duck Ugly, Help I'm a Fish and *Carnivale.*

Earth Horizon Productions `TV`

13 Windsor Place, Dublin 2
℡ 01 661 7475 🖷 01 662 0337

info@earthhorizon.ie
Production Manager *Marcus Stewart*

Has produced documentaries such as *Eco Eye,
About the House* and *Return to Chernobyl.*
 Does not accept unsolicited material.

Element Pictures Ltd `MOVIE` `TV`

21 Mespil Road, Dublin 4
℡ 01 618 5032
info@elementpictures.ie
www.elementpictures.ie
Office Manager *Hilary Barrett*
Development Assistant *Yvonne Donohoe*

Leading producers of film and television drama.
UK office recently opened. Production credits
include *Garage, Death of a President, The Wind
that Shakes the Barley, The League of Gentlemen's
Apocalypse, Omagh* and *The Magdalene Sisters.*
 Does not welcome unsolicited approaches.
Approaches should be through an agent.
Interested in investigating writers who have had
at least one production on radio, film or theatre,
or a novel published

EO Teilifís `TV`

An Chuasnóg, Baile Ard, Spiddal, County Galway
℡ 091 558400 🖷 091 558470
beartla@eoteilifis.ie
www.eoteilifis.ie
Contact *Beartla O Flatharta*

Television production company and facilities
provider established in the 1990s. Specializes
in drama, factual programming and children's
programming. Production credits include the
long-running *Ros na Rún* for TG4, the fantasy
drama *Géibheann*, the documentaries *The
Rescuers* and *Dúiche* and the children's productions
Míre Mara and *Mise agus Pangúr Bán.*

Esperanza Productions `MOVIE` `TV`

54 Hampton Road, Booterstown, County Dublin
℡ 01 278 0003
info@esperanza.ie
www.esperanza.ie

Producers of television and film documentaries,
as well as interactive media. Focus on human
rights issues. Productions include *When
Happiness is a Place for Your Child, We Still Want
You But..., Dropping the Number 10 for Dili* and
Invisible Movement.

Fantastic Films `MOVIE` `TV`

3 Clare Street, Dublin 2
℡ 087 255 1666
info@fantasticfilms.ie
www.fantasticfilms.ie

Formed in 2000 by John McDonnell to produce and develop movies and television drama. Prides itself on high production values and works with emerging and established writers and directors. *Timbuktu* was chosen to open the Dublin International Film Festival in 2004. Other productions include *The Making of a Prodigy*, *Burn the Bed*, *Six Shooter* and *Invisible State*.

Fastnet Films   MOVIE TV

1st Floor, 75–76 Camden Street Lwr, Dublin 2
T 01 478 9566 F 01 478 9567
enquiries@fastnetfilms.com
www.fastnetfilms.com

Film and television production company, working in feature films, television drama series and documentaries. Since 2000, Fastnet has produced over 160 hours of Irish network programming. Also active in international co-productions. Film credits include *An Teanga Runda*, *The Wonderful Story of Kelvin Kind*, *The Halo Effect* and *Eat The Peach*. Television works include *September*, *Lord Haw Haw* and *Bang You're Dead*.

Feenish Productions   MOVIE TV

26 South Frederick Street, Dublin 2
T 01 671 1166
info@feenish.com
www.feenish.com

A film and media production company founded in 2001. Has produced a wide-ranging body of film work, with commissions and funding from TG4, RTE, BCI, the Arts Council, the Irish Film Board and a variety of independent clients. Has worked on documentary films, animations, promotional, corporate and educational films.

Film-makers with an idea for a project are encouraged to make contact. All correspondence handled in the strictest of confidence.

Ferndale Films   MOVIE TV

Ardmore Studios, Bray, County Wicklow
T 01 276 9350 F 01 276 9557
info@ferndalefilms.com
www.ferndalefilms.com

Founded in 1987 by Noel Pearson as a film and theatre production company. Feature film credits include the Oscar-winning *My Left Foot*, *Tara Road*, *The Field*, *Dancing at Lughnasa* and *Frankie Starlight*. Has also produced documentaries including *Bram Stoker's Dracula* and *Brian Friel*.

Frontier Films   TV

2 Northbrook Road, Ranelagh, Dublin 6
T 01 497 7077 F 01 497 7731

info@frontierfilms.ie
www.frontierfilms.ie

An independent television production company, set up in 1986. Works across a range of genres including music programming, factual series and documentaries, including *Big Boys Don't Cry* and *Remember Me*.

Gallowglass Pictures   TV

5 Upper Baggot Street, Dublin 4
T 01 6677050 F 01 6677051
eamon@gallowglasspictures.ie
www.gallowglasspictures.ie

International television company established in 1998 by Eamon McElwee and Tom Clinch. Makes international documentaries, series and corporate videos. Also experienced in multimedia. Output includes *From Clare to Here*, *Dublin City Life*, *Solo in South-East Asia* and *Wolves*.

Glowworm Media   MOVIE TV

Cornmarket Square, Limerick
T 061 446044
kieran@glowworm.ie
www.glowworm.ie

Produces factual programming and documentaries. Aims to treat subjects in a fair and sympathetic way, presenting their stories objectively and truthfully.

Grand Pictures Ltd   MOVIE

44 Fontenoy Street, Dublin 7
T 01 860 2290 F 01 860 2096
info@grandpictures.ie
www.grandpictures.ie

Established in 2000 by the producers Michael Garland and Paul Donovan. Recipients of multiple project development funding from the Irish Film Board and media slate funding. Currently developing a mixture of projects with emerging and experienced talent. Production credits include *Puffball*, *Val Falvey TD*, *Dead Long Enough*, *Spin The Bottle*, *Fergus' Wedding* and *Paths To Freedom*.

Does not accept unsolicited scripts but happy to view brief (one page) synopsis of an idea.

Great Western Films   MOVIE TV

28 Gardiner Place, Dublin 1
T 01 889 8040 F 01 872 8280
info@greatwesternfilms.com
www.greatwesternfilms.com
Development Producer *Niamh Fagan*
Managing Director *Eoin Holmes*

Founded in 2001, Great Western has produced shows such as *Trouble in Paradise*, *The Last Furlong* and *Camera Cafe*.

Willing to accept unsolicited manuscripts but prefers to be contacted with an initial email.

Hawkeye Films ▢MOVIE

Killina, Gort, County Galway
☎ 091 638219 ℻ 091 638048
info@hawkeyefilms.com
www.hawkeyefilms.com
Managing Director *Donal R. Haughey*

Founded in 1994, a small film production company focusing on documentaries, especially social history and the arts. Credits include *Into The Valley*, *An Taibhdhearc*, *Deireadh le hAiocht*, *SLOTS*, *Bothar na Tra* and *Children Of Allah*.

Hell's Kitchen International Ltd ▢MOVIE

21 Mespil Road, Dublin 4
☎ 01 679 5065 ℻ 01 664 3737
info@hellskitcheninternational.com
www.hellskitcheninternational.com

Since 2003 has acted as a co-producer, principally on US films shooting in Ireland. Previous films include *Laws of Attraction* and *The Honeymooners*.

Icebox Films ▢MOVIE ▢TV

56 Temple Road, Blackrock, County Dublin
☎ 01 210 8501 ℻ 01 210 8528
info@iceboxfilms.ie
www.iceboxfilms.ie

TV and film company, established in 2000 by Clíona Ní Bhuachalla and Charlie McCarthy. Dedicated to creating quality contemporary productions. Credits include the television drama *Legend*.

Illusion Animated Productions ▢TV

The Studio 46, Quinn's Road, Shankill, County Dublin
☎ 01 282 1458 ℻ 01 282 1458
info@illusionanimation.com
www.illusionanimation.com

Established in 1999, specializing in the development and creation of pre-production packages for animated TV series, including scripting and storyboarding. Always interested in talking with potential co-production partners for future projects.

JAM Media ▢MOVIE ▢TV

40 Kevin Street Lower, Dublin 8
☎ 01 405 3484 ℻ 01 478 9376

info@jammedia.ie
www.jammedia.ie

Involved in the creation of entertainment brands across genres and platforms. Has been involved in creating animated shorts and TV series. Credits include the children's series, *Picme* and the short film, *Boiled Eggs + Beer*.

Kavaleer Productions Ltd ▢TV

The Digital Hub, Unit A, 2nd Floor, 4 St Catherine's Lane West, Roe Lane, Dublin 8
☎ 01 488 5873 ℻ 01 488 5801
info@kavaleer.com
www.kavaleer.com
CEO *Andrew Kavanagh*
Director of Business Development *Gary Timpson*

Founded in 2001, specializing in original children's animation. Has produced programmes such as *Garth* and *Dad the Impaler*.

Does not accept unsolicited manuscripts. Will look at letters of application briefly outlining past experience and suitability.

Kite Entertainment Ltd ▢TV

4 Windmill Lane, Dublin 2
☎ 01 617 4744
info@kiteentertainment.com
www.kiteentertainment.com

A television production company with a talent management wing. Specializes in comedy, with credits includeing *Just for Laughs* and *Anonymous*.

Like it Love It Productions ▢TV

3/5 Carysfort Avenue, Blackrock, County Dublin
☎ 01 283 44 90 ℻ 01 283 64 20
andy.ruane@likeitloveit.com
www.likeitloveit.com
Managing Director *Andy Ruane*

Independent production company, whose own-format programmes are shown in over twenty countries. Creates programming for Irish and international markets. Credits include *The Lyrics Board*, *Perfect Match* and *Pop TV*.

Little Bird ▢MOVIE ▢TV

13 Merrion Square, Dublin 2
☎ 01 613 1710 ℻ 01 662 4647
info@littlebird.ie
www.littlebird.ie

Independent film and television production company established in 1982. Has additional offices in London and Johannesburg. Credits include *December Bride*, *Into The West*, *A Man of No Importance*, *Croupier*, *Bridget Jones's Diary*, *Trauma* and *Marie and Bruce*.

LKF Productions [MOVIE]

Production Office, 89 Cherry Garth Rivervalley, Swords, County Dublin
℡ 01 840 0561 ℻ 01 840 0561
info@lkf.ie
www.lkf.ie
Contact *Sé Merry Doyle*

Founded in 2005 by experienced producer, Liz Kenny. Aims to develop projects and emerging talent from Ireland and to work with European co-producers to make high-quality low-budget documentaries and feature films for the international market. Credits include the feature *WC* and the documentaries *Unspoken Children* and *In Living Hell*.

Loophead Studio [MOVIE]

Cross, Carrigaholt, County Clare
℡ 065 905 8309
naomi@loopheadstudio.com
www.loopheadstudio.com
Contact *Naomi Wilson*
Contact *Brian Doyle*

Founded in 2000, specializing in the production of animated films. Credits include *Among Strangers*, *Twilight* and *Rehy Fox*. Won the award for best animation at the 2006 Aubagne International Film Festival.
    Does not accept unsolicited manuscripts.

Loopline Film [MOVIE] [TV]

106 Baggot Lane, Dublin 4
℡ 01 667 6498 ℻ 01 667 6604
info@loopline.com
www.loopline.com
Managing Director *Sé Merry Doyle*

Founded in 1992, specializing in high-quality documentaries for domestic and international markets. Also has a training wing, in association with Screen Training Ireland, which runs a residential course in 'Creative Documentary for Directors'. Examples of output include *Hidden Treasures*, *James Gandon - A Life*, *Essie's Last Stand* and *Ahead Of The Class*.

Lunah Productions [MOVIE]

Old Fintra Road, Killybegs, County Donegal
074 973 1379
info@lunah-productions.com
www.lunah-productions.com

Established by the Hannigan family as a creative vehicle by which to focus their extensive experience of working in the media. Works on features, documentaries and corporate productions. Director Declan Hannigan has created work for the International Space Station.

Magma Films [MOVIE] [TV]

16 Merchants Road, Galway
℡ 091 569142 ℻ 091 569148
info@magmaworld.com
www.magmaworld.com

Independent production company, with subsidiaries in Munich and Hamburg. Specializes in animated features, series and high-concept programmes. Also works on live action productions, including films, children's drama, event television and light entertainment. Animation credits include *Norman Normal*, *Pigs Next Door* and *The World of Tosh*. Among its live action credits are *Arte*, *Bus Driver*, *10 Years After* and *The Falcon Thieves*.

McCamley Entertainment [TV]

103 The Woodlands, Ratoath, County Meath
℡ 01 825 7841 ℻ 01 825 7841
dmccamley@eircom.net
www.davidmccamley.com
Director/Producer/Writer *David McCamley*

Founded in 2000, McCamley Entertainment has produced the animated television series *The Island of Inis Cool*, which was nominated for an award at the Cartoons on the Bay international festival in 2005.
    Does not accept unsolicited materials.

Media Platform [MOVIE] [TV]

The Basement, 6 Clare Street, Dublin 2
℡ 01 639 4940
info@mediaplatform.ie
www.mediaplatform.ie
Managing Director *Al Butler*

Founded in 2005, producing for cinema and television.

Meem Productions [MOVIE] [TV]

17a Greenmount Lawns, Terenure, Dublin 6
℡ 01 492 8156
szaidi@oceanfree.net
www.BollywoodIreland.com
Producer *Siraj Zaidi*

Founded in 1989, Meem has produced two TV series and made nine film scripts acquisitions and developments. It has also set up an Indian Film distribution network.
    Particular focus on drama, documentaries and reality TV.
    Does not accept unsolicited manuscripts. Approaches should be via a solicitor or agent.

Midas Productions   (TV)

34 Lower Baggot Street, Dublin 2
☎ 01 661 1384 🖷 01 676 8250
mike@midasproductions.ie
www.midasproductions.ie

Foundeded in 1986, producing for television and corporates. The TV division specializes in English- and Irish-language documentaries, light entertainment and format-driven programming for both the English and Irish speaking markets. Output includes *Close Encounters with Keith Barry* and *The Liffey Laugh*.

Mind The Gap Films   (TV)

6 Wilton Place, Dublin 2
☎ 01 662 4742 🖷 01 662 4758
jennifer@mindthegapfilms.com
www.mindthegapfilms.com

Independent TV production company set up by Bill Hughes and Bernadine Carraher in 2001. Specializes in documentaries, entertainment, arts and music programmes. Recent credits include *Happy Birthday Oscar Wilde*, *Last One Standing* and *The Songs of Mama Cass*.

Mint   (MOVIE) (TV)

205 Lower Rathmines Road, Dublin 6
☎ 01 491 3333 🖷 01 491 3334
info@mint.ie
www.mint.ie

Documentary production company, specializing in historical and observational films. Output includes *Junior Doctors*, *Haughey*, *Our Lady's* and *Who Kidnapped Shergar*. Also has an office in Belfast (13 Fitzwilliam Street, Belfast, BT9 6AW; 028 9024 0555).

Moondance Productions   (MOVIE) (TV)

3 Reilly's Terrace Cork Sreet, Dublin 8
☎ 01 473 4599 🖷 01 473 4598
info@moondance.tv
www.moondance.tv

Established in 1994. Offers a complete service from concept to finished product for a range of corporate and broadcast clients. Production credits include *Sculpting Life*, featuring the artist Rowan Gillespie.

Nemeton Television Productions   (TV)

An Rinn, Dungarvan, County Waterford
☎ 058 46499 🖷 058 46208
geraldine@nemeton.ie
www.nemeton.ie
Head of Development *Geraldine Heffernan*

Founded in 1993. Specializes in factual documentary (singles and series), dramatized documentaries, Irish language programmes, sports coverage and corporate work. Credits include *Laochra Gael*, *Health Squad* and *Micheál O'Hehir*. Twice nominated for IFTAs for *The Brothers* (2006). Particularly interested in co-productions with the UK, Canada and Australia. Nemeton also runs a course in production for Irish speakers with the Waterford Institute of Technology and has an active area of development for Irish language programmes with TG4.

Principally develops ideas in-house. Ideas are welcome but contact via email in first instance or via an agent/professional representative before sending full proposal. Unsolicited material will not be read, copied or stored in any way.

New Decade Film & Television   (MOVIE) (TV)

'Ullord', Brickfield Lane, Killarney Road, Bray, County Wicklow
☎ 01 276 5071
info@newdecade.ie
www.newdecade.ie

Over ten years' experience in film, TV and corporate production. Production credits include *Capital Letters*, *Maybe if You* , *Tales from the Big House* and *So, This is Dyoublong?*

Newgrange Pictures   (MOVIE)

49–50 Berystede, Leeson Park, Dublin 6
☎ 01 498 8028 🖷 01 496 6128
info@newgrangepictures.com
www.newgrangepictures.com
Producer *Jackie Larkin*
Producer *Lesley McKimm*
Development Executive *Aislinn Ni Chuinneagain*

Founded in 2005, Newgrange has produced films such as *Two Wrongs*, *Kings* and *The Year of the Dog*.

Does not accept unsolicited manuscripts or ideas.

One Productions   (MOVIE) (TV)

3 Clare Street, Dublin 2
☎ 01 678 4077 🖷 01 678 4070
enquiries@oneproductions.com
www.oneproductions.com
Director *Tom Hopkins*
Development Executive *Caroline Skinnader*
Director *Christina Brosnan*

Established in 2000, producing films including *Money, Fear & Justice* and *Close*, which were finalists at the Venice Film Festival. *Close* also received the DMA award for best Irish short in 2006.

Welcomes unsolicited manuscripts and prefers to be contacted via email.

Paradox Pictures   [MOVIE] [TV]

26 South Frederick Street, Dublin 2
Ⓣ 01 670 6883 Ⓕ 01 670 6889
paradoxp@iol.ie

Founded in 1993, producing theatrical films,
documentaries, shorts and animated films. Titles
include *How Harry Became A Tree*, *Estella* (a
documentary about the painter Estella Solomons),
Northern Lights and *Pete's Meteor*.

Promedia TV   [MOVIE] [TV]

10 Herbert Place, Dublin 2
Ⓣ 01 662 2500 Ⓕ 01 662 2531
siobhan@promediatv.ie
Line Producer *Siobhan O'Brien*
Producer/Director *Michael O' Connell*
Director *Noduag Houlihan*

Established in 1985, producing films,
documentaries and lifestyle programmes.
Output includes the film *Joe my Friend*, lifestyle
series *Christmas Cracked* and the documentary
Rainbows in their Lives. Received the Crystal Bear
award, Gold Camera award, New Year Film Festival
gold medal and the Jacobs television award.

Will accept unsolicited material but contacted
via phone or email in first instance.

Red Pepper Productions   [TV]

Shamrock Chambers, 1–2 Eustace Street,
Dublin 2
Ⓣ 01 670 7277 Ⓕ 01 670 7278
office@redpepper.ie
www.redpepper.ie

Formed in 1990, producing commercials,
corporates and TV programmes. TV credits
include *Only Fools Buy Horses*, *The Millionaire*, *The
Builders & the Shantytown* and *Hollywood Trials*.

ROSG   [MOVIE] [TV]

An Spidéal, Co. na Gaillimhe
Ⓣ 091 553951 Ⓕ 091 558491
eolas@rosg.ie
www.rosg.ie

A film and television production company
established in 1998 by Ciarán Ó Cofaigh and
Darach Ó Scolaí. Examples of output include
Cosa Nite, *An Leabhar* and *Fíor Scéal*.

RTÉ Independent Productions   [TV]

Stage 7 RTÉ Donnybrook, Dublin 4
Ⓣ 01 208 2743
ipu@rte.ie
www.rte.ie

RTÉ is Ireland's public service broadcaster.
Commissions hundreds of hours of
independently-produced programmes each year.
Submitting instructions available at www.rte.
ie/commissioning/guidelines.html.

Samson Films   [MOVIE] [TV]

The Barracks, 76 Irishtown Road, Dublin 4
Ⓣ 01 667 0533 Ⓕ 01 667 0537
info@samsonfilms.com
www.samsonfilms.com
Managing Director *David Collins*
Producer *Martina Niland*
Head of Department *Neal Rowland*

Founded in 1984, specializing in feature length
drama. Produced *Once*, which was written/
directed by John Carney (winner of World Cinema
Audience Award, Sundance 2007). Also produced
True North, written/directed by Steve Hudson.

Always interested in new writing and welcomes
unsolicited manuscripts, sent via the website.
Over the past few years they have nurtured new
writers, assisting with shorts and developing and
producing first features.

Scannain Lugh   [MOVIE] [TV]

Pier Head, Tory Island, County Donegal
Ⓣ 074 910 0905 Ⓕ 074 910 0905
info@lughfilm.com
www.lughfilm.com
Producer and Director *Loic Jourdain*
Producer and Director *Anne Marie Nic Ruaidhri*

Founded in 2004, producing films and
documentaries. Output includes *Fear na nOileán*.
Materials should be submitted via email.

Screentime ShinAwiL   [MOVIE] [TV]

83–87 Ranelagh Village, Dublin 6
Ⓣ 01 4066 433 Ⓕ 01 4066 434
info@shinawil.ie
www.shinawil.ie

Independent production company, established in
1999 and specializing in entertainment, factual
entertainment and live event programming.

Guarantees to read all proposals but does not
commit to respond within a set time frame. Send
sae if submission is to be returned. Each project
submitted is subject to the Screentime ShinAwiL
development policy process.

Shadowhawk Films (Eire)
International   [MOVIE] [TV]

26 Cromlech Court, Dane Road, Poppintree,
Dublin 11
Ⓣ 01 862 7978/ 01 440 6017 Ⓕ 01 862 7978
shadow@shadowhawkfilms.com
www.shadowhawkfilms.com
President & COO *Andrew O'Reilly*

President & CCO - Production *Martin White*
Producer *Anthony Whelan*

A film and television production company founded in 1996. Recent productions include the films *Love's Delusion* and *Rules of the Game*, and the television shows *Redemption* and *Mind Games*.

Unsolicited materials should be directed to the acquisitions department via email: operations@shadowhawkfilms.com.

Shortt Comedy Theatre/Warehouse TV [TV]

Unit D1A, Eastway Business Park, Ballysimon Road, Limerick
maria@patshortt.com
www.patshortt.com

Founded in 2002, Short Comedy has produced television comedy series such as *Killinaskully* (series 1, 2 and 3).

Does not accept unsolicited materials.

Soilsiú Films [MOVIE] [TV]

Meenderry Falcarragh, Letterkenny, County Donegal
T 074 9180730 F 074 9180732
info@soilsiu.com
www.soilsiu.com

Founded in 1996, developing, scripting and producing feature films, documentaries, and broadcast television programmes. Originally called Vinegar Hill Productions. Production credits include the documentaries *Chiapas, Gods, Faeries and Misty Mountains* and *Saighdiúirí Beaga Gaelacha*, the animated works *Dick Terrapin* and *Sir Gawain and the Green Knight*, and the dramas *This Little Piggy* and *It's All in the Jeans*.

SOL Productions Ltd [TV]

Quarantine Hill, Wicklow Town, County Wicklow
T 0404 68645 F 0404 67153
sol@eircom.net
Marketing Director *Veronica O'Reilly*

Established in 1985, producing shorts, documentaries and dramas. Output includes *A Spiritual Journey*.

Does not accept unsolicited manuscripts.

SP Films [MOVIE]

Unit F5, Riverview Business Park, Nangor Road, Dublin 12
T 01 460 4760 F 01 460 4770
spfilms@eircom.net
www.spfilms.ie

Production company of feature films and short movies. Previous titles include *The Ten Steps, The*

Honourable Scaffolder, Innocence and *The Church of Acceptance*.

Speers Film [MOVIE]

24 Fitzwilliam Street Upper, Dublin 2
T 01 6621130
jonny@speers.ie
www.speers.ie
Contact *Jonny Speers*

Produces commercials and features and offers range of production services. Credits include the film *Adam & Paul*.

Stoney Road Films [MOVIE] [TV]

T 01 677 6681
info@stoneyroadfilms.com
www.stoneyroadfilms.com

Produces drama and documentary for the cinema and TV. Also has a distribution arm. Credits include *Wordweaver - The Legend of Benedict Kiely*, *Convention* and the horror series, *Shiver*.

Telegael Media Group [MOVIE] [TV]

Telegael Spiddal, County Galway
T 091 553460 F 091 553464
info@telegael.com
www.telegael.com

Specialists in animated productions but also develops live action, factual and entertainment programmes.

TG4 [TV]

Baile na hAbhann, Co. na Gaillimhe
T 091 505050 F 091 505021
eolas@tg4.ie
www.tg4.ie

Established in 1996. Has a 3.5% share of the national television market. Core service is its Irish language programme.

Welcomes submissions in the areas of documentaries, traditional music and song, comedy, drama, docu-soaps, lifestyle, travel and the arts. Submissions will only be accepted on TG4 submission forms and should be with TG4 by 16 January, 1 May or the 1 October. Unsuccessful applications will not be retained or returned. See the website for more details.

Tile Films [MOVIE] [TV]

13 Windsor Place, Dublin 2
T 01 611 4646 F 01 661 4647
info@tilefilms.ie
www.tilefilms.ie

Produces high-end factual documentaries with international appeal. Output includes *Ireland's*

Nazis, The Ghosts Of Duffy's Cut and *The Lost Gods*. Also has an office in Manchester.

Send submissions to Rachel@tilefilms.ie

Timesnap Limited   MOVIE TV

15 Luttrell Park Drive, Castleknock, Dublin 15
T 353 86 127 2740
mail@timesnap.com
www.timesnap.com
Director *Declan Cassidy*
Producer *Bill Tyson*

Founded in 2005, producing documentaries including *War on Waste* and *The Crew Cut*. Television production includes the 5-part series *Dancing Dublin*. Films include *Mrs Friendly* and *Whatever Turns You On*.

Does not accept unsolicited scripts.

Totality Pictures Ltd   MOVIE TV

12 Castleview, Artane, Dublin 5
info@totalitypictures.com
www.totalitypictures.com

Film production company offering range of services from script-writing to post-production.

Tyrone Productions   TV

27 Lower Hatch Street, Dublin 2
T 01 662 7200 F 01 662 7217
pcarroll@tyrone-productions.ie
www.tyrone-productions.ie
General Manager *Patricia Carroll*

An independent television production company founded in 1987 and working across genres including drama, documentary and entertainment. Production credits include *Riverdance*, the dramas *Beckett on Film* and *Ros na Rún*, and the documentaries *The Land of Sex and Sinners* and *The Pope's Children*.

Submission form available from website.

Venom   MOVIE

9 Sallymont Gardens, Dublin 6
T 01 4911954 F 01 4911954
info@venom.ie
www.venom.ie

Film production company founded in 1994, developing work for broadcast, internet and theatrical distribution. Credits include the award-winning short, *Undressing my Mother*.

Vico Films   MOVIE

Unit D, Glencormack Business Park, Kilmacanogue, County Wicklow
T 087 6460406 F 01 2014999
info@vicofilms.com
www.vicofilms.com

Independent production company set up in 2003 with the aim of producing innovative and critically acclaimed films for the world market. Focus on developing feature projects, two of which have so far received development funding from the Irish Film Board. Recent titles include *CCW*, *Des Smyth*, *The Lump* and *Takeover*.

Wide Eye Films   MOVIE

30 Herbert Street, Dublin 2
T 01 678 7930 F 01 678 7930
mail@wideeyefilms.com
Producer *Nathalie Lichtenthaeler*

A multi-award-winning feature film production company established in 1999. Recent feature films produced include *Cowboys & Angels* and *The Front Line*.

Any screenplay submissions should be preceded by an initial contact (preferably e-mail).

Wider Vision   TV

West Clare
T 086 260 7822
info@widervision.ie
www.widervision.ie

Formed in 2003, a television production company specializing in documentaries, dramas and corporate productions. Also provides camera crews and services to national and international broadcasters.

Wildfire Films & Television Productions Limited   MOVIE TV

Olympia House, Suite 2B, 61–63 Dame Street, Dublin 2
T 01 672 5553 F 01 672 5573
info@wildfirefilms.net
www.wildfirefilms.net

Produces feature films, television drama and documentaries. Output includes *The Plays of John M. Synge*, *Middletown*, *Red Mist*, *Undertakers*, *The Sack Em Ups*, *When Pigs Carry Sticks* and *Sunday in Dublin*.

World 2000 Productions   MOVIE TV

Ardmore Studios, Herbert Road, Bray, County Wicklow
T 01 276 9672 F 01 286 6810
info@world2000.ie
www.world2000.ie

Established in 1994, developing, producing and distributing feature and television entertainment for the global market. Utilizing the Irish tax incentive programme, finances and provides production services for works including *King Arthur*, *Veronica Guerin*, *Ella Enchanted*, *The*

Count of Monte Cristo, Braveheart and other Irish co-produced projects.

Xstream Pictures Limited `MOVIE` `TV`

Isolde's Tower, 4 Essex Quay, Dublin 2
℡ 01 6799925
info@xstream-pictures.com
www.xstream-pictures.com

Specializes in 'complex and innovative storytelling through a range of media.' Produces documentaries, short dramas, commercials and music videos. Television output includes the documentaries *Fair City, Behind the Scenes* and *RTE Philharmonic Choir.* Also the short film, *The Hall.*

Young Irish Film Makers `MOVIE`

St. Joseph's Studios, Waterford Road, Kilkenny,
℡ 056 776 4677 ℉ 056 775 1405
info@yifm.com
www.yifm.com

A film training and production company set up in 1991 to help young people aged 13 to 20 make digital feature films. Established the National Youth Film School in 2002 to allow young people to spend five weeks shooting a major feature film for television. Also runs film workshops. YIFM's first feature was *Under the Hawthorn Tree* for Channel 4 and RTE in 1998.

Zanzibar Films `MOVIE`

12 Magennis Place, Dublin 2
℡ 01 671 9480 ℉ 01 671 9481
zanzibarfilms@eircom.net
www.zanzibarfilms.net

Founded in 1998, Zanzibar's most notable success is the film *Headrush,* which won the Miramax script award as well as four other international awards.
Willing to look at unsolicited treatments.

Australia

3monkeyfilms Pty Ltd `MOVIE`

5f/26 Wellington Street, Collingwood, VIC 3066
℡ 03 9486 0868 ℉ 03 9486 0867
monkey1@3monkeyfilms.com
www.3monkeyfilms.com
Contact *Michael Robinson Milwright*

Production company established in 2001, focusing on feature film development and short film production. Past credits include *The Opposite of Velocity, The Hedge, A Kind of Hush* and *The Good Oil.*
Welcomes unsolicited manuscripts.

Agenda Film Productions `MOVIE` `TV`

74 Watson Street, Bondi, NSW 2026
℡ 02 9389 4851 ℉ 02 9389 4861
agendafilm@bigpond.com
Director *Ian Iveson*

Founded in 1983, film and television production company specializing in drama. Credits include *Lost Things* and *Kideo!*
Welcomes unsolicited manuscripts by email.

Arenafilm `MOVIE`

Level 2, 270 Devonshire Street, Surry Hills, NSW 2010
℡ 02 9319 7011 ℉ 02 9319 6906
mail@arenafilm.com.au
www.arenafilm.com.au

Founded in 1987, producing feature films that explore social and political themes. Produced the films *Romulus, My Father* (starring Eric Bana), *Three Dollars, The Bank* and *The Boys.*

Associated Creative Talents `MOVIE` `TV`

PO Box 328, Hawthorn, VIC 3122
℡ 03 9855 1571
theproducer@associatedcreativetalents.net
www.associatedcreativetalents.net
Producer/Director *G D Bruny*
Producer *Thana Ed*

Founded in 1983, production company with credits including *Flynn's Flying Doctors, Sixty and Over* and *Rescue.* Currently producing 2-part docu-drama mini-series for TV.
Unsolicited manuscripts are welcomed. 'The first few pages will tell us a lot.'

Aus-TV International `MOVIE` `TV`

PO Box 423, Woollahra, NSW 1350
℡ 02 9929 9300
austv@tvaustralia.com
www.tvaustralia.com
Contact *Bill Payne*

Established in 1986, producing film and TV dramas and documentaries.
Welcomes unsolicited manuscripts. Cautious policy on new writing, 'but open to new concepts and ideas.'

Australian Children's Television Foundation `TV`

3rd Floor, 145 Smith Street, Fitzroy, VIC 3065
℡ 03 9419 8800 ℉ 03 9419 0660
info@actf.com.au
www.actf.com.au
Program Assistant *Melinda Abbey*

A national non-profit organization, committed to providing Australian children with entertaining media made especially for them. Has created over 150 hours of high quality children's content, screened in over 100 countries and winning over 95 local and international awards. Titles include *Round the Twist* and *Crash Zone*. Also acts as a funding body for other children's television producers, offering both script development funding and production investment.

Australian International Pictures Pty Ltd (MOVIE) (TV)

4/305 North Terrace, Adelaide, SA 5000
T 08 8227 0681
austinpic@hotmail.com
www.austinpic.com
Producer *Wayne Grom*

Established in 1981, production company focusing on feature films, television and documentaries. Recent productions include *Maslin Beach* and *The Dreaming*, which was nominated for an Australian Film Institute award.
   Welcomes unsolicited manuscripts and ideas.

Avalon Film Corporation (MOVIE) (TV)

Ultima 6/30 Garfield Terrace, Surfers Paradise, QLD 4217
T 07 5526 7932 F 07 5526 7932
avalonfilms@bigpond.com
Director *Phil Avalon*

Founded in 1976, producing 'good stories that can transfer to film or television'. Past credits include *Liquid Bridge* and *The Pact*.
   Submissions only accepted through agents.

Banksia Productions (TV)

L62–80 Wellington Square, North Adelaide, SA 5006
T 08 8334 5300 F 08 8334 5334
info@banksia.com
www.banksia.com

Founded in 1987, producing children's programmes and documentaries. Credits include *Humphrey*, the series *Great Australian Train Journeys* and the documentary *Pub Crawl With Altitude*. Has received numerous industry awards including the Prix Jeunesse (Germany) and several Logies.

Becker Entertainment (MOVIE) (TV)

Level 1, 11 Waltham Street, Artarmon, NSW 2064
T 02 9438 3377 F 02 9439 1827
info@beckers.com.au
www.beckers.com.au

Integrated screen entertainment company, founded in 1965 by Russell Becker, one of the pioneers of commercial television in Australia. A leader in television and film production, distribution and exhibition. Produces TV programming across a range of genres including documentaries, entertainment shows, children's programmes, live sports and drama. Produces local language programs for Asian markets, operating out of Singapore and Jakarta. Also has a long track record of producing feature films in Australia and the US and is a major independent film distributor in Australia and New Zealand.

Big and Little Films Pty Ltd (MOVIE) (TV)

PO Box 1271, St Kilda South, VIC 3182
T 03 9527 8299 F 03 9527 8266
info@bigandlittlefilms.com
www.bigandlittlefilms.com

Award-winning independent production company working in feature films, documentary and television drama. Develops and finances a slate of projects for domestic and international markets, working with a range of writers and directors.

Blink Films Pty Ltd (MOVIE) (TV)

PO Box 1315, Crows Nest, NSW 1585
T 02 9439 9900 F 02 9439 8099
mail@blinkfilms.com
www.blinkfilms.com
Managing Director *Michael Bourchier*

Television and film production company established in 1984, specializing in drama for children and adults. Recent output includes *The Upside Down Show* for childrens' TV and a feature film, *Lucky Miles*.
   Does not welcome unsolicited manuscripts, but writers can 'pitch a genre and broad idea by email'.

Richard Bradley Productions Pty Ltd (MOVIE) (TV)

PO Box 1417, Bondi Junction, NSW 1355
T 02 9959 3588
rbproductions@bigpond.com
CEO *Richard Bradley*

Founded in 1981, film and television company that has since produced over 100 films including documentaries, TV specials, TV drama and features.
   Welcomes unsolicited approaches by email.

Burberry Productions Pty Ltd (TV)

Level 1, 462 City Road, South Melbourne, VIC 3205
T 03 9693 0600 F 03 9693 0633

info@burberry.com.au
www.burberry.com.au

Founded in 2000, producing television drama series and children's programmes. Output includes *Eugénie Sandler PI*, *The Farm*, *Short Cuts* and *Last Man Standing*.

CAAMA (Central Australian Aboriginal Media Association) MOVIE TV

101 Todd Street, Alice Springs, NT 0870
T 08 8951 9777 F 08 8951 9717
productions@caama.com.au
www.caama.com.au
Executive Producer *Rachel Clements*

Founded in 1980, indigenous media company producing drama and documentaries. Past titles include *Double Trouble* and *Yellow Fella*. Produced the first aboriginal documentary entered at Cannes Film Festival.

Welcomes unsolicited manuscripts by email, but stories must have indigenous content.

Capitol Productions MOVIE

PO Box 734, North Sydney, NSW 2059
T 02 9966 0422 F 02 9966 0522
info@capitolproductions.com
www.capitolproductions.com
Executive Producer *Donna Svanberg*

Founded in 1992, specializing in films and commercials.

Matt Carroll Films Pty Ltd MOVIE TV

12 Sloane Street, Newtown, NSW 2042
T 02 9516 2400 F 02 9516 2099
mcfilms@bigpond.com.au
Contact *Matt Carroll*

Film and TV production company focusing on feature films, TV movies and drama series. Founded in 1995, the company produced *Murder in the Outback* for ITV.

Does not welcome unsolicited manuscripts - submissions 'only via agent or log lines'.

Cascade Films MOVIE

117 Rouse Street, Port Melbourne, VIC 3207
T 03 9646 4022 F 03 9646 6336
info@cascadefilms.com.au
www.cascadefilms.com.au

Production company founded in 1983 and owned by filmmakers Nadia Tass and David Parker. Has produced a series of acclaimed feature films, starting with *Malcolm* in 1985, which won 8 Australian Film Institute awards including Best Picture, Best Script and Best Director. While continuing to produce films by the Tass/Parker

writing/directing team, the company is also interested in engaging writers and directors who will develop their projects under the Cascade Films banner. Also owns and operates the Melbourne Film Studio. More recent credits include *Samantha: An American Girl Holiday*, *Undercover Christmas* and *The Lion, The Witch And The Wardrobe*.

Centaur Enterprises Pty Ltd MOVIE TV

PO Box 126, 4 Mona Vale, Sydney, NSW 1660
T 04 1032 4911
centaurfilms@bigpond.com
CEO *John Meagher*

Founded in 1975, involved in writing, producing and directing feature films and TV programmes, specializing in drama and documentaries. Output includes *Fantasy Man*, *The Rival* and *Australian Cotton*, which won a Gold Medal at the New York Film and TV Festival.

Welcomes new submissions but prefers introductory email first.

Jan Chapman Films Pty Ltd MOVIE TV

PO Box 476, Woollahra, NSW 2025
T 02 9331 2666 F 02 9331 2011
chapman@optusnet.com.au
Producer *Jan Chapman*
Producer's Assistant *Karen Colston*

Film and TV production company founded in 1989 by producer Jan Chapman. Past credits include *The Piano*, *Holy Smoke* and *Somersault*.

Does not welcome unsolicited manuscripts.

Cinetel Productions Pty Ltd MOVIE TV

15 Fifth Avenue, Cremorne, Sydney, NSW 2090
T 02 9953 8071
cinetel@bigpond.net.au
www.cinetel.com.au
Contact *Frank Heimans*

Founded in 1976, producing documentaries and some drama for television and theatrical release. Has won three Gold awards from the International Film & TV Festival of New York, two Blue Ribbons from the American Film & Video Festival, and a Gold award at Berlin. Credits include the documentary, *Battle for Sydney Harbour*.

Welcomes new writing but not unsolicited approaches. Contact by phone in first instance.

Crackerjack TV

75 Chandos Street, St Leonards, NSW 2065
T 02 9438 4255 F 02 9438 4275
info@crackerjack.com.au
www.crackerjack.com.au

A leading boutique production company, owned by FreMantle. Preference for original ideas and formats. Specializes in light entertainment, comedy, factual and reality.

Crawford Productions Pty Ltd (MOVIE) (TV)

Level 10, 575 Bourke Street, Melbourne, VIC 3000
T 03 8613 8250 F 03 8613 8262
admin@crawfords.com.au
www.crawfords.com.au

Founded in 1945 as a radio production facility, it has produced a string of domestic prime time drama successes such as *Consider Your Verdict*, *The Henderson Kids* and *The Violent Earth*.

Eaton Enterprises Pty Ltd (MOVIE) (TV)

18 Plateau Road, Avalon, Sydney, NSW 2107
T 02 9918 7722
eatent@aapt.net.au
Producer/Writer *Barry Eaton*
Producer/Writer *Matthew Eaton*

Founded in 1980, specializing in drama, documentary, internet programmes and corporate/training films. A feature, *Direct Action*, currently in production.

Happy to look at ideas via email. Prefers to develop new ideas and concepts 'from the ground up'.

Electric Pictures (MOVIE)

33 Canning Highway, East Fremantle, WA 6158
T 08 9339 1133 F 08 9339 1183
enquiries@electricpictures.com.au
www.electricpictures.com.au

Founded in 1992, producing award-winning documentaries on a range of themes including history, science, human interest, international affairs, arts and adventure. Output includes *Bom Bali*, *The Black Road* and *Tug of Love*.

Welcomes written submission of strong documentary concepts with international market potential. As a minimum, a submission should consist of a one-page synopsis outlining the idea and any supporting documentation such as character or background notes.

Embryo Films (MOVIE) (TV)

PO Box 300, Artarmon, Sydney, NSW 1570
T 02 8011 3533 F 02 8915 1533
prod@embryo-films.com
www.embryo-films.com

Produces drama, comedy, sci-fi and thrillers written or set in Australia or New Zealand, targeting the international TV and film markets.

Welcomes contact with new writers but submissions accepted only via the website

submission form with no exceptions. See website for further details.

Empire Action Pty Ltd (MOVIE) (TV)

Suite 110, 184 Blues Point Road, McMahons Point, Sydney, NSW 2060
T 02 9655 1067
mickhodge@empireaction.com.au
www.empireaction.com.au
Writer/Director *Mick Hodge*

Established in 2000, film and television production company creating shorts, documentaries and feature films.

Welcomes unsolicited manuscripts. Responds to new writing with 'straightforward honesty which can sometimes appear to be frightfully harsh'.

Enchanter Pty Ltd (MOVIE) (TV) (VG)

PO Box 1027, Darlinghurst, NSW 1300
T 02 9699 4341
info@enchanter.com.au
www.enchanter.com.au
Producer *Matt Carter*

Established in 1998, film, television and media production company focusing on science-fiction and fantasy genres. Limited involvement with computer games. Aims to produce films for the worldwide market utilizing local talent. Recent output includes *Perdition* and *Memory of Breathing*.

Welcomes unsolicited manuscripts by email. Especially interested in low- to medium-budget features and TV sci-fi and fantasy genres.

Essential Viewing (MOVIE) (TV)

PO Box 283, Annandale, NSW 2038
T 02 8568 3100 F 02 9519 2326
www.rbfilms.com.au

Established in 2005 by leading independent Australian feature film and television producers including Rosemary Blight and Chris Hilton (joint CEOs), Kylie du Fresne, Sonja Armstrong, Ben Grant and Ian Collie. Works across dramas, documentaries, drama-docs and factual entertainment and has first look development deals with London-based RDF International and Channel Four International. Essential Pictures is the company's feature films production arm.

Featherstone Productions (TV)

68 Denison Street, Bondi Junction, NSW 2022
T 02 9389 1199 F 02 9369 1432
donaldinho.donald@gmail.com
www.featherstoneproductions.com

Founded in 1985. Has produced over forty documentaries across genres including history, music and arts, science, sports and current affairs. Credits include *The Beach*, *An Imaginary Life* and the seminal TV drama, *Babakuieria*. Notable achievements include Best Documentary (Hot Docs), a Banff Rockie, Best Arts Documentary (San Francisco), Special Jury Prize (San Francisco), a UN Media Peace Prize, a BAFTA nomination and an International Emmy nomination.

Frontier Films Pty Ltd   [MOVIE]

PO Box 294, Harbord, NSW 2096
T 02 9938 5762 F 02 9938 5762
frontier@oceanguard.com
Contact *Frank Shields*

Founded in 1970, specializing in drama, animation and documentary. Credits include *Hurrah* and *The Finder*. Has shown extensively at festivals throughout the world, picking up several awards.

Approach with idea or single-page synopsis in first instance, via mail or email. If interested enough, the company will approach the sender for further material (a treatment or screenplay). At this stage, sender will be required to sign a submission release document.

Gecko Films Pty Ltd   [MOVIE]

PO Box 1320, North Fitzroy, VIC 3068
geckofilms@ozemail.com.au
Screenwriter/Director *Sue Brooks*
Screenwriter/Producer *Alison Tilson*

Founded in 1992, film production company specializing in drama. Recent credits include *Road to Nhill* and *Japanese Story*, which was selected for 'Un Certain Regard' category at the Cannes Film Festival.

Not actively seeking manuscripts or new writing; generates scripts within the company.

Generation Films   [VG] [MOVIE] [TV]

367 Beaconsfield Parade, St Kilda, VIC 3182
T 03 9537 1963 F 03 9923 6388
bob@weisfilms.com
www.weisfilms.com
CEO *Bob Weis*

Established in 1981, production company working across a range of media including film, television and computer games. Winner of awards including the UN Media Peace Prize and AFI awards.
Output includes *Women of the Sun: 25 Years Later*.

Wants 'subject driven', not 'market driven', films. Welcomes unsolicited manuscripts but prefers synopsis first.

Granada Productions   [TV]

Fox Studios Australia, Level 1, Building #61, Driver Avenue, Moore Park, NSW 1363

T 02 9383 4360 F 02 8353 3494
www.granadaproductions.com.au

Founded in 1998, producing shows including *Australia's Next Top Model, An Aussie Goes Barmy, Teen Fat Camp* and the Logie award-winning show *Dancing with the Stars*.

Great Western Entertainment Pty Ltd   [TV] [VG]

140 Stirling Highway, North Fremantle, WA 6159
T 08 9433 6899 F 08 9433 6922
annem@gwe.net.au
Producer *Paul Barron*

Established in 1983, producing for television and starting to develop computer games. Particular focus on children's television but currently developing adult drama and documentaries.

Credits include *Wormwood, Parallax* and *Streetsmartz*, which was awarded a Screen Producers' Association of Australia award for children's projects.

Always on the lookout for fresh new talent. Email a synopsis only in first instance.

Grundy   [TV]

FremantleMedia Australia, 110–112 Christie Street, St Leonards, Sydney, NSW 2065
T 02 9434 0666 F 02 9434 0700
www.fremantlemedia.com

Formed out of Grundy Television, the company founded by George Grundy in 1959. Went on to become Australia's most important producer and distributor, with titles including *Sons and Daughters, Prisoner Cell Block H* and *Neighbours*. Was acquired by FreMantle (then called Pearson Television) in 1995.

Hoodlum Active Pty Ltd   [MOVIE] [TV] [VG]

PO Box 38, Paddington, QLD 4064
T 07 3367 2965
info@hoodlum.com.au
www.hoodlumactive.com
CEO *Tracey Robertson*

Founded in 2000, producing multi-platform drama and documentary projects. Output includes *Fat Cow Motel* and *Emmerdale Online*, an online narrative synchronised with the television serial.

Unsolicited manuscripts welcome. 'We prefer people to have an understanding and experience of the industry, not necessarily as a writer.'

Instinct Entertainment   [MOVIE]

Level 1, 111 Nott Street, Port Melbourne, VIC 3207
T 03 9646 0955 F 03 9646 1588

admin@instinctentertainment.com.au
www.instinctentertainment.com.au

Founded in 1999, producing documentaries, shorts and feature films. Produced documentaries such as *To be Frank* and *SALUTE - The Peter Norman Story*. Feature film output includes *Torn*, *Strange Bedfellows* and *Takeaway*.

In order to submit material for consideration, include a completed application form together with a signed submission agreement, a one-page synopsis of the project and a CV.

Jigsaw Entertainment   (TV)

Development Office, 1st Floor, 129 Cathedral Street, Woolloomooloo, NSW 2011
T 02 9326 9922 F 02 9326 9277
thefolks@jigsaw.tv
www.jigsaw.tv

Established in 1999, an indie production company specializing in light entertainment, comedy and drama production. Produced the *BlackJack Trilogy* (telemovie) and the sitcom *Welcher & Welcher*.

Company policy not to accept unsolicited material.

Kapow Pictures   (MOVIE) (TV)

Studio 33, Technopark 6–8 Herbert Street, St Leonards, Sydney, NSW 2065
T 02 9439 0399 F 02 9439 0398
info@kapowpictures.com.au
www.kapowpictures.com.au

Independent animation company founded in 1997. Works on feature films, TV series, long form TV productions and short films. Titles include *CJ the DJ* and *Here Comes Peter Cottontail*.

Kojo Pictures   (MOVIE) (TV)

31 Fullarton Road, Kent Town, SA 5067
T 08 8363 8300 F 08 8363 8329
dean@kojo.com.au
www.kojo.com.au

Formed in 2004 to develop and produce feature length movies, short films and television drama. Full service from script development through to full production. Outputs include *Elise*, *2:37* and *Wolf Creek*.

Lifesize Film and Television Pty Ltd   (TV)

212 Punt Road, Prahran, Melbourne, VIC 3181
T 03 9510 3027 F 03 9524 2777
lipscombejames@hotmail.com
Director *James Lipscombe*
Director *Maxine Lipscombe*

Founded in 1993, an independent production company specializing in documentaries and lifestyle programmes.

Happy to look at ideas via email.

Lonely Planet Television   (TV)

90 Maribyrnong Street, Footscray, Melbourne, VIC 3011
T 03 8379 8000 F 03 8379 8111
info@lonelyplanet.tv
www.lonelyplanet.tv
Executive Producer *Laurence Billiet*
Head of Content Development *David Collins*

The television production section of Lonely Planet Publishing, producing travel documentaries and factual entertainment. Titles include *The Sport Traveller* and *Going Bush*.

Welcomes one-page submissions by email.

J McElroy Holdings Pty Ltd   (MOVIE)

FSA#48, Fox Studios Australia, Driver Avenue, Moore Park, NSW 1363
T 02 9383 4475 F 02 9383 4471
jmcelroy@msn.com.au
Director *Jim McElroy*
Director *Marta McElroy*

Film production company established in 1991, producing feature films in a wide range of genres. Credits include *The Year of Living Dangerously*, *Mr Reliable* and *Picnic at Hanging Rock*.

Unsolicited manuscripts are welcomed by mail but prefers to read synopsis first.

Media World Pictures   (MOVIE) (TV)

PO Box 90, Carlton South, VIC 3053
T 03 9329 3252
frontdoor@mediaworld.com.au
www.mediaworld.com.au

Founded in 1982. Has produced the television programmes *The Circuit* and *Dogstar*. Films include *Zone 39*, *Beware of Greeks Bearing Guns* and the award-winning children's feature, *The Silver Brumby* (starring Russell Crowe).

Has a company policy not to accept unsolicited material.

Melodrama Pictures Pty Ltd   (MOVIE)

179 Johnston Street, Fitzroy, VIC 3065
T 03 9416 3566 F 03 9417 7336
info@melodramapictures.com
www.melodramapictures.com
Producer/Company Director *Melanie Coombs*

Film production company specializing in features and animation. Founded in 1999, output includes *The Glenmoore Job* and *Harvie Krumpet*, winner of Oscar for Best Animated Short.

Does not welcome unsolicited manuscripts, but 'open to genuine collaboration'.

Midas Films (David Douglas Productions Pty Ltd) (MOVIE) (TV)

Farm Dogs Bite, 13B Lytton Road, Moss Vale, NSW 2577
T 02 4869 5150
midaslectures@bigpond.com
Producer/MD *David Douglas*

Founded in 1978, specializing in documentaries, commercials and features for domestic and international markets.
Does not welcome unsolicited approaches.

Moving Targets (MOVIE) (TV)

14/41 Broughton Road, Artarmon, NSW 2064
T 02 9413 3932 F 02 9413 3931
rodhay@bigpond.com.au
Writer/Producer/Director *Rod Hay*

Established in 1986, film and television production company focusing on documentaries and feature films. Output includes TV series *Height of Passion*. Documentaries are often a mixture of drama and sport.
Welcomes unsolicited manuscripts; prefers to be approached first by telephone. 'Must be able to understand the theme and characterisation as soon as possible, but also must push the envelope.'

Mushroom Pictures (MOVIE) (TV)

135 Forbes Street, Woolloomooloo, Sydney, NSW 2011
T 02 9360 6255 F 02 9360 7307
info@mushroompictures.com.au
www.mushroompictures.com.au

Produces television shows such as *Tribal Voice* (based on the Australian band Yothu Yindi) and the series *Greatest Australian Albums*. Co-produced the film *Chopper*, and produced the films *Cut*, *Gettin' Square* and *Wolf Creek*.
Does not accept unsolicited scripts.

Music Arts Dance Films Pty Ltd (MOVIE) (TV)

PO Box 7, Elwood, VIC 3184
T 03 9596 9999
info@musicartsdance.com
www.musicartsdance.com
Director/Executive Producer *Kevin Lucas*
Managing Director/Producer *Aanya Whitehead*

Founded in 1986, independent production company making drama, features and documentaries for TV and film. Selections in Sundance, Berlin, London and many other film festivals. Projects focus particularly on arts in all areas, including music, dance and theatre.

Welcomes unsolicited ideas but prefers to receive a letter and synopsis first.

New Music Theatre (MOVIE) (TV)

PO Box 128, Petersham, NSW 2049
T 02 9558 3645 F 02 9558 3645
dstrahan@revolve.com.au
www.revolve.com.au
Director *Derek Strahan*

Founded in 1972. Has provided scripts for projects produced by the company and third parties. Also involved in production of music. Credits include *Leonora*, *Fantasy* and the *Inspector Shanahan Mysteries*.
Does not welcome unsolicited approaches.

Nomad Films International Pty Ltd (MOVIE) (TV)

PO Box 176, Prahran, Melbourne, VIC 3181
T 03 9819 3350 F 03 9819 3395
nomfil@rabbit.com.au
www.nomadfilmsinternational.com
Executive Producer *Douglas Stanley*
Producer *Kate Faulkner*

Established in 1976, film and TV production company mostly concentrating on feature films and TV drama. Past credits include a drama miniseries, *Shadow of the Osprey*, and *The Promise*, a feature film.
Welcomes unsolicited manuscripts but prefers a treatment of the story first. Encourages new writing but not funding new scripts at this stage.

Perpetual Motion Pictures (TV)

PO Box 1161, Crows Nest, NSW 1585
T 02 8333 4167 F 02 8333 3344
jo@perpetualmotionpictures.tv
www.perpetualmotionpictures.tv
Producer *Jo Cadman*
Producer/Editor *Maurice Todman*

Television production company founded in 1990, specializing in documentary and lifestyle programmes.
Happy to read all new writing, particularly within the company's genre.

Porchlight Films (MOVIE)

Suite 31, 94 Oxford Street, Darlinghurst, NSW 2010
T 02 9326 9916 F 02 9357 1479
mail@porchlightfilms.com.au
www.porchlightfilms.com.au

Founded in 1996. Has produced award-winning short films, documentaries, television programmes and feature films including the independent box office hit *Mullet*, *Walking on Water* (winner of five AFI Awards), the critically-acclaimed *Little Fish* (starring Cate Blanchett, Sam Neill and Hugo Weaving), and *The Home Song Stories* (winner of eight AFI Awards).

Radhart Pictures Pty Ltd MOVIE TV

Level 1, 479 Crown Street, Surry Hills, NSW 2034
T 02 9699 7622
jason@radhart.com
Producer *Jason Harty*

Film and television production company founded in 2001. Recent credits include *Clutch* and *Charmed Robbery*.
Welcomes unsolicited manuscripts by email.

Red Carpet Productions Pty Ltd MOVIE

PO Box 1199, Potts Point, NSW 1335
T 02 9356 8677 F 02 9360 2421
info@redcarpetfilms.com.au
www.redcarpetfilms.com.au

Independent feature film production company, whose debut feature was *Somersault*. Is developing several more features and has a catalogue of award-winning short films. Offers additional services including script editing and script analysis.

Seven Dimensions TV

Unit 19, 156 Beaconsfield Parade, Albert Park, VIC 3206
T 03 9686 9677
info@7dimensions.com.au
www.7dimensions.com.au
CEO *Eve Cash*

Founded in 1979, with a back catalogue of over 500 training, business and educational films and over 140 awards. Produces for television, DVD and electronic delivery. Drama, documentary and interview-based.
Examples of titles include the *Take Away Training Series*, *Feedback Solutions*, *People Skills Series* and *Boomerang*.
No real capacity to accept unsolicited ideas or material. Normally uses internal team of writers.

John Sexton Productions MOVIE

The Pavilion 3, Courallie Road, Northbridge, NSW 2063
T 02 9967 2222 F 02 9967 2234
sextonfilms@spin.net.au
Managing Director *John Sexton*

Founded in 1972. Has produced 8 feature films focusing primarily on drama.
Submissions accepted only via recognized literary agents.

Shooting Star Picture Company Pty Ltd TV

Suite 60, Upper Deck, Jones Bay Wharf, Pirrama Road, Pyrmont, NSW 2009
T 02 9660 6969 F 02 9660 8989
info@shootingstar.com.au
www.shootingstar.com.au
Executive Producer *Janelle Mason*
Executive Producer *Peter Skillman*

Established in 1992, television production company specializing in drama and entertainment shows. Recent output includes police drama *Stingers*.
Welcomes unsolicited ideas in relevant genres.

Southern Star Entertainment TV

Level 9, 8 West Street, North Sydney, NSW 2060
T 02 9202 8555 F 02 9955 8302
general@sstar.com.au
www.sstar.com.au

One of Australia's largest independent creators and producers of television programming, including prime-time drama (*Blue Heelers*, *The Secret Life of Us*, *Water Rats*); children's and family television programming (*The Sleepover Club*, *Tracey McBean*, *The Adventures of Bottle Top Bill*) and factual (*Forensic Investigators*). Endemol Southern Star is a joint venture with Europe's Endemol, working on programmes such as *Big Brother* and *Deal or No Deal*.

Taxi Film Production TV MOVIE VG

332 Montague Road, West End, Brisbane, QLD 4101
T 07 3334 4200 F 07 3334 4250
shoot@taxifilm.tv
www.taxifilm.tv
Executive Producer *Andrew Wareham*

Founded in 2001. A multi-disciplinary pool of directors allows Taxi to take on projects of any form and genre. TVC's and music videos are core business. Also has office in Sydney.

Stacey Testro International Production MOVIE TV

26a Dow Street, South Melbourne, VIC 3205
T 03 9690 0099 F 03 9696 5110
melb@sti.com.au
www.sti.com.au

Production arm of Stacey Testro International, producing film, theatre and television for a global market.

TriMax Films MOVIE

Unit 35, Mockridge Avenue, Newington, NSW 2127

Ⓣ 04 1435 2410 Ⓕ 02 8572 6088
info@trimaxfilms.com
Producer/Director *Lester Crombie*

Established in 1982, film production company focusing on thrillers, action films and drama features.

Does not welcome unsolicited manuscripts.

Village Roadshow Limited (MOVIE)

Level 1, 500 Chapel Street, South Yarra, VIC 3141
Ⓣ 03 9281 1000 Ⓕ 03 9660 1764
www.villageroadshow.com.au

A leading international media and entertainment company. Incorporates a major movie production business operating out of Los Angeles.

Vitascope Filmed Entertainment (MOVIE) (TV)

GPO Box 1193, Sydney, NSW 2001
Ⓣ 02 9590 7666 Ⓕ 02 6337 5239
sydney@vfe.com.au
vfe.com.au

Independent company with a remit to develop, finance, produce and distribute feature films and television for the domestic and international markets. Work spans a range of genres from documentary and adventurous art house releases to international commercial features. Credits include *Hemispheres*, *Luv Sux*, *Romance of the Rose*, *The Grandfather Chest*, *The Song They Sing in Heaven* and *Wild Country*.

Zoot Film Tasmania (TV) (VG)

PO Box 3105, West Hobartt, Tasmania, TAS 7000
Ⓣ 04 1856 8086
andy@zoot.net.au
www.zoot.net.au
Producer *Andrew Wilson*

Founded in 2004, providing casting, crewing and location management services. Currently extending into computer game production. Examples of recent credits include *Race Through Time* for the BBC, the docu-drama *Exile in Hell*, *Kaleidoscope*, *The Barn*, *Dark Decisions* and *Cable*.

Happy to receive unsolicited submissions but reserves the right not to reply or accept any material offered. Approach by email with all relevant contact details and a one-page synopsis. Full manuscripts will not be considered. Accepts scripts of all genres but especially interested in art house, contemporary stories. Particualrly interested in new media, cross media and trans media stories that have life beyond the box office and are suited to online distribution.

New Zealand

AJ Films (MOVIE) (TV)

PO Box 37024, Halswell, Christchurch
Ⓣ 03 322 9279 Ⓕ 03 322 9280
info@ajfilms.co.nz
www.ajfilms.co.nz

Recently established, working particularly on short films, television drama and commercials. Examples of output include the short films *Closer*, *A Quiet Night* and *Kicken*, an action adventure for Dutch television.

George Andrews Productions Ltd (MOVIE) (TV)

5H Westminster Court, 5 Parliament Street, Auckland
Ⓣ 09 307 9196
george@gaproductions.co.nz
www.gaproductions.co.nz

Produces documentaries and feature films. Examples of output includes *The Game of Our Lives*, *Out of the Dark*, *Nuclear Reaction* and *Nga Tohu Signatures*.

Bitesize Productions Ltd (TV)

62 Rame Road, Greenhithe, Auckland
Ⓣ 09 413 9791
Laurence@bitesize.co.nz
www.bitesize.co.nz
Contact *Laurence Belcher*

Producers specializing in documentaries and series for television. Past titles include *Taste NZ*, *Garden Show*, *5.30 With Jude* and *Need for Speed*.

Blueskin Films Ltd (MOVIE)

PO Box 27261, Marion Square, Wellington
Ⓣ 04 8017176
catherinef@blueskinfilms.co.nz
www.blueskinfilms.co.nz
Contact *Catherine Fitzgerald*

Film producer established in 2002, with particular focus on short drama. Also offers script consultancy services for emerging writers for stage and screen. Examples of output include *Kerosene Creek*, *The Little Things*, *Two Cars, One Night* and *Turangawaewae: a Place to Stand*.

Borderless Productions Limited (TV)

Studio 4, 2A New North Road, Eden Terrace, Auckland
Ⓣ 09 302 3103
qiujing@borderlessproductions.com
www.borderlessproductions.com
Principal *Qiujing Wong*

Produces programmes for a range of clients including not-for-profits, television and internet broadcasters, corporations and individuals. Specialists in documentaries and factual programming. Output includes *Postcards from China, Nepal: Caught up in the people's war* and *A World of Kindness.*

Conbrio Media Ltd [MOVIE] [TV]

PO Box 34367, Birkenhead, Auckland 0746
☎ 027 470 6895
christina@conbrio.co.nz
www.conbrio.co.nz

Company run by producer/writer Christina Milligan and post-producer/editor Roger Grant. Currently developing several projects for cinema, television and mobile phones. Also working with actor/writer Rawiri Paratene on short films under the banner Shorts Conbrio. Production credits include *Pet Detectives* (a children's TV drama series) and the documentary *Te Whanau a Putiputi.*

Daybreak Pacific [MOVIE]

PO Box 91642, Auckland 1030
☎ 09 580 0167 ℻ 09 580 0402
dale@daybreakpacific.com
www.daybreakpacific.com

Feature film production company whose credits include *Treasure Island Kids 1, 2* and *3, Terror Peak, Cupid's Prey* and *Ozzie.* Also has offices in London and Los Angeles.

Diva Productions [MOVIE] [TV]

PO Box 5986, Wellesley Street, Auckland 1141
☎ 09 360 9852
arani@divaproductions.co.nz
www.divaproductions.co.nz

Production company with focus on film, documentaries and television comedy. Examples of previous productions include three series of *Topp Twins* and *Mr and Mrs*, co-produced with Ninox Films.

Drum Productions [MOVIE] [TV]

PO Box 91782, AMC, Auckland
☎ 09 376 2919
stan@drumproductions.co.nz

Works across range of genres include films (features and shorts) and dramas, comedy and documentaries for television.

Emmanuel Productions [MOVIE]

PO Box 25736, St Heliers, Auckland
☎ 09 575 3030
emmanuelproductions@xtra.co.nz

Produces short films as well as corporate videos. Its short film, *Rice*, won several awards at the 2003 Wanganui Film Festival.

Evoke Pictures [MOVIE]

info@evokepictures.co.nz
www.evokepictures.co.nz

Independent film company with strong focus on the art of story-telling. Credits include *Death to the Premonition* and *Make Evelyn Smile.* Currently 'in hibernation, as we look to develop our stories and establish international contacts in the film industry'.

Execam Television and Video Production [TV]

Level 2, Katipo House, 195 Victoria Street,
PO Box 6577, Wellington
☎ 04 801 5600 ℻ 04 801 5558
info@execam.co.nz
www.execam.co.nz

Works across a range of genres from television programmes to corporate video to streaming video for the internet. Involved in every stage of television production from concept through to finished show. Has worked with all three major New Zealand networks. Credits include the documentary series, *Extraordinary Kiwis.*

Exposure [MOVIE] [TV]

PO Box 99350, Newmarket, Auckland,
☎ 09 302 4031 ℻ 09 302 4037
kevin@exposure.org
www.exposure.org

A communications company making human interest programmes for commercial, documentary and humanitarian purposes. Titles include *Journey of Hope* and *Tragedy of Child Labour.*

Eyeworks Touchdown [MOVIE] [TV]

PO Box 90018, Auckland Mail Centre,
Auckland 1142
☎ 09 379 7867 ℻ 09 379 7868
nz@eyeworks.tv
www.touchdowntv.com

A leading entertainment television production company founded in 1991. The country's market leader in broad appeal entertainment programmes, specializing in entertainment, lifestyle, sport and factual series. Examples of output include *Mountain Dew On The Edge, Police Stop!, In The Face of Fear, Profilers, Going Straight* and *Game of Two Halves.* Is also expanding into drama and film.

To submit an idea, see the online ideas submission form on the website. Automatically respects individual's intellectual property rights but has policy not to sign non-disclosure or confidentiality agreements.

Flux Animation Studio Ltd [MOVIE] [TV]

Private Bag MBE P280, Auckland
T 09 360 6003 F 09 360 6004
flux@fluxmedia.co.nz
www.fluxmedia.co.nz

Award-winning character animation specialist able to take a project from script to completion. Founded in 1997.

Front Page Limited [TV]

PO Box 90 361, Auckland Mail Centre, Auckland
T 09 377 4433 F 09 377 4434
harman@frontpage.co.nz
www.frontpage.co.nz
Executive Producer *Richard Harman*

A leading independent producer of television current affairs and documentaries, working in both broadcast and non broadcast TV. Titles include *Agenda* and the *TV One Insight* debates.

Rachel Gardner [MOVIE] [TV]

21 Picton Street, Ponsonby
T 021 765 405
Rachel@rachelgardner.co.nz

Pursues production interests across feature films, short films, television dramas, fact-based series and documentaries. Credits include *The Pretender, Fog, Russia's Forgotten Children, Truant* and *The Lion Man*.

The Gibson Group Limited [MOVIE] [TV]

PO Box 6185, Te Aro, Wellington
T 04 384 7789 F 04 384 4727
info@gibson.co.nz
www.gibson.co.nz

Founded by Dave Gibson in 1977 and now among New Zealand's most significant independent film and television production companies. Specializes in high-end television drama for both primetime and children's audiences, along with documentary, comedy, arts magazine and factual programming. Programmes are shown in 80 countries around the world. Also works in new media. Recent credits include *The Insiders Guide to Love, The Insiders Guide to Happiness, The Strip* and *Holly's Heroes*.

Godzone Pictures Ltd [MOVIE] [TV]

PO Box 78–186, Grey Lynn, Auckland
T 025 974 025 F 09 378 7333

liz@godzone.co.nz
www.godzone.co.nz

Produces films, TV series and documentaries. Output includes *The Creakers, Playing Possum* and *Letters about the Weather*.

Green Stone Pictures [MOVIE] [TV]

Private Bag 56 909, Dominion Road, Auckland
T 09 630 7333 F 09 623 7764
www.greenstonepictures.com

Award-winning film and television production company, established in 1994 and specializing in factual series and documentaries. Also produces some entertainment and studio shows and children's drama. Credits include the children's drama series *Secret Agent Men Series II*, the feature film *Bella* and documentaries, *Cave Creek - The full story of a national tragedy, Back from the Dead - the Saga of the Rose Noelle* and *Whanau*.

Hawke Films [TV]

PO Box 505, Nelson
T 03 547 8262
khawke@paradise.net.nz
www.valiantpoint.com

Produces television series, documentaries and programmes for children.

Huntaway Films [MOVIE] [TV]

Level 1, Steamer Wharf, Lower Beach Street, Queenstown
T 03 441 1441 F 03 441 1451
cassells@huntawayfilms.co.nz
www.huntawayfilms.co.nz
Contact *Jay Cassells*

Operates in New Zealand and Australia, and has an office in Melbourne. Develops and produces content for film, television, DVD and new media.

Works with New Zealand and Australian writers and filmmakers in a bid 'to record and tell good New Zealand and Australian yarns'.

Isola Productions Ltd [MOVIE] [TV]

PO Box 90–781, AMC, Auckland
T 09 815 1225 F 09 815 1224
info@isola-productions.co.nz
www.isola-productions.co.nz

Established by Rachel Jean in 2001, specializing in drama and documentary. Credits include the documentaries *Bird Flu, High Times* and *Dawn Raids*, the films *Karma* and *Minefield*, and the television series, *The Market*.

Joyride Films Ltd (MOVIE) (TV)

120 Williamson Avenue, Grey Lynn,
Auckland 1021
☏ 09 360 5380 🖷 09 360 5386
info@joyridefilms.com
www.joyridefilms.com
Executive Producer *Anzak Tindall*
Production Manager *Kerry Prendeville*

Founded in 2003 & with office in central
Auckland. Offers production support services in
New Zealand to clients from all over the world.
Have completed short film called 'Aphrodite's
Farm' and has a feature film and documentaries in
development.
  Happy to explore new opportunities. Email in
first instance.

Kiwa Media (MOVIE) (TV)

PO Box 41136, St Lukes, Auckland
☏ 09 377 7674 🖷 09 375 2868
chelsea@kiwafilm.com
www.kiwafilm.com
Producer/Director *Chelsea Winstanley*

Produces feature films, shorts and documentaries
for domestic and international television and film
markets. Specialists in dubbing programmes into
indigenous languages. Production credits include
Buzz and Poppy, *Stickmen*, *Pakeha Maori*, *The
Natives of New Zealand* and *Gang Kids*.
  Welcomes ideas from novice, intermediate
and experienced screenwriters. Assesses
all unsolicited scripts, novels, treatments,
screenplays and pitches sent in. See website for a
submissions guide.

La Luna Studios Ltd (MOVIE) (TV) (VG)

PO Box 258, Silverdale 0944, North Shore,
Auckland
☏ 09 974 2283
cris@lalunastudios.com
www.lalunastudios.co.nz
Contact *Christina Casares*

Film and TV production company founded
in 2005. Extends to computer games. Focuses
on family and children's entertainment. Work
includes the 3D animated short film *Kea* and
upcoming 3D animated feature *The Magic Shoes*.
  Welcomes unsolicited material by mail. Keen
to see entertaining writing with a good moral
message.

Legge Work (TV)

PO Box 32 576, Devonport
☏ 09 445 0110
gordon@leggework.co.nz
www.leggework.co.nz

Over ten years' experience of producing
sports-based programmes. Titles include
Speedweek, *World of Motorsport* and *Coast to
Coast*.

Little Ed Films Ltd (MOVIE) (TV)

PO Box 41152, Eastbourne, Wellington
☏ 04 384 6385
angelalittlejohn@xtra.co.nz

Produces films as well as dramas, children's
programmes and documentaries for television.
Production credits include *Redial*, *They're
Gone*, *They Ain't No Moa* and *Not Only But
Always*.

Livingstone Productions Ltd (TV)

PO Box 7896, Symonds Street, Auckland 1150
☏ 09 636 5826 🖷 09 636 5825
info@livingstoneprods.co.nz
www.livingstoneprods.co.nz

Television production company founded in
1989 by John A Givins. Develops programmes
from ideas generated internally and also for
network operators and funded scriptwriters.
Works across genres including documentaries,
situation-comedy series, entertainment series,
drama and television specials. Has worked with
the Maori Television Service on projects in
both Maori and English. Output includes *B&B*,
Queer Nation, *Captain's Log* and *Nga Wahine
Mauri Ora*.

M F Films Ltd (MOVIE) (TV)

PO Box 91695, AMSC, Auckland
☏ 09 376 0876 🖷 09 376 9675
micheledriscoll@mffilms.co.nz
www.mffilms.co.nz
Assistant Producer *Michele Driscoll*

Produces features, TV dramas, documentaries
and commercials. Past credits include *One
of Them*, *When Love Comes*, *50 Ways of
Saying Fabulous*, *The Big OE* and *God, Sreenu
and Me*.

Natural History New Zealand (TV)

8 Dowling Street, PO Box 474, Dunedin 9016
☏ 03 479 9799 🖷 03 479 9917
ideas@nhnz.tv
www.nhnz.tv

A leading producer of factual television, creating
over 60 hours of programmes each year on the
subjects of the natural world, health, science,
adventure and people. Also has offices in Beijing
and Washington, D.C. Examples of recent output
include *Spider Power*, *World's Biggest Baddest
Bugs*, *Nature's Warzone*, *Equator 2: Rivers of the*

Sun, Wild Horses - Return to China and *The Curse of the Elephant Man.*

New Zealand Greenroom Productions Limited (TV)

68 Harbour Village Drive, Gulf Harbour, Whangaparaoa, Auckland
T 021 888 665
info@nzgreen.tv
www.nzgreen.tv

Television company founded in 1997. Leading player in surf/snow/alpine extreme sports and adventure programming. Output includes *NZ National Surf Champs* and *Arctic Challenge.*

Ninox Television Ltd (TV)

PO Box 9839, Marion Square, Wellington 6141
T 04 801 6546 F 04 801 6573
ninox@ninox.co.nz
www.ninoxtv.com
Managing Director *David Harry Baldock*
Finance Director *Pauline Downie*
Production Manager *Peter Baldock*

Television production company formed in 1988. Over 300 hours of broadcast programming to date. Focuses on documentary, reality and dramatized documentary formats. Past credits include *Sensing Murder*, *Dream Home* and *Location, Location, Location.*

Open to unsolicited material via email with letter of introduction.

Origin One Films (MOVIE)

PO Box 33–223, Petone, Lower Hutt, Wellington 5046
T 04 970 9039 F 04 970 9039
originone@paradise.net.nz
Producer *Brendon Hornell*

Wellington-based production company focused on drama, bio-pic and comedy film. Founded in 1999. has produced four short films to date and one digital feature film (2001).

Encourages unsolicited submissions by email. Keen to find an inspiring script from any writer, unknown or not. Especially potential UK/NZ co-prods.

Point of View Productions (MOVIE) (TV)

PO Box 78084, Grey Lynn, Auckland
T 09 360 1044
info@pointofview.co.nz
www.pointofview.co.nz
Contact *Shirley Horrocks*

Producing short films and documentaries. Examples of output include *Flip and Two Twisters* (a documentary about Len Lye), the short film, *Stay in Touch* and the drama, *Managing Diversity.*

Gaylene Preston Productions (MOVIE) (TV)

59 Austin Street, Mount Victoria, Wellington
T 04 384 4242
the1@gaylenepreston.com
www.gaylenepreston.com

Gaylene Preston is a leading New Zealand filmmaker and in 2001 was made New Zealand's first Filmmaker Laureate by the New Zealand Arts Foundation. Output includes the films *Perfect Strangers* and *Bread and Roses* and the mini-series, *Ruby and Rata.* Documentaries include *Earthquake!* , *Coffee Tea or Me?*, *Titless Wonders* and *Lands of our Fathers.*

Raconteur (MOVIE) (TV)

PO Box 1051, Christchurch
T 03 3777 266 F 03 3777 268
info@raconteur.co.nz
www.raconteur.co.nz

Founded in 1996 by Veronica McCarthy and Bill de Friez. Produces high-quality film and television productions across genres including documentaries, children's series and drama. Output includes *The Crossing* (drama), *Mum Can I Drive?* and *Animation Station* (children's series), and *Between the Lines - Dennis Glover* and *Madame Morison* (documentaries).

Has an active development slate and always willing to consider new ideas. Send an email to development@raconteur.co.nz or contact Kim on 03 3777 268.

Raynbird Productions (MOVIE) (TV)

PO Box 19–817, Woolston, Christchurch
T 03 384 3500
gaylene@raynbird.com
www.raynbird.com

Works across media including films and documentaries, as well as educational and corporate projects. Involved at all stages from inception to completion.

Severe Features (MOVIE) (TV)

44 Francis Street, Grey Lynn, Auckland 1245
T 021 764198
severe@paradise.net.nz
Company Director *Leanne Saunders*

Founded in 2000, a television and film production company, specializing in drama.

Examples of output include *Christmas* (2003) and *Nature's Way* (2006).

Welcomes unsolicited manuscripts and ideas. Approach by email with a brief synopsis only.

Shooting Stars Productions [MOVIE]

Apartment 8, 20 Central Road, Kingsland, Auckland 1021
T 09 815 8277 F 09 815 8377
gavin.butler@xtra.co.nz
Director *Gavin Butler*

Film production company founded in 2001. Develops screenplays and directs; specialized interest in drama.

Always keen to look over new material. Unsolicited submissions via email.

SHOTZ Film and Video Production Ltd [TV]

49 Archmillen Avenue, Pakuranga, Auckland 2010
T 09 576 8890 F 09 577 5343
ronel@shotzproductions.co.uk
www.shotzproductions.co.nz
Producer *Ronel Schodt*
Director of Production *Karl Schodt*

Television production company founded in October 2002. Specializes in TVC and documentary.

Welcomes unsolicited manuscripts and ideas via email.

South Pacific Pictures [MOVIE] [TV]

PO Box 104–124, Lincoln North, Henderson, Auckland
T 09 839 0999 F 09 839 0990
jjohnson@spp.co.nz
www.spp.co.nz

Over the last 20 years South Pacific Pictures has produced over 3,000 hours of programming, both for television and cinema. Encompasses drama, entertainment, reality programming and documentaries. Has undertaken work for broadcasters in New Zealand, Australia, Canada and the UK. Believes that 'great writing is the foundation of good television [and] great execution is the realisation of that writing.' Recent production credits include the films *We're Here to Help* and *Sione's Wedding*, the documentary series *Cook - Obsession and Betrayal in the New World*, a telefeature *The Man Who Lost His Head* and the drama series *Shortland Street*, *Outrageous Fortune* and *Interrogation*.

Development department reviews all scripts, novels, screenplays and treatments submitted. Commissions writers to work on original film and television projects. See the website for the submissions guide and to download a submission release form.

Sprout Productions [TV]

62 Church Street, Onehunga, Auckland
T 021 775 224
pclews@hotmail.com

Specializing in TV comedy, drama, documentary and corporate productions. Credits include *House Call 2*, *Last Laugh* and *Going Straight*.

Sticky Pictures [MOVIE] [TV]

PO Box 27440, Wellington
T 04 8025511 F 04 8025522
mhairead@stickypictures.co.nz
www.stickypictures.co.nz
Development Executive

Established in 2000, an independent producer of film and TV programmes. Produces across a range of genres including animation, documentary, drama and sports. Has a second office in Auckland. Credits include the arts show *The Living Room*, the documentaries *War of the Words* (about spelling bees), *The Magical World of Misery* (about a graffiti artist) and the short film *Dead End*.

Te Aratai Film & TV Production [TV]

PO Box 3717, Shortland Street, Auckland
T 09 378 7833
paora.tahi@xtra.co.nz

Produces programmes for television, specializing in documentaries and children's. Credits include *Tikitiki*, *Tu Te Puehu*, *Ka Hao te Rangatahi* and *Mika Live*.

Timeline Productions [MOVIE] [TV]

PO Box 90943, AMSC, Auckland 1142
T 09 309 2613 F 09 309 4048
info@timelineprods.com
www.timelineprods.com

Specializes in short and feature documentaries for broadcasters and film festivals around the world.

Top Shelf Productions Ltd [MOVIE] [TV]

PO Box 9101, Wellington
T 04 382 8364 F 04 801 6920
topshelf@top-shelf.co.nz
www.topshelfproductions.co.nz
Contact *Trevor Haysom*

Founded in 1988 by Vincent Burke, producing for film and television. Specializes in documentaries, drama-documentaries, factual, information and education, arts programmes, television drama, features and short films. Has co-produced and co-financed ventures with companies from Australia, North America and Europe.

The TV Set Ltd   MOVIE TV

PO Box 91189, AMC, Auckland
T 09 360 3214 F 09 360 3218
megan@thetvset.co.nz
www.thetvset.co.nz

An independent documentary production company established in 2000. Company aims to tell strong personal stories that explore the human condition. Titles include *3 Chord's & The Truth: The Anika Moa Story*, *Back for Good*, *Big Tahuna*, *Death on the Beach* and *One for the Road: The Michael Utting Story*.

Representation

Just because someone writes fabulous TV drama does not mean that they can handle movies – or vice versa. There are great TV writers out there, and there are great movie writers out there. But for every Allan Ball, who can write a movie so filmic and ambitious as American Beauty and then can take that skill and adapt it to an amazing series like Six Feet Under, there is a Dennis Potter whose work simply came alive on the small screen but refused to work in the cinema.

Nick Moorcroft in 'Film, television... it's all the same, isn't it?'

Representation

Agents sometimes fail to acknowledge, respond to or even read unsolicited submissions. It is always worth contacting a company before sending in any material to find out their current policy on new writing.

USA

Above the Line Agency

468 N. Camden Drive, Suite 200, Beverly Hills, CA 90210
Ⓣ 310 859 6115 Ⓕ 310 859 6119
www.abovethelineagency.com

Represents writers and directors, working primarily in motion pictures but also providing services in television, books and other media. Writers must complete an online questionnaire in first instance.

Abrams Artists

275 Seventh Avenue, 26th Floor, New York, NY 10001
Ⓣ 646 486 4600 Ⓕ 646 486 2358
www.abramsartists.com

Operating since 1977, with offices in New York and Los Angeles. The literary division represents writers and directors for film, TV and theatre.

Acme Talent and Literary Agency

4727 Wilshire Boulevard, Suite 333, Los Angeles, CA 90010
Ⓣ 323 954 2263
www.acmetalentagents.com

A full service talent and literary agency. Includes a literary division representing writers of feature films, television series, TV movies and novels.
  Submission accepted by mail only. Recommendations highly advised. Do not send original material and include sae for a reply and a submission release form (if necessary).

Agency for the Performing Arts (APA)

405 S. Beverly Drive, Beverly Hills, CA 90212
Ⓣ 310 888 4200 Ⓕ 310 888 4242
www.apa-agency.com

The agency's literary department has an expansive roster of established screenwriters, directors, show creators and novelists. Clients work across both cinema and television. Also offices in Nashville and New York.

The Alpern Group

15645 Royal Oak Road, Encino, CA 91436
Ⓣ 818 528 1111 Ⓕ 818 528 1110

Represents writers for both film and television.

Anonymous Content Talent and Literary Management

8522 National Boulevard, Suite 101, Culver City, CA 90232
Ⓣ 310 558 6031
info@anonymouscontent.com
www.anonymouscontent.com

Representing writers, actors and directors, with 13 managers working to develop film and television projects.

Artist International

9595 Wilshire Boulevard, 9th Floor, Beverly Hills, CA 90212
Ⓣ 310 358 9239
www.artistint.com

Provides professional representation to actors, models, writers, directors, athletes, music performers, artists, and other entertainment properties. Divided into four divisions, of which one is literary. Also offices in New York, Miami and Dubai.

The Artists Agency

1180 S Beverly Drive, Suite 400, Los Angeles, CA 90035
Ⓣ 310 277 7779 Ⓕ 310 785 9338

Specializes in representing screenwriters and directors for film and television. Has affiliated contacts in New York and the UK.

Ballistic Talent Management

talent@ballistic-media.com
www.ballistic-media.com

Represents writers, directors, performers and producers.
Do not call or send unsolicited material.

Bicoastal Talent

Literary, 210 North Pass Avenue, Suite 204, Burbank, CA 91505
T 818 845 0150 F 818 845 0152
literary@bicoastaltalent.com
www.bicoastaltalent.com

Formed in 2001, with a second office in Orlando, Florida. Represents actors and writers for film, television and theatre. Preference is given writers in the Los Angeles area. Will look at all queries but will only respond if interested.
Send query letter in first instance. Unsolicited materials will not be reviewed. Does not represent authors of self- or unpublished manuscripts, treatments, concepts or TV pilots. A minimum of three completed screenplays must be available for review upon request. Will also consider new writers with strong placements in industry-recognized competitions or by industry referral. Queries should include a list of completed screenplays with title, genre and a one-paragraph synopsis.

The Bohrman Agency

8899 Beverly Boulevard, West Hollywood, CA 90048
T 310 550 5444

Offers representation to screenwriters.

Brant Rose Agency

6671 Sunset Boulevard, Suite 1584 B, Los Angeles, CA 90028
T 323 460 6464

Provides representation for screenwriters.

Brillstein-Grey Entertainment

9150 Wilshire Boulevard, Ste 350, Beverly Hills, CA 90212
T 310 275 6135 F 310 275 6180

Established by Bernie Brillstein in 1969. Manages a number of major stars along with a roster of high profile writers and producers.

Don Buchwald and Associates, Inc.

10 East 44th Street, New York, NY 10017
info@buchwald.com
www.buchwald.com

A full service talent agency, founded in 1977. Offers representation across all sectors of the entertainment industry, including a literary department and television and film packaging departments.

Caliber Talent Group

244 Fifth Avenue, Suite 254F, New York, NY 10001
T 212 252 4600 F 212 591 6082
info@calibertalent.com
www.calibertalent.com
Manager of Motion Pictures *Erik D. Parks*
Manager of Theatre *Rashad V. Chambers, Esq.*

Currently has 44 clients active in TV, films, and on- and off-Broadway. Particularly interested in documentary, drama and thrillers.

Cambridge Literary Associates

135 Beach Road, Unit C-3, Salisbury, MA 01952
T 978 499 0374
www.cambridgeliterary.com

Represents books and screenplays.
Approach by mail (including a synopsis and an sae) before sending in a manuscript. Only interested in 'well-published authors'. No reading fees.

Maria Carvainis Agency, Inc.

1270 Avenue of the Americas, Suite 2320, New York, NY 10020
T 212 245 6365 F 212 245 7196
mca@mariacarvainisagency.com
President *Maria Carvainis*

Founded 1977. Handles fiction: literary and mainstream, contemporary women's, mystery, suspense, historical, young adult novels; non-fiction: business, women's issues, memoirs, health, biography, medicine. No film scripts unless from writers with established credits. No science fiction. Commission: Domestic & Dramatic 15%
No faxed or emailed queries. No unsolicited manuscripts; they will be returned unread. Queries only, with IRCs for response. No reading fee.

The Chasin Agency, Inc.

8899 Beverly Boulevard, Suite 716, Los Angeles, CA 90048
T 323 278 7505

Offers representation to screenwriters.

Concept Entertainment

334 1/2 North Sierra Bonita Avenue, Los Angeles, CA 90036
enquiries@conceptentertainment.biz
www.conceptentertainment.biz

A production and management company, representing screenwriters.

Only able to respond to ideas of particular interest. Company will make contact with a release form. Will not accept unsolicited mail.

Creative Artists Agency

2000 Avenue of the Stars, Los Angeles, CA 90067
⊤ 424 288 2000 ⒡ 424 288 2900
www.caa.com

Founded in 1975, one of Hollywood's most prestigious talent and literary agencies. Also offices in New York, Beijing, Calgary, Kansas City, London, Nashville, St. Louis and Stockholm.

Creative Convergence

4055 Tujunga Avenue, Suite 200, Studio City, CA 91604
⊤ 818 508 8052 ⒡ 818 508 8052
infor@creative-convergence.com
www.creative-convergence.com
Partner *Philippa Burgess*
Literary Manager/Partner *Bradley Kushner*

Established in 2004. Currently has a list of 25 writers for film and TV. Particualr emphasis on action, adventure, family, comedy and science fiction. Not interested in drama or romance but always seeking published books for film and television. Clients include Gary Sinyor (*The Bachelor*), Rod Johnson (*Queen Size*), Gary Boulton Brown (*Trinity*), Chuck Austen (*Tripping the Rift*), Ann Austen (*Power Rangers*) and Peter Manus (*The Hive*).

Happy to recieve unsolicited manuscripts and ideas. Contact via query@creative-convergence.com. Commission charged at 10%. No reading fee.

Curtis Brown Ltd

10 Astor Place, New York, NY 10003
⊤ 212 473 5400 ⒡ 212 598 0917
www.curtisbrown.com
Film & TV Rights *Edwin Wintle*
CEO *Timothy Knowlton*

Founded 1914. Handles general fiction and non-fiction. Also some scripts for film, TV and theatre. Representatives in all major foreign countries.

No unsolicited manuscripts; queries only, with IRCs for reply. No reading fee.

Diverse Talent Group

1875 Century Park East, Suite 2250, Los Angeles, CA 90067
⊤ 310 201 6565 ⒡ 310 201 6572
www.diversetalentgroup.com

Talent and literary agency, founded and run by Christopher Nassif.

The Endeavor Talent Agency

9601 Wilshire Boulevard, 10th Floor, Beverly Hills, CA 90210
⊤ 310 248 2000 ⒡ 310 248 2020

Talent and literary agency, formed in 1995. Provides representation for writers, directors and actors, working across TV, film and video games. Also has an office in New York.

Energy Entertainment

999 N. Doheny Drive, Suite 711, Los Angeles, CA 90069
⊤ 310 274 3440
info@energyentertainment.net
www.energyentertainment.net

Launched in 2001, a boutique agency representing writers actors and directors.
Unsolicited calls or mail will not be returned.

Evatopia

400 S. Beverly Drive, Suite 214, Beverly Hills, CA 90212
⊤ 310 270 3868 ⒡ 310 601 7702
submissions@evatopia.com
www.evatopia.com
Principal *Margery Walshaw*

Founded in 2004, with a list of 15 clients. Interested in film scripts, especially drama, comedy, sci-fi, children's and horror. No erotica. A well-written story is the most important criteria.

Will review unsolicited queries but not manuscripts. If interested, the writer will then be approached for a manuscript. Use online submission form on website (under the 'For our consideration' link). Commission is 15% in the USA and 20% abroad. No reading fee. Although based in Beverly Hills, also has a presence in the UK and welcomes submissions from overseas.

Robert A. Freedman Dramatic Agency, Inc.

Suite 2310, 1501 Broadway, New York, NY 10036
⊤ 212 840 5760
rfreedmanagent@aol.com

Founded 1928 as Brandt & Brandt Dramatic Department, Inc. Took its present name in 1984. Works mostly with established authors. Send letter of enquiry first with sae. Specializes in plays, film and TV scripts. Commission: Dramatic 10%.

Unsolicited mss not read.

Gallagher Literary

26500 W. Agoura Road, Suite 102-221,
Los Angeles, CA 91302
℡ 818 880 0490
dealmakerx@aol.com
www.robgallagher.freeservers.com

A boutique management company representing
writers and directors. Clients include Dan
O'Bannon. Only completed specs may be pitched,
via online web form.

The Geddes Agency

8430 Santa Monica Boulevard, Ste 200,
West Hollywood, CA 90069
℡ 323 848 2700
www.geddes.net

Represents writers, actors and voice-over artists.
(Also has an office at 1633 North Halsted Street,
Ste 300, Chicago, Illinois 60614. T 312 787 8333)

Submissions accepted by mail only. In first
instance, send letter of inquiry with a brief
description of project (either to address above or
to lit@geddes.net). Will only respond if there is
an interest in further investigating the project for
representation.

The Gersh Agency (TGA)

41 Madison Avenue, Floor 33, New York,
NY 10010
℡ 212 997 1818 ℱ 212 997 1978
info@gershla.com
www.gershagency.com

Established as the Phil Gersh Agency in the 1950s
and now represents many of the world's leading
writing, directing and acting talents. Includes a
dedicated TV literary and packaging division and
a feature literary division (led by David Gersh,
Richard Arlook and Abram Nalibotsky). Also has
an LA office (232 North Canon Drive, Suite 201,
Beverly Hills, CA 90210; T 310 274 6611).

Does not accept or consider unsolicited
material, ideas or suggestions of any nature
whatsoever.

The Charlotte Gusay Literary Agency

10532 Blythe Avenue, Los Angeles,
CA 90064
℡ 310 559 0831 ℱ 310 559 2639
gusay1@ca.rr.com (queries only)
www.gusay.com
Owner/President *Charlotte Gusay*

Founded 1989. Handles fiction, non-fiction, young
adult/teen and screenplays (feature scripts/plays
to film). Commission Home 15%; Foreign &
Translation 25%.

Does not accept unsolicited manuscripts; send
one-page query letter and sae in the first instance.
No reading fee.

Guy Walks into a Bar Management

7421 Beverly Boulevard, No. 4, Los Angeles,
CA 90036
℡ 323 930 9935
info@guywalks.com
www.guywalks.com

Associated with a television production company,
the management division represents motion
picture writers and directors.

John Hawkins & Associates, Inc.

71 West 23rd Street, Suite 1600, New York,
NY 10010
℡ 212 807 7040 ℱ 212 807 9555
www.jhalit.com

Founded 1893. Handles film and TV rights.
Commission Apply for rates.

No unsolicited manuscripts. Send queries
with 1–3-page outline and one-page CV. IRCs
necessary for response. No reading fee.

ICM (International Creative Management)

10250 Constellation Boulevard, Los Angeles,
CA 90067
℡ 310 550 4000
www.icmtalent.com

A wide-ranging talent and literary agency.
Represents directors and writers responsible for
many major motion pictures and televison series.
Recently acquired boutique agency Broder Webb
Chervin Silbermann. Also has offices in New York
and London.

Will not accept or consider any unsolicited
material, ideas or suggestions of any nature
whatsoever. Unsolicited materials will not be
forwarded to or discussed with any third parties.

Carolyn Jenks Agency

24 Concord Avenue, Suite 412, Cambridge,
MA 02138
℡ 617 354 5099
cbjenks@att.net

Represents screenwriters but do not send
unsolicited manuscripts. Approach first
by email or letter with outline of project, a short
author biography and an sac. Does not charge
reading fee.

Kaplan Stahler Gumer Braun

8383 Wilshire Boulevard, Suite 923, Beverly Hills,
CA 90211

Ⓣ 323 653 4483 Ⓕ 323 653 4506
info@ksgbagency.com
www.ksgbagency.com

Established in 1981 by Elliot Stahler and Mitch Kaplan. Now one of Hollywood's most prestigious agencies, representing writers, directors and producers for cinema and television.

The Jon Klane Agency

120 El Camino Drive, Suite 112, Beverly Hills, CA 90212
Ⓣ 310 278 0178
query@klaneagency.com
www.klaneagency.com

Represents writers, directors and producers to the film and television industries.

Before submitting material, send an email with the title of your work in the subject portion of the email form and your letter and/or synopsis pasted in the body of the email form. Do not send attachments. The agency will make contact via phone or email if interested in reviewing material.

Kneerim & Williams

c/o Fish & Richardson PC, 225 Franklin Street, Boston, MA 02110-2804
info@fr.com
www.fr.com

Founded in 1990 to deal with books, screenplays, and film and TV rights.

Before sending in manuscript, contact in writing with cover letter, outline of project, a two-page synopsis, a CV and an sae. No fees charged. Not currently seeking new original screenplays.

Lenhoff & Lenhoff

830 Palm Avenue, West Hollywood, CA 90069
Ⓣ 310 855 2411 Ⓕ 310 855 2412
charles@lenhoff.com
www.lenhoff.com

Management company founded in 1991. Represents writers, directors, producers and cinematographers.

Unable to accept unsolicited materials.

Management 819

39 East 12th Street, Suite 709, New York, NY 10003
gillian@management819.com
www.newrootfilms.com/management819.html

A full service talent management company and sister company to New Root Films. Represents writers, directors, producers and actors.

Metropolitan Talent Agency

4500 Wilshire Boulevard, Second Floor, Los Angeles, CA 90010
Ⓣ 323 857 4500
www.mta.com

Well-established talent agency, representing numerous screenwriters.

Monteiro Rose Agency

17514 Ventura Boulevard, Suite 205, Encino, CA 91316
Ⓣ 818 501 1177 Ⓕ 818 501 1194
monrose@monteiro-rose.com
www.monteiro-rose.com

Over fifteen-years experience representing live-action and animation writers for children's television, features, home video and interactive markets.

William Morris Agency (WMA)

1325 Avenue of the Americas, New York, NY 10019
Ⓣ 212 586 5100 Ⓕ 212 246 3583
www.wma.com

One of the world's most famous talent and literary agencies, founded in 1898. Works across all areas of the entertainment industry, including movies and TV. Principal offices in New York, Beverly Hills, Nashville, London, Miami Beach and Shanghai.

Henry Morrison Inc.

105 S. Bedford Road, Mt. Kisco, New York, NY 10549
Ⓣ 914 666 3500 Ⓕ 914 241 7846
hmorrison1@aol.com

Handles fiction, non-fiction and screenplays.

Happy to read unsolicited manuscripts but send letter, project outline and sae in first instance. Does not charge a reading fee.

B.K. Nelson Literary Agency

84 Woodland Road, Pleasantville, NY 10570
Ⓣ 914 741 1322 Ⓕ 914 741 1324
bknelson4@cs.com
www.bknelson.com
President *Bonita K. Nelson*
Vice President *Leonard 'Chip' Ashbach*
Editorial Director *John W. Benson*

Founded 1979. Specializes in novels, business, self-help, how-to, political, autobiography, celebrity biography. Major motion picture and TV documentary success. Commission 20%.

No unsolicited manuscripts. Send letter of inquiry in first instance. Reading fee charged.

New Wave Talent Management

2660 West Olive Avenue, Burbank, CA 91505
T 818 295 5000
www.nwe.com

Represents a wide-ranging list of writers, directors, actors and comedians.

The Orange Grove Group, Inc.

12178 Ventura Boulevard, Suite 205, Studio City, CA 91604
T 818 762 7498 F 818 762 7499
agent@orangegrovegroup.com
www.orangegrovegroup.com

Established in 1986 as a management consultancy to small businesses. Also optioned literary material for development into feature film and television projects and in 1993 became a full-service talent and literary agency for actors, writers and directors.
By referral only.

Pangea Management

T 310 309 6155
www.pangeamg.tv

Formed in 2006 as part of the RDF Media Group to exploit the intellectual property rights of British and European producers of scripted and non-scripted formats. Based in Santa Monica, California.

Paradigm

360 North Crescent Drive, North Building, Beverly Hills, CA 90210
T 310 288 8000 F 310 288 2000
www.paradigmagency.com

Established in 1992, representing writers, producers, directors and actors for motion pictures and television. Also has offices in New York, Monterey and Nashville.

Stephen Pevner Inc

382 Lafayette Street, Suite 8, New York, NY 10003
T 212 674 8403

Established in 1991, with clients including screenwriters, producers and directors.
Send a note, project synopsis and sae before sending in script. No reading fees.

PMA Literary & Film Management, Inc.

PO Box 1817, Old Chelsea Station, New York, NY 10013
T 212 929 1222 F 212 206 0238
queries@pmalitfilm.com
www.pmalitfilm.com
President *Peter Miller*

Founded 1976. Commercial fiction and nonfiction. Specializes in books with motion picture and television potential, and in true crime. No poetry, pornography, non-commercial or academic. Commission: Home 15%; Dramatic 10–15%; Foreign 20–25%.
No unsolicited manuscripts. Approach by letter with one-page synopsis.

Result Talent Group

468 N. Camden Drive, Suite 300J, Beverly Hills, CA 90210
T 310 601 3186 F 310 388 3121
www.rtgtalent.com

Specializes in the representation of actors, screenwriters and directors.

Rosenstone/Wender

38 East 29th Street, 10th Floor, New York, NY 10016
T 212 725 9445 F 212 725 9447
rosenstone@aol.com

Founded 1981. Handles fiction, non-fiction, children's, and scripts for film, TV and theatre. No material for radio. Commission: Home 15%; Dramatic 10%; Foreign 20%.
No unsolicited manuscripts. send letter outlining the project, credits, etc. No reading fee.

Jack Scagnetti Talent & Literary Agency

5118 Vineland Avenue, Suite 106, North Hollywood, CA 91601
T 818 762 3871
Contact *Jack Scagnetti*

Founded 1974. Works mostly with established/published authors. Handles non-fiction, fiction, film and TV scripts. No reading fee. Commission: Home & Dramatic 10% (scripts), 15% (books); Foreign 15%.

SMA, LLC

8950 W Olympic Boulevard, PMB #380, Beverly Hills, CA 90211
T 310 203 8787
www.smaagency.com
Agent/Manager *Sara Margoshes*

Represents screenplays and story outlines to producers and development executives.
Takes clients by referral only. Do not send query letter. A release form signed by the author is required before accepting a submission.

Philip G. Spitzer Literary Agency, Inc.

50 Talmage Farm Lane, East Hampton, NY 11937
Ⓣ 631 329 3650 Ⓕ 631 329 3651
Contact *Philip Spitzer*

Founded 1969. Works mostly with established/published authors. Specializes in general non-fiction and fiction – thrillers. Commission: Home & Dramatic 15%; Foreign 20%.
No reading fee for outlines.

The Stein Agency

5125 Oakdale Avenue, Woodland Hills, CA 91364
Ⓣ 818 594 8990 Ⓕ 818 594 8998
mail@thesteinagency.com
www.thesteinagency.com

Provides representation for writers for cinema and television.

Sterling Lord Literistic, Inc.

65 Bleecker Street, New York, NY 10012
Ⓣ 212 780 6050 Ⓕ 212 780 6095
info@sll.com
www.sll.com

Founded 1979. Handles all genres, fiction and non-fiction. Commission: Home 15%; UK & Translation 20%.
Unsolicited manuscripts by regular mail only. Prefers letter outlining all non-fiction. Does not accept screenplays. Screen deals involve established clients only (ie book-to-screen deals). No reading fee.

Gloria Stern Agency

12535 Chandler Boulevard, Suite 3, North Hollywood, CA 91607
Ⓣ 818 508 6296 Ⓕ 818 508 6296
Contact *Gloria Stern*

Founded 1984. Handles film scripts, genre (romance, detective, thriller and sci-fi) and mainstream fiction; electronic media. Accepts interactive material, games and electronic data. Commission: Home 15%; Offshore 20%.
Currently not accepting unsolicited material. Fee required for critique.

Stone Manners Agency

6500 Wilshire Boulevard, Suite 550, Los Angeles, CA 90048
Ⓣ 323 655 1313

Represents some screenwriters.

Suite A Management Talent & Literary Agency

120 El Camino Drive, Suite 202, Beverly Hills, CA 90212
Ⓣ 310 278 0801 Ⓕ 310 278 0807
suite-a@earthlink.net
Agent/Principal *Lloyd D. Robinson*

Agency dedicated to the needs of writers, producers, directors and actors. For writers, the agency specializes in representing screenplays for development as made-for-television movies and low- to mid-budget features. Special focus on novels and plays for adaptation to film or television. Also represents writers for hire or attachment.

Summit Talent & Literary Agency

9454 Wilshire Boulevard, Suite 203, Beverly Hills, CA 90212
Ⓣ 310 205 9730

Represents screenwriters but references required. Send letter of inquiry in first instance.

Sweet 180

141 West 28th Street, Suite 300, New York, NY 10001
Ⓣ 212 541 4443 Ⓕ 212 563 9655
www.sweet180.com

Looks after a stable of actors, writers and directors.

TalentScout Management

1484 1/2 South Robertson Boulevard, Los Angeles, CA 90035
Ⓣ 310 276 5160 Ⓕ 310 276 5134
contact@atalentscout.com
www.atalentscout.com

Un-produced and un-published writers are encouraged to enter one of TalentScout's contests. Produced writers and published novelists should send a query letter or email in the first instance, stating writing credits as well as a synopsis of the proposed project.
Established in 1991, representing writers, directors and actors internationally.

Trancas International Management

1875 Century Park East, Suite 1145, Los Angeles, CA 90067
Ⓣ 310 553 5599 Ⓕ 310 553 0536
info@trancasfilms.com
www.trancasfilms.com

Also operates from London. Represents screenwriters, producers, directors and actors across sectors including film and television.

Treasure Entertainment, Inc.

468 N. Camden Drive, Suite 200, Beverly Hills,
CA 90210
ⓣ 310 860 7490 ⓕ 310 943 1488
info@treasureentertainment.net
www.treasureentertainment.net

A division of the production company of the
same name, managing screenwriters, television
writers, feature film directors and television
directors.

The Tudor Group

2921 Cavendish Drive, Los Angeles,
CA 90064-4615
ⓣ 510 876 5214 ⓕ 510 876 5215
www.tudorgroup.com

Specialists in personal management in
the entertainment arena. Roster includes
screenwriters.

Underground Management

447 S. Highland Avenue, Los Angeles,
CA 90036
ⓣ 323 930 2588 ⓕ 323 930 2334
submissions@undergroundfilms.net
www.undergroundfilms.net

Interested in developing and representing new
writers and filmmakers.

United Talent Agency

9560 Wilshire Boulevard, Suite 500,
Beverly Hills, CA 90212-2401
ⓣ 310 273 6700 ⓕ 310 247 1111
www.unitedtalent.com

A major talent and literary agency founded in
1991. Represents people across all sectors of the
creative arts, including screenwriters, filmmakers
and television writers.

Warden, White & Associates

8444 Wilshire Boulevard, 4th Floor,
Beverly Hills, CA 90211
ⓣ 323 852 1028

Represents writers and directors. Clients' credits
include *Sleepless in Seattle, Meet Joe Black* and
Enemy of the State.

Irene Webb Literary Film Representation & Publishing

1112 Montana Avenue, Suite 294, Santa Monica,
CA 90403
ⓣ 310 394 9024
webblit@verizon.net
www.irenewebb.com

Established in 2003, with a strong track record
of selling film and television rights for authors.
Among properties sold are *The Alienist, Fletch,
Hitchhikers Guide to the Galaxy, Men in
Black, Patriot Games, Postcards from the Edge,
Striptease* and *The Witches of Eastwick.*

Email query letter initially. Do not send
unsolicited manuscripts.

Working Artists

ⓣ 818 907 1122 ⓕ 818 891 1293
info@workingartists.net
www.workingartists.net

Los Angeles-based production and management
company founded in 1994. Represents film,
long-form television and documentary projects.
Using its contacts in the independent and studio
markets, aims to facilitate relationships between
writers, directors and producers and appropriate
production companies.

Ann Wright Representatives

165 West, 46th Street, Suite 1105, New York,
NY 10036–2501
ⓣ 212 764 6770 ⓕ 212 764 5125
Contact *Dan Wright*

Founded 1961. Specializes in material with strong
film potential. Handles screenplays and novels,
drama and fiction. Commission on screenplays:
10% of gross.

Approach by letter; no reply without IRCs.
Include outline and credits only. 'Has reputation
for encouraging new writers.' No reading fee.
Signatory to the Writers Guild of America
Agreement.

UK

A & B Personal Management Ltd

Linden Hall, 162-168 Regent Street, Suite 330,
London, W1B 5TD
ⓣ 020 7434 4262

Deals in full-length manuscripts for TV, theatre
and cinema. Does not welcome unsolicited
manuscripts. In first instance send letter with
return postage. No reading fee for synopses or
screenplays.

The Agency (London) Ltd

24 Pottery Lane, London, W11 4LZ
ⓣ 020 7727 1346 ⓕ 020 7727 9037
info@theagency.co.uk
www.theagency.co.uk

Founded 1995. Deals with writers and rights for
TV, film, theatre, radio scripts and children's

fiction. Also handles directors. Only existing clients for adult fiction or non-fiction. Commission: Home 10%; US by arrangement.

Send letter with sae. No unsolicited manuscripts. No reading fee.

The Ampersand Agency Ltd

Ryman's Cottages, Little Tew, OX7 4JJ
℡ 01608 683677
peter@theampersandagency.co.uk
www.theampersandagency.co.uk
Contact *Peter Buckman*
Contact *Anne-Marie Doulton*

Founded 2003. Handles literary and commercial fiction and non-fiction; contemporary and historical novels, crime, thrillers, biography, women's fiction, history, memoirs. No scripts unless by existing clients.

Unsolicited manuscripts, synopses and ideas welcome, as are e-mail enquiries, but send sample chapters and synopsis by post (sae required if material is to be returned). No reading fee.

Darley Anderson Literary, TV & Film Agency

Estelle House, 11 Eustace Road, London, SW6 1JB
℡ 020 7385 6652 ℻ 020 7386 5571
enquiries@darleyanderson.com
www.darleyanderson.com/
www.darleyandersonchildrens.com
Agency Assistant *Ella Andrews*

Founded 1988 and run by an ex-publisher. Handles commercial fiction and non-fiction, children's fiction and non-fiction; also selected scripts for film and TV. Special fiction interests: all types of thrillers and crime (American/hard boiled/cosy/historical); women's fiction (sagas, chick-lit, contemporary, love stories, 'tear jerkers', women in jeopardy) and all types of American and Irish novels.

Overseas associates APA Talent and Literary Agency (LA/Hollywood); Liza Dawson Literary Agency (New York); and leading foreign agents throughout the world. Send letter, synopsis and first three chapters; plus sae for return. No reading fee.

Author Literary Agents

53 Talbot Road, London, N6 4QX
℡ 020 8341 0442 ℻ 020 8341 0442
agile@authors.co.uk
Contact *John Havergal*

Founded 1997. 'We put to leading publishers and producers strong new novels, thrillers and graphic media ideas.'

Send sae with first chapter, scene or section writing sample, plus half-to-one page outline. Include graphics samples, if applicable. No reading fee.

Alan Brodie Representation Ltd

6th Floor, Fairgate House, 78 New Oxford Street, London, WC1A 1HB
℡ 020 7079 7990 ℻ 020 7079 7999
info@alanbrodie.com
www.alanbrodie.com
Contact *Alan Brodie*
Contact *Sarah McNair*
Contact *Lisa Foster*

Founded 1989. Handles theatre, film and TV scripts. No books. Commission: Home 10%; Overseas 15%.

Preliminary letter plus professional recommendation and CV essential. No reading fee but sae required.

Brie Burkeman

14 Neville Court, Abbey Road, London, NW8 9DD
℡ 0870 199 5002 ℻ 0870 199 1029
brie.burkeman@mail.com
Contact *Brie Burkeman*
Contact *Isabel White*

Begun in 2000 and now representing an average of 10–25 clients. Clients include Philip Gawthorne of *Dream Team* and Jean Pasley of *How About You*. Interested primarily in film scripts. Does not deal in short film or musicals. Also associated with Serafina Clarke Ltd and independent film/TV consultant to literary agents.

Sample material accepted by post only with return postage. Email attachments will be deleted. No reading fee. 15% commission charged.

Capel & Land Ltd

29 Wardour Street, London, W1D 6PS
℡ 020 7734 2414 ℻ 020 7734 8101
rosie@capelland.co.uk
www.capelland.com
Contact *Georgina Capel*

Handles fiction and non-fiction. Also film and TV. Clients include Kunal Basu, John Gimlette, Andrew Greig, Dr Tristram Hunt, Liz Jones, Andrew Roberts, Simon Sebag Montefiore, Stella Rimington, Diana Souhami, Louis Theroux, Fay Weldon. Commission: Home, US & Translation 15%.

Send sample chapters and synopsis with covering letter and sae (if return required) in the first instance. No reading fee.

Casarotto Ramsay and Associates Ltd

Waverley House, 7–12 Noel Street, London,
W1F 8GQ

ⓉT 020 7287 4450 ⓕF 020 7287 9128
agents@casarotto.co.uk
www.casarotto.uk.com

Handles scripts for TV, theatre, film and radio.
Clients include: (film) Laura Jones, Neil Jordan, Nick
Hornby, Shane Meadows, Purvis & Wade, Lynne
Ramsay; (TV) Howard Brenton, Amy Jenkins, Susan
Nickson, Jessica Stevenson. Commission: Home
10%. Overseas associates worldwide.

No unsolicited material without preliminary
letter.

Mic Cheetham Literary Agency

50 Albemarle Street, London, W1S 4BD
ⓉT 020 7495 2002 ⓕF 020 7495 8961
info@miccheetham.com
www.miccheetham.com
Contact *Mic Cheetham*

Established 1994. Handles general and literary
fiction, crime and science fiction, and some
specific non-fiction. Film/TV scripts from
existing clients only. Clients include Iain Banks,
Simon Beckett, Carol Birch, Anita Burgh, Laurie
Graham, M. John Harrison, Toby Litt, Ken
MacLeod, China Miéville, Antony Sher, Janette
Turner Hospital. Commission: Home 15%;
US & Translation 20%. Works with The Marsh
Agency for all translation rights.

No unsolicited manuscripts. Approach in
writing with publishing history, first two chapters
and return postage. No reading fee.

Mary Clemmey Literary Agency

6 Dunollie Road, London, NW5 2XP
ⓉT 020 7267 1290 ⓕF 020 7482 7360
Contact *Mary Clemmey*

Founded 1992. Handles fiction and
non-fiction – high-quality work with an
international market. TV, film, radio and theatre
scripts from existing clients only. US clients:
Frederick Hill Bonnie Nadell Inc., Lynn C.
Franklin Associates Ltd, The Miller Agency,
Roslyn Targ, Weingel- Fidel Agency Inc, Betsy
Amster Literary. Commission: Home 10%; US
& Translation 20%. Overseas associate Elaine
Markson Literary Agency, New York.

No unsolicited manuscripts. Approach by letter
only in the first instance giving a description of
the work (include sae). No reading fee.

Jonathan Clowes Ltd

10 Iron Bridge House, Bridge Approach, London,
NW1 8BD

ⓉT 020 7722 7674 ⓕF 020 7722 7677
Contact *Ann Evans*
Contact *Lisa Whadcock*

Founded 1960. Pronounced 'clewes'. Now one of
the biggest fish in the pond, and not really for the
untried unless they are true high-flyers. Fiction
and non-fiction, plus scripts. Special interests:
situation comedy, film and television rights.
Clients include David Bellamy, Len Deighton,
Elizabeth Jane Howard, Doris Lessing, David
Nobbs, Gillian White and the estate of Kingsley
Amis. Commission: Home & US 15%; Translation
19%. Overseas associates Andrew Nurnberg
Associates; Sane Töregard Agency.

No unsolicited manuscripts; authors come by
recommendation or by successful follow-ups to
preliminary letters.

Elspeth Cochrane Agency

16 Trinity Close The Pavement, London,
SW4 0JD
ⓉT 020 7622 3566
elspeth@elspethcochrane.co.uk

Founded in 1960, representing some
screenwriters. Commission is a negotiable 12.5%.

Rosica Colin Ltd

1 Clareville Grove Mews, London, SW7 5AH
ⓉT 020 7370 1080 ⓕF 020 7244 6441
Contact *Joanna Marston*

Founded 1949. Handles all full-length
manuscripts, plus theatre, film, television and
sound broadcasting but few new writers being
accepted. Commission: Home 10%;
US & Translation 20%.

Preliminary letter with return postage essential;
writers should outline their writing credits and
whether their manuscripts have previously been
submitted elsewhere. May take 3–4 months to
consider full manuscripts; synopsis preferred in
the first instance. No reading fee.

Creative Media Management

3b Walpole Court, Ealing Studios, London,
W5 5ED
ⓉT 020 8584 5363 ⓕF 020 8566 5554
enquiries@creativemediamanagement.com
www.creativemediamanagement.com
Contact *Chris Calitz*

Formed in July 1999 by Jacqui Fincham.
Represents film, television and radio writers
specializing in drama and comedy. Clients
include David Freedman, Kieran Galvin,
Stuart Grieve, Jon Groom, Adam Howe, Maggie
Innes, Katya Jezzard, Paul Mari and Claire
Wilson.

Not currently accepting any new submissions or spec scripts from writers. See website for detailed submissions guidelines.

Curtis Brown Group

5th Floor, Haymarket House, 28-29 Haymarket, London, SW1Y 4SP
☎ 020 7393 4400
media@curtisbrown.co.uk
www.curtisbrown.co.uk
Film/TV/Theatre *Nick Marston*

Established in 1899 and one of Europe's oldest literary agencies. Interested in all forms of fiction; rarely deals in factual formats.
  Accepts unsolicited material only by post. Requires synopsis, covering letter and CV. Return postage essential. No reading fee.

Judy Daish Associates Ltd

2 St Charles Place, London, W10 6EG
☎ 020 8964 8811 🖷 020 8964 8966
Contact *Judy Daish*
Contact *Tracey Elliston*
Contact *Howard Gooding*

Founded 1978. Theatrical literary agent. Handles scripts for film, TV, theatre and radio. No books.
  Preliminary letter essential. No unsolicited manuscripts.

Felix de Wolfe

Kingsway House, 103 Kingsway, London, WC2B 6QX
☎ 020 7242 5066 🖷 020 7242 8119
info@felixdewolfe.com
www.felixdewolfe.com
Contact *Felix de Wolfe*

Founded 1938. Handles quality fiction only, and scripts. Clients include Jan Butlin, Jeff Dowson, Brian Glover, Sheila Goff, Aileen Gonsalves, John Kershaw, Ray Kilby, Bill MacIlwraith, Angus Mackay, Gerard McLarnon, Malcolm Taylor, David Thompson, Paul Todd, Dolores Walshe. Commission: Home 12½%; US 20%.
  No unsolicited mss. No reading fee.

The Dench Arnold Agency

10 Newburgh Street, London, W1F 7RN
☎ 020 7437 4551 🖷 020 7439 1355
contact@dencharnold.com
www.dencharnold.com
Contact *Elizabeth Dench*
Contact *Michelle Arnold*

Founded 1972. Handles scripts for TV and film. Clients include Peter Chelsom. Commission: Home 10–15%. Overseas associates include

William Morris/Sanford Gross and C.A.A., Los Angeles.
  Unsolicited manuscripts will be read, but a letter with sample of work and CV (plus sae) is required.

Bryan Drew Ltd

Quadrant House, 80-82 Regent Street, London, W1B 5AU
☎ 020 7437 2293 🖷 020 7437 0561
bryan@bryandrewltd.com
www.bryandrewltd.com

Represents all types of authors, with specialism in representing writers for big and small screen.

Janet Fillingham Associates

52 Lowther Road, London, SW13 9NU
☎ 020 8748 5594 🖷 020 8748 7374
info@janetfillingham.com
www.janetfillingham.com

Founded in 1992, representing a select list of writers and directors working in television and feature films in the adult and family markets.
  Not currently seeking new clients but will accept CVs sent by email. Promises to read every CV but any other correspondence will not be read.

Film Rights Ltd/Laurence Fitch Ltd

Mezzanine, Quadrant House, 80–82 Regent Street, London, W1B 5AU
☎ 020 7734 9911 🖷 020 7734 0044
information@filmrights.ltd.uk
www.filmrights.ltd.uk

Agency established in 1932. Now representing around 20 clients such as Ray Coovey and John Chapman. Handles most genres of film and TV projects, particularly drama. Little interest in historical scripts.
  Reads new material mostly on recommendation only. Commission charged: 10% domestic, 15% overseas.

Jill Foster Ltd

9 Barb Mews, Brook Green, London, W6 7PA
☎ 020 7602 1263 🖷 020 7602 9336
agents@jflagency.com
www.jflagency.com
Managing Director *Jill Foster*
Agent *Alison Finch*
Agent *Dominic Lord*
Agent *Gary Wild*
Agent *Simon Williamson*

Established in 1976 to represent film and television screenwriters. Specialists in comedy and drama. Accepts no unsolicited manuscripts. Charges 12.5% domestic commission, 15% overseas.

FRA

91 St Leonard's Road, London, SW14 7BL
ⓣ 020 8255 7755 ⓕ 020 8286 4860
suzanne@futermanrose.co.uk
www.futermanrose.co.uk
Biography/Non-Fiction/Screenplays *Guy Rose*
TV and Screenplays *Suzanne Cowie*

Formerly Futerman, Rose & Associates, founded 1984. Handles fiction; scripts for film and TV; biography; show business; current affairs and teenage fiction. Clients include Larry Barker, Christian Piers Betley, Ginger Baker, Tom Conti, Philip Dart, Lynsey de Paul, Iain Duncan Smith, Elyssa Edmonstone, Royston Ellis, Sir Martin Ewans, Yvette Fielding, Susan George, Stephen Griffin, Brian Harvey, Paul Hendy, Russell Warren Howe, Keith R. Lindsay, Stephen Lowe, Eric MacInnes, Paul Marx, Valerie Mendes, Max Morgan-Witts, Ciarán O'Keeffe, Erin Pizzey, John Repsch, Liz Rettig, Peter Sallis, Pat Silver-Lasky, Paul Stinchcombe, Gordon Thomas, Bill Tidy, Mark White, Toyah Willcox, Simon Woodham, Tappy Wright, Allen Zeleski.

No unsolicited manuscripts. Send preliminary letter with brief resumé, detailed synopsis, first 20 pages (approx.) and sae.

French's

78 Loudoun Road, London, NW8 0NA
ⓣ 020 7483 4269

Founded in 1973. Handles scripts for all media with particular focus on screenplays. Write in first instance before sending in manuscript. Charged reading service for first-time writers. Commission charged at 10% for home market.

Blake Friedmann Literary Agency

122 Arlington Road, London, NW1 7HP
ⓣ 020 7284 0408 ⓕ 020 7284 0442
info@blakefriedmann.co.uk
www.blakefriedmann.co.uk
Film/TV *Julian Friedman*
Original Scripts/Radio *Conrad Williams*

Founded 1977. Handles all kinds of fiction from genre to literary; a varied range of specialist and general non-fiction, plus scripts for TV, radio and film. No poetry, science fiction or short stories (unless from existing clients). Clients include Gilbert Adair, Jane Asher, Edward Carey, Elizabeth Chadwick, Anne de Courcy, Anna Davis, Barbara Erskine, Ann Granger, Ken Hom,

Billy Hopkins, Peter James, Glenn Meade, Deon Meyer, Lawrence Norfolk, Gregory Norminton, Joseph O'Connor, Sheila O'Flanagan, Siân Rees, Michael Ridpath, Craig Russell, Tess Stimson, Michael White. Commission Books: Home 15%; US & Translation 20%. Radio/TV/Film: 15%. Overseas associates: 24 worldwide.

Unsolicited manuscripts welcome but initial letter with synopsis and first two chapters preferred. Letters should contain as much information as possible on previous writing experience, aims for the future, etc. No reading fee.

Noel Gay

19 Denmark Street, London, WC2H 8NA
ⓣ 020 7836 3941
info@noelgay.com
www.noelgay.com
CEO *Alex Armitage*

Agency representing some of the UK's leading writers, broadcasters, journalists and presenters. Principal business of the Noel Gay Organisation, which includes Noel Gay Television.

Eric Glass Ltd

25 Ladbroke Crescent, London, W11 1PS
ⓣ 020 7229 9500 ⓕ 020 7229 6220
eglassltd@aol.com
Contact *Janet Glass*

Founded 1934. Handles fiction, non-fiction and scripts for publication or production in all media. Overseas associates in the US, Australia, Czech Republic, France, Germany, Greece, Holland, Italy, Japan, Poland, Scandinavia, Slovakia, South Africa, Spain.

No unsolicited manuscripts. Return postage required. No reading fee.

The Rod Hall Agency Limited

6th Floor, Fairgate House, 78 New Oxford Street, London, W1A 1HB
ⓣ 020 7079 7987 ⓕ 020 7079 7988
office@rodhallagency.com
www.rodhallagency.com
Contact *Charlotte Knight*

Founded 1997. Handles drama for film, TV and theatre and writers-directors. Does not represent writers of episodes for TV series where the format is provided but represents originators of series. Clients include Simon Beaufoy (*The Full Monty*), Jeremy Brock (*Mrs Brown*), Liz Lochhead (*Perfect Days*), Martin McDonagh (*The Pillowman*), Simon Nye (*Men Behaving Badly*). Commission: Home 10%; US & Translation 15%. No reading fee.

Roger Hancock Ltd

4 Water Lane, London, NW1 8NZ
☎ 020 7267 4418 ℻ 020 7267 0705
info@rogerhancock.com
www.rogerhancock.com

Founded 1960. Special interests: comedy drama and light entertainment. Commission: Home 10%; Overseas 15%.

Unsolicited manuscripts not welcome. Initial phone call required.

David Higham Associates Ltd

5–8 Lower John Street, Golden Square, London, W1F 9HA
☎ 020 7434 5900 ℻ 020 7437 1072
dha@davidhigham.co.uk
www.davidhigham.co.uk
Contact *Nicky Lund*
Contact *Georgina Ruffhead*

Founded 1935. Handles fiction, general non-fiction (biography, history, current affairs, etc.) and children's books. Also scripts. Clients include John le Carré, J.M. Coetzee, Stephen Fry, Jane Green, James Herbert, Alexander McCall Smith, Lynne Truss, Jacqueline Wilson. Commission: Home 15%; US & Translation 20%; Scripts 10%.

Preliminary letter with synopsis essential in first instance. No email submissions. No reading fee. See website for further information.

Valerie Hoskins Associates Limited

20 Charlotte Street, London, W1T 2NA
☎ 020 7637 4490 ℻ 020 7637 4493
vha@vhassociates.co.uk
Contact *Valerie Hoskins*

Founded 1983. Handles scripts for film, TV and radio. Special interests: feature films, animation and TV. Commission: Home 12½%; US 20% (maximum).

No unsolicited scripts; preliminary letter of introduction essential. No reading fee.

ICM London

4-6 Soho Square, London, W1D 3PZ
☎ 020 7432 0800
www.icmtalent.com

The London wing of the American firm, International Creative Management.

Strict policy not to accept or look at unsolicited submissions.

Independent Talent Group

Oxford House, 76 Oxford Street, London, W1D 1BS

☎ 020 7636 6565 ℻ 020 7323 0101
www.independenttalent.com

Founded 1973. Formerly known as ICM. Specializes in scripts for film, theatre, TV and radio. No reading fee.

International Scripts

1A Kidbrooke Park Road, London, SE3 0LR
☎ 020 8319 8666 ℻ 020 8319 0801
internationalscripts@btinternet.com
Contact *Bob Tanner*
Contact *Pat Hornsey*
Contact *Jill Lawson*

Founded 1979 by Bob Tanner. Handles most types of books (non-fiction and fiction) and scripts for most media. No poetry, articles or short stories. Clients include Jane Adams, Zita Adamson, Ashleigh Bingham, Simon Clark, Ann Cliff, Dr James Fleming, June Gadsby, Julie Harris, Robert A. Heinlein, Anna Jacobs, Anne Jones, Richard Laymon, Trevor Lummis, Margaret Muir, Nick Oldham, Chris Pascoe, Christine Poulson, John and Anne Spencer, Janet Woods. Commission: Home 15%; US & Translation 20%. Overseas associates include Ralph Vicinanza, USA; Thomas Schlück, Germany.

Preliminary letter, one-page synopsis, wordage, plus sae required. No unsolicited manuscripts by post or email accepted.

Michelle Kass Associates

85 Charing Cross Road, London, WC2H 0AA
☎ 020 7439 1624 ℻ 020 7734 3394
office@michellekass.co.uk

Founded 1991. Handles literary fiction, film and television primarily. Commission: Home 10%; US & Translation 15–20%.

No manuscripts accepted without preliminary phone call. No reading fee. No email submissions.

LAW

14 Vernon Street, London, W14 0RJ
☎ 020 7471 7900 ℻ 020 7471 7910
www.lawagency.co.uk

Founded 1996. Handles full-length commercial and literary fiction, non-fiction and children's books. Film and TV scripts handled for established clients only. Commission: Home 15%; US & Translation 20%. Overseas associates worldwide.

Unsolicited manuscripts considered; send brief covering letter, short synopsis and two sample chapters. Sae essential. No emailed or disk submissions.

The Leo Media & Entertainment Group

150 Minories, London, EC3N 1LS
ⓣ 020 8905 5191 ⓕ 08701 330 258
info@leomediagroup.com
www.leomediagroup.com
Managing Director *Alexander Sullivan*

Founded in 2003, a 'one-stop shop', able to secure finance and distribution for films and provide casting, secure directors and production crew, edit screenplays, help set up websites and provide legal advice. Represents writers for feature films and television, across genres. Clients include Buddy Bregman, EM Lake, Marino Colmano, Nik Goldman, Dr Mia Sperber, Eden Yorke, Philippe Bouin, CTP Productions, Karyn & Jay Milner, Chuck Kelly Powers, Giles Ecott & Roger Cayless, Michael Leavy, Jim Curtis, Fountain of Life Productions, Lisa Munsterhjelm, Matthew Lewis, Jill D'Angelo Powers, Nigel Pascoe and Leon Gildin.

Happy to look at unsolicited manuscripts or ideas. Approach by email in first instance. No reading fee. Commission varies from project to project.

The Christopher Little Literary Agency

Eel Brook Studios, 125 Moore Park Road, London, SW6 4PS
ⓣ 020 7736 4455 ⓕ 020 7736 4490
info@christopherlittle.net or firstname@ www.christopherlittle.net
Contact *Christopher Little*

Founded 1979. Handles commercial and literary full-length fiction and non-fiction. Film scripts for established clients only. Authors include Steve Barlow and Steve Skidmore, Paul Bajoria, Andrew Butcher, Janet Gleeson, Carol Hughes, Alastair MacNeill, Robert Mawson, Haydn Middleton, Andrew Quinnell, Christopher Matthew, Robert Radcliffe, J.K. Rowling, Darren Shan, Wladyslaw Szpilman, John Watson, Pip Vaughan-Hughes, Gorillaz, Christopher Hale, Peter Howells, Gen. Sir Mike Jackson, Lauren Liebenberg, Shiromi Pinto, Dr Nicholas Reeves, Shayne Ward, Angela Woolfe, Anne Zouroudi. Commission: Home 15%; US, Canada, Translation, Audio & Motion Picture 20%.

Send detailed preliminary letter in the first instance with synopsis, first 2–3 chapters and sae. No reading fee.

Andrew Mann Ltd

Third Floor, 1 Old Compton Street, London, W1D 5JA
ⓣ 020 7734 4751 ⓕ 020 7287 9264
info@manuscript.co.uk
www.andrewmann.co.uk
Literary Agent *Tina Betts*

Literary Agent *Anne Dewg*
Literary Agent *Louise Burns*

Agency founded in 1968, now representing about 80 clients. Handles film and television comedy and drama. Also interested in children's scripts. Does not deal in horror, sci-fi, fantasy or sitcoms.

Unsolicited material accepted as a letter, short synopsis and 30 pages of text. Sae essential. No reading fee. 15% commission charged (US and Translation, 20%).

Marjacq Scripts Ltd

34 Devonshire Place, London, W1G 6JW
ⓣ 020 7935 9499 ⓕ 020 7935 9115
subs@marjacq.com
www.marjacq.com
Books *Philip Patterson*
Film/TV *Luke Speed*

Handles fiction and non-fiction, literary and commercial as well as film, TV, radio scripts. No poetry. Commission: Home 10%; Overseas 20%.

New work welcome; send brief letter, synopsis and approx. first 50 pages plus sae. No reading fee.

MBA Literary Agents Ltd

62 Grafton Way, London, W1T 5DW
ⓣ 020 7387 2076 ⓕ 020 7387 2042
diana@mbalit.co.uk
www.mbalit.co.uk
Contact *Diana Tyler*
Contact *Meg Davis*
Contact *Jean Kitson*

Founded in 1971. Handles fiction and non-fiction, TV, film, radio and theatre scripts.

Submission guidelines for unsolicited material on the website. 10% commission charged for television, and 10–20% for film.

McKernan Agency

5 Gayfield Square, Edinburgh, EH1 3NW
ⓣ 0131 557 1771
maggie@mckernanagency.co.uk
www.mckernanagency.co.uk
Contact *Maggie McKernan*

Founded 2005. Works in conjunction with Capel & Land Ltd. Handles fiction, both literary and commercial; also TV, film, radio and theatre scripts. Commission: Home 15%; US & Translation 15%.

Send sae for return of manuscripts. No reading fee.

Bill McLean Personal Management

23b Deodar Road, Putney, London, SW15 2NP
ⓣ 020 8789 8191
Contact *Bill McLean*

Established in 1972 to represent talent from film, TV, radio and stage. Clients include Dwynwen Berry, Phil Clark, Tony Jordan, Jeffrey Segal, Frank Vickery, Laura Watson and Mark Wheatley. No unsolicited material.

William Morris Agency (UK) Ltd

CentrePoint Tower, 103 Oxford Street, London, WC1A 1DD
☏ 020 7534 6800 ℻ 020 7534 6900
www.wma.com
Head of TV *Holly Pye*
Film *Lucinda Prain*

London office founded in 1965. Worldwide talent and literary agency with offices in New York, Beverly Hills, Nashville, Miami and Shanghai. Handles fiction, general non-fiction, TV and film scripts. Commission: TV & Film 10%; UK Books 15%; US Books & Translation 20%.
Accepts email submissions only. Send synopsis and three sample chapters (50 pages or less) to ldnsubmissions@wma.com. No reading fee.

The Narrow Road Company

182 Brighton Road, Coulsdon, Surrey, CR5 2NF
☏ 020 8763 9895 ℻ 020 8763 2558
richardireson@narrowroad.co.uk
Managing Director *Richard Ireson*

Agency founded in 1986, dealing mainly in drama across theatre, film, TV and radio.
No unsolicited manuscripts. Approach by letter with CV and return postage. The agency advises that first time writers need not apply.

PFD

Drury House, 34–43 Russell Street, London, WC2B 5HA
☏ 020 7344 1000 ℻ 020 7836 9539/9541
postmaster@pfd.co.uk
www.pfd.co.uk
Managing Director *Caroline Michel*
Agent *Lesley Davey*
Agent *Gemma Hirst*
Agent *Michelle Archer*
Agent *Jessica Cooper*

PFD represents authors of fiction and non-fiction, children's writers, screenwriters, playwrights, documentary makers, technicians, presenters and actors throughout the world.
Consult the website for submission guidelines.

Pollinger Limited

9 Staple Inn, Holborn, London, WC1V 7QH
☏ 020 7404 0342 ℻ 020 7242 5737

info@pollingerltd.com
www.pollingerltd.com
Managing Director *Lesley Pollinger*
Authors Agent *Tim Bates*
Dramatic Rights Agent *Ruth Needham*

Established in 1935, interested in scripts for film and television, predominantly drama. Existing clients are in the main novelists but currently developing the dramatic rights and screenwriters lists.

RDF Management

3-6 Kenrick Place, London, W1U 6HD
☏ 020 7317 2230
info@rdfmanagement.com
www.rdfmanagement.com

A specialist talent agency representing actors, writers, presenters and stand-up comedians.

Real Creatives Worldwide

14 Dean Street, London, W1D 3RS
☏ 020 7437 4188
business@realcreatives.com
www.realcreatives.com
Contact *Mark Maco*

Founded in 1984 and now has 420 clients.
Interested in scripts for film and television, especially dramas, documentaries and action films.
Email in first instance, before mailing or posting manuscript/idea. Charges commission of 10-20% but no reading fee.

Redhammer Management Ltd

186 Bickenhall Mansions, Bickenhall Street, London, W1U 6BX
☏ 020 7486 3465 ℻ 020 7000 1249
info@redhammer.info
www.redhammer.info
Vice President *Peter Cox*

'Provides in-depth management for a small number of highly talented authors.' Willing to take on unpublished authors who are professional in their approach and who have major international potential, ideally for, book, film and/or TV. Clients include Martin Bell OBE, John Brindley, Brian Clegg, Joe Donnelly, Audrey Eyton, Maria Harris, Senator Orrin Hatch, Amanda Lees, David McIntee, Hon. Nicholas Monson, Michelle Paver, Carolyn Soutar, Carole Stone, Donald Trelford, Justin Wintle, David Yelland.
Submissions considered only if the guidelines given on the website have been followed. Do not send unsolicited manuscripts by post. No radio or theatre scripts. No reading fee.

Sayle Screen Ltd

11 Jubilee Place, London, SW3 3TD
T 020 7823 3883
info@saylescreen.com
www.saylescreen.com
Agent *Jane Villiers*
Agent *Toby Moorcroft*
Agent *Matthew Bates*
Agent's Assistant *Hannah Barker*

Agency specializing in writers and directors for film and television. Particularly interested in drama, documentary and comedy writing. Represents around 60–70 clients including Andrea Arnold and Marc Evans.

No unsolicited material without preliminary letter. No email submissions. No reading fee.

The Sharland Organisation Ltd

The Manor House, Manor Street, Raunds, Northants, NN9 6JW
T 01933 626 600 F 01933 624 860
info@sharlandorganisation.co.uk
www.sharlandorganisation.co.uk
Director *Mike Sharland*

Founded in 1988. Specializes in national and international film and TV negotiations. Also negotiates multimedia, interactive TV deals and computer game contracts. Handles scripts for film, TV, radio and theatre; also non-fiction. Markets books for film and handles stage, radio, film and TV rights for authors. Commission: Home 15%; UK and Translation 20%.

No unsolicited manuscripts. Preliminary enquiry by letter or phone essential. No reading fee.

Sheil Land Associates

52 Doughty Street, London, WC1N 2LS
T 020 7405 9351 F 020 7831 2127
info@sheilland.co.uk
Film, TV and Theatrical Agent *Sophie Janson*
Film, TV and Theatrical Agent *Emily Hayward*

Film, television and theatrical agents interested to see scripts of all genres. Contact with reception advised before sending letters and sample writing. No reading fee. 15% commission charged.

Elaine Steel

110 Gloucester Avenue, London, NW1 8HX
T 020 7483 2681 F 020 7483 4541
ecmsteel@aol.com
Contact *Elaine Steel*

Founded 1986. Handles scripts, screenplays and books. No technical or academic. Clients include Les Blair, Anna Campion, Michael Eaton, Pearse Elliott, Gwyneth Hughes, Brian Keenan, Troy

Kennedy Martin, James Lovelock, Rob Ritchie, Albie Sachs, Ben Steiner. Commission: Home 10%; US & Translation 20%.

Initial phone call preferred.

Micheline Steinberg Associates

104 Great Portland Street, London, W1W 6PE
T 020 7631 1310
info@steinplays.com
www.steinplays.com
Agent *Micheline Steinberg*
Agent *Matt Connell*
Assistant *Helen Macauley*

Founded in 1988, dealing in drama and comedy for film, television, theatre and radio. Represents writers for film and TV rights in fiction and non-fiction on behalf of book agents.

Does not welcome unsolicited approaches. Industry recommendation preferred. Commission charged at 10% for home and 10–20% for overseas.

Rochelle Stevens & Co.

2 Terrets Place, Upper Street, Islington, London, N1
T 020 7359 3900 F 020 7354 5729
info@rochellestevens.com
www.rochellestevens.com

Represents writers and directors for stage and screen.

The Tennyson Agency

10 Cleveland Avenue, Wimbledon Chase, London, SW20 9EW
T 020 8543 5939
submissions@tenagy.co.uk
www.tenagy.co.uk
Partner *Christopher Oxford*
Partner *Adam Sheldon*

Agency founded in 2002 with 16 clients including Julian Howell, Tony Bagley and Steve MacGregor. Interested in any fictional material for both film and television except fantasy, science fiction, children's animation or foreign language projects.

Letters of introduction encouraged at first by post or email. No reading fee. 15% domestic commission charged, 20% overseas.

United Agents Ltd

12-26 Lexington Street, London, W1F 0LE
info@unitedagents.co.uk
www.unitedagents.co.uk

Formed in 2007, with department for TV, film, radio and theatre scripts. Roster includes many high-profile television and film writers, from the award-winning to the emerging.

Unsolicited material welcome; see submissions policy on the website. No reading fee.

Cecily Ware Literary Agents

19C John Spencer Square, Canonbury, London, N1 2LZ

ⓣ 020 7359 3787 ⓕ 020 7226 9828
info@cecilyware.com
www.cecilyware.com
Contact *Cecily Ware*
Contact *Gilly Schuster*
Contact *Warren Sherman*

Founded 1972. Primarily a film and TV script agency representing work in all areas: drama, children's, series/serials, adaptations, comedies, etc. Commission: Home 10%; US 10–20% by arrangement.

No unsolicited manuscripts or phone calls. Approach in writing only. No reading fee.

AP Watt Ltd

20 John Street, London, WC1N 2DR
ⓣ 020 7405 6774 ⓕ 020 7831 2154
apw@apwatt.co.uk
www.apwatt.co.uk

Agency founded in 1875. Film and television specialists. Interested in drama, comedy and documentary.

No unsolicited material. Approach via agent by letter FAO Media Department.

Canada

Agence Omada

451 Rue Ste-Catherine Ouest, bur. 201, Montréal, Québec, H3B 1B1
ⓣ 514 287 1246 ⓕ 514 287 7694
info@omada.ca
www.omada.ca

Represents variety of 'behind the camera' artists including screenwriters and directors.

The Alpern Group

585 Bloor Street West, 2nd Floor, Toronto, Ontario, M6G 1K5
ⓣ 416 907 5845 ⓕ 416 531 4961

Represents writers for both film and television.

Ambition

439 Wellington Street West, Suite 204, Toronto, Ontario, M5V 1E7
ⓣ 416 916 8340 ⓕ 647 435 4865
info@ambitiontalent.com
www.ambitiontalent.com
Owner *David Ritchie*

Founded in 2006. Represents talent from the film and television industries. Interested in all types of material.

Material read only upon recommendation. No unsolicited emails or calls. Commission: 10-15% for clients.

Aurora Artists

19 Wroxeter Avenue, Toronto, Ontario, M4K 1J5
ⓣ 416 463 4634
aurora.artists@sympatico.ca

Represents writers for film and TV.

Christopher Banks & Associates

Suite 410, 6 Adelaide Street East, Toronto, Ontario, M5C 1H6
ⓣ 416 214 1155 ⓕ 416 214 1150
info@chrisbanks.com
www.chrisbanks.com

Founded in 1983, a talent agency for professionals in the film, television and performing arts.

Accepts submissions for representation by referral only.

The Characters

8 Elm Street, Toronto, Ontario, M5G 1G7
ⓣ 416 964 8522 ⓕ 416 964 8206
www.thecharacters.com

Founded in 1969, a full service agency with offices in Toronto and Vancouver. Represents editors and writers, among others.

Submit a package including a resume, approximately 30-word logline, and a 1-page synopsis or a resume and demo reel or portfolio. Label "Attn: Literary Department" and include sae for return of materials. Will not consider unsolicited scripts.

Great North Artists Management Inc.

350 Dupont Street, Toronto, Ontario, M5R 1V9
ⓣ 416 925 2051 ⓕ 416 925 3904
info@gnaminc.com
President *Ralph Zimmerman*
Literary Agent *Rena Zimmerman*
Literary Assistant *Abigail Algar*

Represents literary talent from film, television and theatre. Founded in 1972.
No unsolicited material.

Green Light Artist Management Inc.

1240 Bay Street, Suite 804, Toronto, Ontario, M5R 2A7
ⓣ 416 920 5110 ⓕ 416 920 4113

info@glam.on.ca
www.glam.on.ca

Represents people working across creative disciplines in the film and television industry, including directors, editors and writers.

Lowenbe Holdings

680 Indian Point Road, Glen Haven, Nova Scotia, B3Z 2TZ
℡ 902 823 1409
jmiller@loweba.ca
www.lowenbe.ca
President *Jan Miller*

Development consultant company founded in 1997. Specializes in drama and documentary projects for film and television. Also produces 'Pitcher Perfect' workshops presented at the Toronto International Film Festival, Hot Docs, the Banff Television Festival, the Atlantic Film Festival and training facilities across Canada.

Lucas Talent

7th Floor, 100 West Pender Street, Vancouver, British Columbia, V6B 1R8
℡ 604 685 0345 ℻ 604 685 0341
info@lucastalent.com
www.lucastalent.com
Literary Agent *Anna Archer*
Literary Agent *Doreen Holmes*

Agency for screen talent only. Founded in 1986. Handles all genres of writing; adults and children's, live action and animation.

Unsolicited material only accepted following introduction via email/phone. Copy read only when requested. Process outlined on website. Only represents writers from western Canada. 10% commission charged to clients.

Mensour Agency

41 Springfield Road, Ottawa, Ontario, K1M 1C8
www.mensour.ca

Roster of clients including writers working in the entertainment and advertising industries.

Meridian Artists Inc.

207-2 College Street, Toronto, Ontario, M5G 1K3
℡ 416 961 2777
info@meridianartists.com
www.meridianartists.com
Agent/Owner *Glenn Cockburn*
Associate *Andrea Glenn*
Assistant *Chris Paré*

Film and television literary agency founded in October 2005. Interested in all genres except documentary.

Welcomes unsolicited submissions via email. No reading fee. 10% commission charged to clients.

Pangaryk Productions, Inc.

Script Services Division, 14885 Alpine Boulevard, Hope, British Columbia, V0X 1L5
℡ 604 202 5400
scriptservices@pangarykproductions.com
www.pangarykproductions.com
President *Tihemme Gagnon*

Creative development and script services company for film and television projects. Provides story editing, script analysis, proofreading and pitch packaging services to production companies, agencies and independent screenwriters on a contractual basis.

Material accepted via emails. Non-disclosure agreements mandatory for all submissions.

Vanguarde Artists Management

119 Spadina Avenue, Suite 5012, Toronto, Ontario, M5V 2L1
info@vanguardeartists.com
www.vanguardeartists.com

A boutique literary film and television agency. Does not accept unsolicited submissions.

Westwood Creative Artists

94 Harbord Street, Toronto, Ontario, M5S 1G6
℡ 416 964 3302 ext. 22 ℻ 416 975 9209
wca-office@wcaltd.com
www.wcaltd.com

Literary agency with a dedicated film and television department.

Ireland

The Lisa Richards Agency

108 Upper Leeson Street, Dublin 4
℡ 01 637 5000 ℻ 01 667 1256
info@lisarichards.ie
www.lisarichards.ie
Contact *Faith O'Grady*

Founded in 1989. Accepts screen and stage plays. No reading fee. Authors should submit first 2–3 scenes, a short synopsis and an sae for return of material if necessary.

Australia

Cameron Creswell Agency

7th Floor, 61 Marlborough Street,
(Locked Bag 848), Surry Hills, NSW 2010
Ⓣ 02 9319 7199 Ⓕ 02 9319 6866
info@cameronsmanagement.com.au
www.cameronsmanagement.com.au

Founded in 1976, with clients including
screenwriters, directors, film editors, playwrights
and a distinguished list of authors.

HLA Management Pty Ltd

87 Pitt Street, Redfern, NSW 2016
Ⓣ 02 9310 4948 Ⓕ 02 9310 4113
hla@hlamgt.com.au
www.hlamanagement.com.au
Managing Director *Kate Richter*

Founded in 1973. Roster of 115 clients including
screenwriters, playwrights and directors. No
reading fee. Commission charged at 15%.

Prefers an initial approach prior to receiving
work for consideration.

Stacey Testro International Management

26a Dow Street, South Melbourne, VIC 3205
Ⓣ 03 9645 9181
melb@sti.com.au
www.sti.com.au

Literary and talent agency representing writers,
directors, producers, composers, directors of
photography and actors.

New Zealand

Johnson & Laird and Rachel Gardner Managements

PO Box 78340, Grey Lynn, Auckland 1002
Ⓣ 09 378 9800 Ⓕ 09 378 9801
info@rachelgardner.co.nz
www.rachelgardner.co.nz

Formed in 2002 representing actors, directors,
presenters, voice-over artists and writers.
Affiliated with agencies in Sydney, London
and LA.

Richards Literary Agency

PO Box 31 240, Milford, North Shore City 074
Ⓣ 09 410 0209 Ⓕ 09 410 0209
rla.richards@clear.net.nz
Partner/Agent *Ray Richards*

Small agency founded in 1977. Currently
representing over 100 clients including novelists
Maurice Gee and Joy Cowley. Not interested in
original screenplays. Only seeks adaptations of
published books.

No unsolicited manuscripts.

Scriptease

PO Box 10251, Dominion Road, Auckland,
jane@scriptease.info
www.scriptease.info

Established in 2000 to represent writers and
assist with script development, editing and
assessing.

Courses

... the screenwriter's relationship with the director is at the very heart of filmmaking. But the screenwriter must learn, and it is sometimes a painful lesson, that he is not an equal partner, indeed he is somewhat subservient. Once the film goes into preparation, the writer is distanced from the centre of things. Some writers, I am told, visit the set and watch the filming. I avoid it like the plague. I would rather check-in at an airport than watch a day's filming.

Ronald Harwood in 'The art of adaptation'

Courses

USA

American Film Institute (AFI)

2021 N. Western Avenue, Los Angeles,
CA 90027-1657

T 323 856 7600 F 323 467 4578
www.afi.com

The AFI Conservatory offers a 2-year Master of
Fine Arts degree. Topics covered in the first year
include: cycle production, screenwriting and
narrative workshops, writing the thriller, and
American approaches to film. In the second year,
issues covered include writing for television and
world approaches to film.

American University

School of Communication, 4400 Massachusetts
Avenue, NW, Washington, D.C. 200

T 202 885 2060 F 202 885 2019
communication@american.edu
www.soc.american.edu

Offers a two-week screenwriting workshop. The
course demonstrates how to develop the right
idea, create interesting characters and manage the
plot in the development of a short dramatic film.

Boston University

1 Sherborn Street, Boston, MA 02215

T 617 353 4636
script@bu.edu
www.bu.edu

The School of Communication offers a 2-year
Master of Fine Arts degree in screenwriting. The
course consists of 64 credit hours of an intensive
combination of writing and film and television
classes. Students are required to write at least
three original feature-length screenplays.

Brooklyn College

2900 Bedford Avenue, 0314 Plaza Building,
Brooklyn, NY 11210

T 718 951 5664 F 718 951 4733
film@brooklyn.cuny.edu
depthome.brooklyn.cuny.edu

The Department offers an undergraduate
Bachelor of Arts in Film with a concentration in
film production, film studies, screenwriting or
film marketing. Also offers a 2-year Certificate
Programme in Film with a concentration in film
production or screenwriting.

California Institute of the Arts

24700 McBean Parkway, Valencia, CA 91355

T 661 255 1050
www.calarts.edu

Offers several courses with screenwriting
elements including MFA programmes in Writing
for Performance, Film Directing and Writing.

Chapman University

Lawrence & Kristina Dodge College of Film
& Media Arts, 1 University Drive, Orange,
CA 92866

T 714 997 6765
dodgecollege@chapman.edu
www.ftv.chapman.edu
Associate Professor and Chair *Joseph Slowensky*

Graduate courses on screenwriting offered at
the Dodge College of Film and Media Arts'
specialist Conservatory of Motion Pictures.
Course includes tuition on writing for both film
and television. The programme is led by David S.
Ward, writer of *The Sting*.

Graduate programme is open to anyone
with a bachelor's degree. Screenwriting course
costs approximately $41,000 for 2 years.

Cogswell Polytechnical College

1175 Bordeaux Drive, Sunnyvale, CA 94089

T 408 541 0100 F 408 747 0764
info@cogswell.edu
www.cogswell.edu

Offers courses in producing motion picture
ideas, scriptwriting and advanced scriptwriting.
The scriptwriting course demonstrates the
fundamentals of writing a script for animation,
television, commercials, films, and digital
games, whilst the advanced course explores
the manipulation of time and space, in-depth

application of conflict and resolution, and the application of alternative writing formats.

Emerson College

Department of Writing Literature & Publishing, 120 Boylston Street, Boston, MA 02116-4624
☎ 617 824 8500
www.emerson.edu

Offers a Master of Fine Arts in creative writing. Courses aimed at students interested in pursuing careers in writing fiction, poetry, nonfiction, plays or screenplays.

Florida State University Film School

University Center, 3100A Tallahassee, FL 32306-2350
☎ 850 644 7728 ⊞ 850 644 2626
mfainfo@film.fsu.edu
film.fsu.edu

Offers a Master of Fine Arts in professional writing. The programme covers screen- and playwriting, screenplay craft, story conceptualization, and adaptation of a novel for both stage and screen.

Hollins University

PO Box 9647, Roanoke, Virginia, VA 24020
☎ 540 362 6308
hugrad@hollins.edu
www.hollins.edu
Director, MA in Screenwriting and Film Studies
Klaus Phillips

Founded in 1842. Offers range of film and scriptwriting courses including: Screenwriting; Advanced Screenwriting; Narrative Theory and Practice for Screenwriters; Writing Short Scripts; Television Sitcom Writing; Incorporating Folklore and Myths into your Screenplays.

Applications invited from students who can demonstrate their suitability for the courses by submission of undergraduate transcripts, writing samples and letters of recommendation. See website for details.

Johns Hopkins University

1717 Massachusetts Avenue, Suite 101, Washington, D.C. 200
☎ 202 452 1940
aapadmissions@jhu.edu
advanced.jhu.edu

Offers an elective Film and Screenwriting course as part of the Master of Arts programme. Students view and analyze classic films, such as *Chinatown* and *Psycho*, that illustrate screenwriting techniques, and use a successful major film script as text to examine storyline and structure from concept to synopsis. The focus is on dialogue, characterization, plot development, pacing, text and subtext, and visual storytelling.

Institute of Screenwriting

1754 Cadiz Road, Wintersville, OH 43953
☎ 888 659 3897
www.inst.org

Founded in 1998, the Institute offers a Screenwriting Diploma Course accredited by the Open and Distance Learning Quality Council.

Los Angeles Film School

6363 Sunset Boulevard, Hollywood, CA 90028
☎ 323 860 0789
info@lafilm.com
www.lafilm.com

Offers a 1-year Immersion Filmmaking Programme, which includes a course on screenwriting.

Louisiana State University

The Department of English, 260 Allen Hall, Baton Rouge, LA 70803
☎ 225 578 4086 ⊞ 225 578 4129
www.english.lsu.edu

The department of English offers a Master of Fine Arts in creative writing, which includes courses on screenwriting. Students usually complete the degree within 2½ to 3 years.

Loyola Marymount University

Film Studies, 1 LMU Drive, Los Angeles, CA 90045-2659
☎ 310 338 3033
SFTV-INFO@lmu.edu
www.lmu.edu

Offers a Bachelor of Arts in screenwriting, and a Master of Fine Arts in screenwriting. In the undergraduate programme, students are introduced to the basic elements of screenwriting, learning about character, dialogue, plotting, visual writing, and classic and alternative structures. In intermediate and advanced courses, students write and rewrite feature length screenplays, study genres, take classes in sitcom and dramatic television writing, and learn how to adapt stories to suit different mediums.

Middlebury College

Film/Video Studies, Wright Theatre, Middlebury, VT 05753
☎ 802 443 3190
fdrexel@middlebury.edu
www.middlebury.edu

Department of Film and Media Studies offers workshops in screenwriting. The first workshop is a creative writing course which explores key dramatic elements such as character, setting, point of view, plot, and theme set in a narrative medium. In the second workshop, students are exposed to the various purposes of film treatments, character backstories, synopses, sequence outlines, master scene scripts, and shooting scripts.

Minnesota State University

Department of Communication Studies, Film Studies and Theatre Arts, 1104 7th Avenue South, Moorhead, MN 56563
☎ 218 477 2126
borchers@mnstate.edu
www.mnstate.edu

The Department offers evening courses in writing for film and television as part of their film studies major programme.

Mount Holyoke College

Film Studies Program, Art Building, South Hadley, ME 01075
☎ 413 538 2200
filmstudies@mtholyoke.edu
www.mtholyoke.edu

Offers an intensive screenwriting course with an emphasis on structure and character, which prepares the student for the step outline of a feature-length film.

New York Film Academy

100 E. 17th Street New York, NY 10003
☎ 212 674 4300 🖷 212 477 1414
film@nyfa.com
www.nyfa.com

Offers a comprehensive range of screenwriting courses from an 8-week (or 12-week evening) workshop to a one-year programme. In addition to writing classes, students also study film craft, acting, pitching, and cinema studies as they apply to screenwriting. They will also write, direct and edit a short digital film or scene from a feature script.

New York University

School of Continuing and Professional Studies Office of Admissions, 145 4th Avenue, Room 201, New York, NY 10003
☎ 212 998 7200
scps.info@nyu.edu
www.scps.nyu.edu

Offers wide range of screenwriting certificates including: Screenwriting I: An Introduction;

Screenwriting II: A Workshop; Master Class for Screenwriting Certificate Students; Adapting True Stories Into Screen Stories; Advanced Storyboarding and Storytelling; Marketing Your Screenplay; Writing a Screenplay in 10 Weeks; Rewriting a Screenplay in 10 Weeks; Screenwriting Technique, Theory, and Practice; The Screenwriter's Craft; Writing for Prime-Time Television; Writing the Sundance Movie.

North Carolina School of the Arts

1533 South Main Street, Winston-Salem, NC 27127-2188
☎ 336 770 3399
admissions@ncarts.edu
www.ncarts.edu

Offers a 4-year Bachelor of Fine Arts degree with the option to specialize in screenwriting. In the first year the course covers topics such as character, conflict, visual storytelling, setting, dialogue, emotional tone, stage directions, and professional format. In the second year students explore genre, climaxes and resolutions, the controlling idea and an introduction to traditional three-act structure.

Ohio University School of Film

Lindley Hall, 378 Athens, OH 45701
☎ 740 593 1323 🖷 740 593 1328
lagraff@ohio.edu
www.finearts.ohio.edu

Offers a Master of Fine Arts degree with courses in screenwriting. The focus is on the narrative screenplay.

Pennsylvania State University

Department of Film/Video and Media Studies, 204 Carnegie Building, University Park, PA 16802
☎ 814 865 1503
jrp5@psu.edu
www.comm.psu.edu

Runs an undergraduate programme which has film-video as a major. It involves the in-depth study of film and video production and incorporates idea development, writing, production craft and production management.

Purchase College

State University of New York Conservatory of Theatre Arts & Film, 735 Anderson Hill Road, Purchase, NY 10577
☎ 914 251 6830
www.purchase.edu

Offers a dramatic writing course as part of the Bachelor of Fine Arts programme. The introduction to screenwriting course involves exercises in writing short stories and scenes.

Regent University

School of Communication and the Arts, 1000 Regent University Drive, Virginia Beach, VA 23464
T 888 777 7729 F 757 226 4394
www.regent.edu

Offers a Master of Fine Arts with the option to major in script and screenwriting. Some of the courses covered include story structure for stage and screen, playwriting, writing for television, feature film scriptwriting, and writing Christian drama for stage and screen.

San Diego State University

5500 Campanile Drive, San Diego, CA 92182-4561
T 619 594 1375 F 619 594 1391
kzaccari@mail.sdsu.edu
www.sdsu.edu

Offers an undergraduate BS degree in Television, Film, and New Media Production (TFM), with courses in screenwriting.

San Francisco State University

1600 Holloway Avenue, San Francisco, CA 94132
T 415 338 1111
cinedept@sfsu.edu
www.cinema.sfsu.edu

Offers a Bachelor of Arts degree in cinema which includes courses in screenwriting. The course involves practice in film writing, emphasizing story and plot dynamics, characterization, narration, dialogue and script forms.

Columbia University School of the Arts

305 Dodge Hall, Mail Code 1808, 2960 Broadway, New York, NY 10027
T 212 854 2134
admissions-arts@columbia.edu
wwwapp.cc.columbia.edu

The Film division offers a Master of Fine Arts degree with concentrations in screenwriting, producing and directing. Aside from the courses in screenwriting, they also offer a course on writing for television which concentrates on dramas primarily based on non-fictional material.

Scottsdale Community College

9000 E. Chaparral Road, Scottsdale, AZ 85256
T 480 423 6000
kate.herbert@sccmail.maricopa.edu
www.scottsdalecc.edu

The Motion Picture/Television department offers Certificates of Completion and Associate

of Applied Science degrees, with the option to specialize in screenwriting (2-year programme). It also offers classes in the hour-long television drama, sitcom writing, and short script writing which includes writing training films, promotional pieces, 30 and 60 second advertising spots, infomercials and other corporate materials.

Syracuse University

College of Visual and Performing Arts, 200 Crouse College, Syracuse, NY 13244-1100
T 315 443 2769
admissu@syr.edu
www.syr.edu

Department of Transmedia offers a 2-year Master of Fine Art in Film, with courses in scriptwriting. The introductory scriptwriting course explores the basic elements of film scripting: dramatic fundamentals, screenplay format, narrative strategies, character creation and dialogue development.

Temple University

School of Communications & Theater Room, 210 Annenberg Hall, 2020 North 13th Street, Philadelphia, PA 19122
T 215 204 3859
dlannon@vm.temple.edu
www.temple.edu

The school offers a Bachelor of Arts in Film and Media Arts. Includes a seminar- and workshop-based 'writing for media' course. Other elements include scene analysis for writers and directors.

UCLA Extension Writers' Program

10995 LeConte Avenue, Suite 440, Los Angeles, CA 90024
T 310 825 9415 F 310 206 7382
writers@uclaextension.edu
uclaextension.edu/writers
Program Representative in Screenwriting
Leigh-Michil George
Program Assistant in Screenwriting *Shannon Corder*

Programme offers almost 150 onsite and online film and television courses each year. Also certificates in feature film and television, an annual Writers' Studio, a screenplay competition and a nine-month 'masterclass'. Alumni include Iris Yamashita (Oscar nominee for *Letters from Iwo Jima*) and Gavin Hood (Oscar winner for *Tsotsi*).

Open admission policy. Courses cost between $95 and $535.

University of California (Berkeley)

Program in Film Studies, 7408 Dwinelle Hall, Suite 2670, Berkeley, CA 94720
☎ 510 642 1415 ℱ 510 642 8881
rfa@berkeley.edu
filmstudies.berkeley.edu

Offers both undergraduate and postgraduate degrees in films studies, with courses in screenwriting.

University of California (Los Angeles)–UCLA

UCLA School of Theatre Film & TV, 102 East Melnitz Hall, Box 951622, Los Angeles, CA 90095-1622
☎ 310 825 5761 ℱ 310 825 3383
info@tft.ucla.edu
www.tft.ucla.edu
Chairman *Richard Walter*
Co-chairman *Hal Ackerman*

Offers courses on screenwriting fundamentals, advanced screenwriting and creating TV pilots. Also runs a professional screenwriting programme. Alumni have been involved in films including *Sideways*, *Little Miss Sunshine*, *War of the Worlds* and *Forrest Gump*.

Applicants should have an undergraduate degree and provide writing samples. Applications 'should be made carefully'.

University of California (Santa Barbara)

Santa Barbara, 1720 Ellison Hall, Santa Barbara, CA 93106-4010
☎ 805 893 2347 ℱ 805 893 8630
admin@filmandmedia.ucsb.edu
www.filmandmedia.ucsb.edu

Offers both undergraduate and graduate Arts degrees in Film and Media Studies. Undergraduate students interested in screenwriting are advised to take on additional screenwriting courses. The graduate course allows students the opportunity to explore the creative process in autobiographical screenplay construction.

University of Central Florida

School of Film & Digital Media, PO Box 163120, Orlando, FL 32816-3120
☎ 407 823 4285 ℱ 407 823 3659
film@mail.ucf.edu
film.ucf.edu

Offers a Bachelor of Fine Arts degree in production. Students take courses in Directing, Editing, Screenwriting and Production Management.

University of Iowa

425 English-Philosophy Building, Iowa City, IA 52242
☎ 319 335 0330
cinema-complit@uiowa.edu

Offers both undergraduate and graduate degrees in film studies. Also short-form and long-form screenwriting courses. In the short-form course, students are involved in exercises and projects in writing, developing, and work-shopping screenplays for short film or video, as well as budgeting, location scouting, and other preproduction activities. The long-form course deals with topics such as visualization, sequencing, dialogue, preparation of treatment, screenplay for fiction film, and script problems.

University of Kansas

The Department of Theatre and Film, 356 Murphy Hall, Lawrence, KS 66045-3102
☎ 785 864 3511
kuthf@ku.edu
www2.ku.edu

Offers undergraduate and graduate degrees in theatre and film studies. In the undergraduate basic screenwriting course the emphasis is on the creation of a treatment and a screenplay, whilst the intermediate course explores genre, character, dialogue, and the development of a personal writing style. In the graduate programme students are exposed to dramatic scriptwriting and explore the problems with basic screenwriting.

University of Miami

Frances L. Wolfson Communication Building 5100 Brunson Drive, Coral Gables, FL 33146
☎ 305 284 2265
communication@miami.edu
www.miami.edu/com

Offers both undergraduate and graduate degrees in their Motion Picture programme. In the undergraduate programme, students are given an introduction to scriptwriting, which involves the creation and formatting of narrative material for motion pictures. Also offers a Master of Fine Arts in screenwriting.

University of Michigan

Department of Screen Arts & Culture, 6525 Haven Hall, 505 S. State Street, Ann Arbor, MI 48109-1045
☎ 734 766 0147 ℱ 734 936 1846
mlouisa@umich.edu
filmvid.info@umich.edu
Program Co-ordinator *Mary Lou Chlipala*

Offering undergraduate screenwriting courses taught in the Department of Screen Arts and Culture. Courses are taught by working screenwriters. Notable alumni include Arthur Miller, Jim Burnstein, Meg Kasdan and Richard Friedenberg.

Tuition for in-state pupils is $9,724 per semester, $29,132 out of state.

University of Nevada (Las Vegas)–UNLV

Film Department, 4505 Maryland Pkwy.,
Las Vegas, NV 89154-5015
Ⓣ 702 895 3547
gradcollege@unlv.edu
www.unlv.edu

Offer a 3-year Master of Fine Arts degree in screenwriting. The degree focuses specifically on the art and craft of writing for screen.

University of New Orleans

Department of Film Theatre & Communications, 2000 Lakeshore Drive, New Orleans, LA 70148
Ⓣ 504 280 6317
www.uno.edu

Offers a Master of Fine Arts degree in film production, with the option to specialize in creative writing (including fiction writing, non-fiction writing, playwriting, poetry writing and screenwriting).

University of North Carolina at Greensboro

Department of Broadcasting and Cinema, 210 Brown Building UNCG, PO Box 26170, Greensboro, NC 27402-6170
Ⓣ 336 334 5360 Ⓕ 336 334 5039
www.uncg.edu

Offers a Master of Fine Arts in drama with a concentration in film and video production, with the option to specialize in screenwriting. Aside from a screenwriting course, they also offer a media writing course which gives the student practice in television script-writing, with emphasis given to development of concepts and proposals for episodic television.

University of Oklahoma

640 Parrington Oval, Norman, OK 73019
Ⓣ 405 325 3020 Ⓕ 405 325 7135
fvs@ou.edu
www.ou.edu/fvs

The College of Arts and Sciences offers an interdisciplinary programme in film and video studies. The feature screenwriting course within the programme is an introduction to writing for the screen which includes a variety of assignments leading up to developing and writing a feature screenplay.

University of Southern California (USC)

USC School of Cinematic Arts, University Park, LUC 301, Los Angeles, CA 90089-2211
Ⓣ 213 740 3303 Ⓕ 213 740 8035
writing@cinema.usc.edu
cinema.usc.edu

The school offers both an undergraduate degree in Fine Arts and a Master of Fine Arts degree. Their philosophy in storytelling is a combination of compelling characters and three-act structure. Also teaches experimental narratives, television pilot writing and writing for growing media like mobisodes, internet series and gaming.

University of Texas

Austin Department of Radio-TV-Film CMA 6.118, Austin, TX 78712-1008
Ⓣ 512 471 4071 Ⓕ 512 471 4077
rtf-advising@austin.utexas.edu
rtf.utexas.edu

Offers a Master of Fine Arts degree in screenwriting. The objective of the 2-year programme is to provide a foundation of skills that will enable writers to achieve success either in features or television and in either Hollywood or independent arenas.

University of Utah

Division of Film Studies, 375 S 1530 E RM 257b, Salt Lake City, UT 84112-0380
Ⓣ 801 581 5127
info@film.utah.edu
www.film.utah.edu

Offers both a Bachelor of Arts and a Master of Fine Arts in Film Studies. The screenwriting course involves the development of a narrative screenplay, script format, character development, dramatic construction, dialogue and other storytelling skills.

Western Michigan University

School of Communication, Kalamazoo,
MI 49008
Ⓣ 269 387 3130 Ⓕ 269 387 3990
eric.mcconnell@wmich.edu
www.wmich.edu

Offers a Bachelor of Arts degree in Film, Video and Media studies, with courses in film and television scripting and playwriting.

Writers Guild of America East Foundation/ Columbia School of the Art Screenwriting Workshop

555 West, 57th Street, Suite 1230, New York, NY 10019

Ⓣ 212 767 7800 Ⓕ 212 582 1909
info@wgaeast.org
www.wgaeast.org

The workshop is an intense 10-week course. The 3-hour Saturday workshops at Columbia University are followed by one-on-one mentoring sessions. In addition, workshop participants attend master classes taught by renowned screenwriters.

UK

Arista Development

11 Wells Mews, London, W1T 3HD
Ⓣ 020 8341 2663
arista@aristotle.co.uk
www.aristadevelopment.co.uk

Created in 1996 by Stephen Cleary, then Head of Development at British Screen, to provide script development skills training for European-based development executives, producers and writers. Provides long term training for writers, such as the Adept 3 programme programme for MA screenwriter graduates, supported by Skillset's Film Skills Fund, The Media Programme of the EU and Arista's own Writer Scholarship Fund. Also offers an international consultancy service to create bespoke training around the globe.

Arvon Foundation

National Administration: 2nd Floor, 42A Buckingham Palace Road, London, SW1W 0RE
Ⓣ 020 7931 7611
london@arvonfoundation.org
www.arvonfoundation.org
National Director *Ariane Koek*

Founded 1968. Offers people of any age (over 16) and any background the opportunity to live and work with professional writers. Four-and-a-half-day residential courses are held throughout the year at Arvon's four centres, covering poetry, fiction, drama, writing for children, songwriting and the performing arts. Bursaries towards the cost of course fees are available for those on low incomes, the unemployed, students and pensioners. Runs a biennial international poetry competition.
Campuses at:
Devon: Totleigh Barton, Sheepwash, Beaworthy EX21 5NS; T 01409 231338; F 01409 231144;

E totleighbarton@arvonfoundation.org
Yorkshire: Lumb Bank, Heptonstall, Hebden Bridge HX7 6DF; T 01422 843714; F 01422 843714; E lumbbank@arvonfoundation.org
Invernessshire: Moniack Mhor, Teavarran, Kiltarlity, Beauly IV4 7HT; T 01463 741675; E moniacmhor@arvonfoundation.org
Shropshire: The Hurst, Clunton, Craven Arms SY7 0JA; T 01588 640658; F 01588 640509; E thehurst@arvonfoundation.org

Bath Spa University

Newton Park, Bath, BA2 9BN
Ⓣ 01225 875573 Ⓕ 01225 875503
r.kerridge@bathspa.ac.uk
www.bathspa.ac.uk

MA in Creative Writing. A course for creative writers wanting to develop their work. Teaching is by published writers in the novel, poetry, short stories and scriptwriting. In recent years, several students from this course have received contracts from publishers for novels, awards for poetry and short stories and have had work produced on BBC Radio.

Bournemouth University

The Media School, Weymouth House, Talbot Campus, Fern Barrow, Poole, BH12 5BB
Ⓣ 01202 966763 Ⓕ 01202 965530
media@bournemouth.ac.uk
media.bournemouth.ac.uk
Programme Administrator *Katrina King*

Three-year, full-time BA (Hons) course in Scriptwriting for Film and Television. Graduates emerge with the ability to write TV and film scripts across genres – movies, sitcoms, TV drama series, soaps, short comedies, TV screenplays, etc. The programme is designed to suit both school leavers and those of mature years wishing to prepare themselves for new career opportunities. Was the UK's first undergraduate screenwriting programme and is one of only two undergraduate screenwriting programmes in the UK to be accredited by Skillset.

Bridgend College

Cowbridge Road, Bridgend, CF31 3DF
Ⓣ 01656 302302 Ⓕ 01656 663912
enquiries@bridgend.ac.uk
www.bridgend.ac.uk

Runs course teaching scriptwriting for TV and film. Aims to develop skills in character analysis, narrative, storytelling and script layout. Areas of script writing covered include drama, factual and documentary. Costs £90 for 30 weeks teaching.

Burton Manor

The Village, Burton, Neston, CH64 5SJ
ⓣ 0151 336 5172 ⓕ 0151 336 6586
enquiry@burtonmanor.com
www.burtonmanor.com

Wide variety of short courses, residential and non-residential, on writing and literature. Full details in brochure.

Cardiff University Centre for Lifelong Learning

Senghennydd Road, Cardiff, CF24 4AG
ⓣ 029 2087 0000 ⓕ 029 2066 8935
learn@cardiff.ac.uk
www.cf.ac.uk
Coordinating Lecturer for English, Media and Creat *Ian Spring*

University founded by royal charter in 1883. Course availability detailed in *Choices* brochure, published twice yearly. Normally a range of 10-20 week courses in Media and Creative Writing. Approximately £65 for a 10-week course (reduced fees available).

The Central School of Speech and Drama

Central School of Speech and Drama, The Embassy Theatre, Eton Avenue, London, NW3 3HY
ⓣ 020 7722 8183
enquiries@cssd.ac.uk
www.csd.ac.uk
Course Leader *Dymphna Callery*

MA in Writing for Stage & Broadcast.

City Lit

Keeley Street, Covent Garden, London, WC2B 4BA
ⓣ 020 7492 2652 ⓕ 020 7492 8256
humanities@citylit.ac.uk
Head of School *Tabby Toussaint*

City Lit's Creative Writing section offers over 120 courses each year on a part-time basis. General screenwriting courses are taught at both basic and more advanced level. Also offers classes in more specialist areas such as: Writing comedy scripts; Scriptwriting for television; Making characters work; Screenwriting: the horror film; Practical sitcom writing; Screenwriting: the short film.

Most courses are open access, the main exception being the more advanced screenwriting courses, where basic knowledge and some experience—equivalent to taking a stage 1 course—is required. An 11-week evening course in 2007–8 comes in at around £90.

City University

Northampton Square, London, EC1 0HB
ⓣ 020 7040 8268 ⓕ 020 7040 8256
ell@city.ac.uk
www.city.ac.uk
Courses Co-ordinator, Writing and Journalism
Alison Burns

Courses for adults offered by the Department of Education and Lifelong Learning include Writing for Children, Writing Comedy, Writing Situation Comedy and Writing TV Drama. Open-access evening classes with expert tutors. Alumni include Catherine Tate and Rachel Zadok.

Community Creative Writing and Publishing

'Sea Winds' 2 St Helens Terrace, Spittal, Berwick-upon-Tweed, Northumberland, TO15 1RJ
ⓣ 01289 305213
mavismaureen@aol.com
Author/Tutor/Moderator *Maureen Raper*

Offering writing courses for all ability levels, including Writing for Radio and Television. Courses run for 10 weeks, with competitions and prizes available at the end.
Costs £2.50 per lesson.

Dartington College of Arts

Totnes, TQ9 6EJ
ⓣ 01803 862224 ⓕ 01803 861666
registry@dartington.ac.uk
www.dartington.ac.uk
Director of Writing *Mark Leahy*

BA (Hons) Writing or Writing (Contemporary Practices) or Writing (Scripted Media); Textual Practices minor award; MA Performance Writing: exploratory approaches to writing as it relates to performance, visual arts, sound arts and contemporary culture. Encourages the interdisciplinary, with minor awards and electives at BA level in arts and cultural management, choreography, music, theatre, visual performance.

De Montfort University

The Gateway, Leicester, LE1 9BH
ⓣ 0116 250 6199 ⓕ 0116 257 7196
huadmiss@dmu.ac.uk
www.dmu.ac.uk

MA offered in television scriptwriting. Offers direct links and networking opportunities within the industry by introducing the students to professional writers, script editors, agents and producers through a regular programme of guest lectures, workshops and location visits. By writing 'shadow' scripts of existing television shows, the students learn the disciplines of writing to a brief and experience the demands of writing for a popular soap opera or multi-episodic television drama. Guest writers have included Jimmy McGovern (*Cracker/Gunpowder Treason & Plot*), Peter Berry (*Silent Witness*), Ellie Barker (*Tracy Beaker*), Rob Gittins (*EastEnders*), Andy Hamilton (*Drop the Dead Donkey/Trevor's World of Sport*), Lizzie Mickery (*Messiah*), Tony Marchant (*Take Me Home/Mark of Cain*) and Sarah Bagshaw (*Emmerdale/Heartbeat/The Royal*).

Documentary Filmmakers Group

4th Floor, Shacklewell Studios, 28 Shacklewell Lane, London, E8 2EZ
℡ 020 7249 6600
info@dfgdocs.com
www.dfgdocs.com

A non-profit, comprehensive resource for documentary filming, established in 2001. Offers training and hosts screenings, festivals and forums.

Edge Hill University

St Helens Road, Ormskirk, Lancashire, L39 4QP
℡ 01695 575171 Ⓕ 01695 579997
daniele.pantano@edgehill.ac.uk
www.edgehill.ac.uk
Contact *Daniele Pantano*

Offers a two-year, part-time MA in Writing Studies. Combines advanced-level writers' workshops with closely related courses in the poetics of writing and contemporary writing in English. A full range of creative writing courses is available at undergraduate level, in poetry, fiction and scriptwriting which may be taken as part of a modular BA.

Jim Eldridge Writing Workshops

Pear Tree Farm, Bowness on Solway, Wigton, Cumbria, CA7 5AF
℡ 01697 352246
jimeldridge@jimeldridge.force9.co.uk
www.jimeldridge.com
Director and Workshop Tutor *Jim Eldridge*

Courses run by a professional screenwriter since 2004, showing how to write for film, TV and radio. Workshops deal with the business side as well as the creative side of writing.

Workshops take place twice a year, in spring and autumn. Costs £29 per workshop.

Euroscript

64 Hemingford Road, London, N1 1DB
℡ 07958 244656
ask@euroscript.co.uk
www.euroscript.co.uk

An independent, script development organization. Aims to facilitate the production of high-quality marketable screenplays for film and television via a process of on-going creative collaboration with writers and producers/directors. The Advanced Screenwriting Workshop costs £595 (concessions) and £950 for writer/producer or writer/director teams.

First Take

13 Hope Street, Liverpool, L37 7DD
℡ 0151 708 5767 Ⓕ 0151 709 2613
all@first-take.org
www.first-take.org
Director *Lynne Harwood*
Director *Jane Farley*
Director *Stephen Barr*

An independent television and film production and training organization. Offers courses such as Distinct Voices: Diverse Lives, a talent development scheme aimed at groups who are under-represented in the industry, and From Scratch to Screen, a short film writing course. Schemes include some of the scripts being produced.

Only open to writers living in the North West. Participation is free.

Goldsmiths, University of London

New Cross, London, SE14 6NW
℡ 020 7919 7060 Ⓕ 020 7919 7509
admissions@gold.ac.uk
www.goldsmiths.ac.uk

Founded in 1891, the institution offers masters courses in Script Writing, Filmmaking, Feature Film, Screen Documentary and Screen Studies. Script Writing course focuses on developing professional skills such as pitching for employment and production funding. During the programme students produce a short fiction script and a feature film or equivalent length TV or radio script.

Applicants to all MA courses should have a first degree or relevant professional experience.

Mark Grindle Associates Ltd

15 Gladstone Square, Ashfield, Dunblane, Perthshire, FK15 0JN
℡ 07515 526 828 Ⓕ 01786 201 067
mga@markgrindle.com
www.markgrindle.com
CEO *Mark Grindle*
MGA Ltd offers offers bespoke training for screenwriters who want to write and develop intellectual properties for the interactive entertainment industry. Training by people with genuine industry experience.

1/ The Gameplay: Screenwriting and Development for Interactive Entertainment. (Full 10 week and Bootcamp Versions available) Individual tuition also available on request. 2/ The Animation Finishing School: Creating and Defining Memorable Characters (Bringing Animators and Screenwriters closer together). 3/ Evolve! A course designed for Producers and Development Personnel from traditional media who want to develop or oversee the development of intellectual properties for interactive entertainment.

Initialize Films

15 Southampton Place, London, WC1V 6QA
℡ 0207 404 7020
info@initialize-films.co.uk
www.initialize-films.co.uk
Co-ordinator *Marion Simon*

Consultancy to producers and film funds that has run courses since 2005 on writing for TV, soaps and features, development money and marketing scripts. Also available are courses for developing new talent and for women screenwriters.

Anyone is eligible to apply. Fees from £70 upwards.

Institute of Screenwriting

Overbrook Business Centre, Poolbridge Road, Blackford, Wedmore, Somerset, BS28 4PA
℡ 0800 781 1715
screenwriting@inst.org
www.inst.org/screenwriting-course
Customer Services Co-ordinator
Alan Asbridge

Institution providing courses run by active screenwriters 'not armchair theorists'. Offering Distance Learning Diploma in Screenwriting. Accredited by the Open and Distance Learning Quality Council.

International Film School Wales

University of Wales Newport, Caerleon Campus, PO Box 101, Newport, NP18 3YG
℡ 01633 432432 ℻ 01633 432046
uic@newport.ac.uk
artschool.newport.ac.uk/filmschool.html

The leading institution for the promotion and development of the audiovisual culture of Wales, accredited as a Screen Academy by Skillset. Currently over 500 undergraduate and postgraduate students. Excellent links with industry. Many courses involve a screenwriting component.

Lancaster University

Department of English and Creative Writing
Bowland College, Lancaster, LA1 4YT
℡ 01524 594169 ℻ 01524 594247
L.Kellett@lancaster.ac.uk
www.lancs.ac.uk
Course Officer in Creative Writing *Lyn Kellett*

Department of English and Creative Writing offers both campus-based and distance learning masters degrees in creative writing.

First degree in any subject is required. Applicants should send samples of their writing along with application form.

Leeds Metropolitan University

School of Film Television & Performing Arts
H505 Civic Quarter, Calverley Street, Leeds, LS1 3HE
℡ 0113 812 3860
c.j.pugh@leedsmet.ac.uk
www.leedsmet.ac.uk
Administrator *Chris Pugh*

Offers a Diploma/MA in Screenwriting (Fiction). Based firmly upon the traditional crafts of storytelling and screen narrative the course emphasizes the development of the student's individual voice and creativity, with originality and innovation at the heart of the course. Graduates will have a thorough understanding of the creative and business processes involved in development, commissioning and writing for both television and film, and will have a strong portfolio of work, and contacts, to enable them to develop their careers in their chosen area/s. Skillset-accredited.

Lighthouse Arts and Training

1 Zone B, 28 Kensington Street, Brighton, East Sussex, BN1 4AJ
℡ 01273 647197
info@lighthouse.org.uk
www.lighthouse.org.uk
Administrator *Emma Bassett*

Founded in 1986, offering professional development courses for aspiring and practicing filmmakers, screenwriters, animators and artists working with digital and moving image media. Courses include Introduction to Screenwriting and The Craft of Writing TV Drama. Offers a wide range of initiatives such as production schemes, masterclasses, networking events and bursaries.

London College of Communication

School of Media, Elephant and Castle, London, SE1 6SB
℡ 0207 514 6858

k.marshall@lcc.arts.ac.uk
www.lcc.arts.ac.uk
Course Director *Kelly Marshall*
Course Administrator *Simeon Paskell*

Since 1992 the college has offered a part-time MA in screenwriting. Develops screenwriting skills through a series of lectures, seminars and tutorials. Students build up a portfolio of work, including a full-length film or television series. The college is part of the University of the Arts.

Requires a first degree or proven experience. Annual fees are £1,584 for EU students and £4,900 overseas.

London Film Academy

The Old Church, 52a Walham Grove, London, SW6 1QR
℡ 020 7386 7711 ℻ 020 7381 6116
info@londonfilmacademy.com
www.londonfilmacademy.com

A non-profit making trust, providing training in key areas of filmmaking. A member of the Skillset Media Academy Network.

1-day courses include "From Script To Screen - The Business of Film" and "Screenplay in a Day".

The London Film School

24 Shelton Street, Covent Garden, London, WC2H 9UB
℡ 020 7836 9642 ℻ 020 7497 3718
b.dunnigan@lfs.org.uk
www.lfs.org.uk
MA Screenwriting Course Leader *Brian Dunnigan*

Founded in 1956, offers Masters courses in Screenwriting and Filmmaking. One-year screenwriting course includes one-to-one mentoring with industry professionals and tuition in small groups. Alumni include Michael Mann, Mike Leigh, Tak Fujimoto and Roger Pratt. Degree courses are validated by London Metropolitan University.

Metropolitan Film School

Ealing Studios, Ealing Green, London, W5 5EP
℡ 020 8280 9119 ℻ 020 8280 9111
info@metfilmschool.co.uk
www.metfilmschool.co.uk

Founded in 2003, courses are taught by a professional script development executive. Offers 6-month part-time Write a Feature Film course, in which students write a synopsis, treatment and first draft of a script.

Admission is open to all. Course costs £1,400.

Middlesex University

Trent Park, Bramley Road, London, N14 4YZ
℡ 020 8411 5000
enquiries@mdx.ac.uk
www.mdx.ac.uk

The UK's longest established writing degree offers a Single or Joint Honours programme in Creative and Media Writing (full or part-time). This modular programme gives an opportunity to explore journalism, poetry, prose fiction and dramatic writing for a wide range of genres and audiences. Option for work experience in the media and publishing industries. Contact Admissions or David Rain (d.rain@mdx.ac.uk). MA in Writing (full-time, part-time; day and evening classes) includes writing workshops; critical seminars; lectures and workshops from established writers; introduction to agents and publishers. Options available: Fiction, Poetry or Scriptwriting. Contact: Sue Gee, 020 8411 5941 (s.gee@mdx.ac.uk). Also offers research degrees M.Phil/PhD in Creative Writing. Contact: Maggie Butt (m.butt@ mdx.ac.uk). The University has a thriving Writing Centre running an annual literary festival, weekly talks, community projects and writers in residence.

Moonstone International Screen Labs

67 George Street, Edinburgh, EH2 2JG
℡ 0131 220 2080 ℻ 0131 220 2081
info@moonstone.org.uk
www.moonstone.org.uk

Founded in 1997 in consultation with the Sundance Institute. Supports the work of European independent screenwriters and directors through a programme of advanced project-based development and training. Two Screenwriters' Labs held each year for a week each, giving participants the opportunity to develop their screenplays with the assistance of leading screenwriters.

The Labs are open to feature film writers and directors from throughout Europe. Applicants must have a proven extensive track record.

Morley College

61 Westminster Bridge Road, London, SE1 7HT
℡ 020 7928 8501 ℻ 020 7928 4074
enquiries@morleycollege.ac.uk
www.morleycollege.ac.uk
Head of Media *Mike Harris*

Offers a one-term screenwriting course. Aims to teach students how to develop feature script outlines and complete a script for a short film. Concentration on mainstream and independent world cinema, with an emphasis on narrative and story structure.

Stage 1 and Stage 2 courses. Stage 2 is a continuation and aims to help students develop their scripts beyond the outline and treatment stage.

For Stage 2 students should have completed a Stage 1 course or already have started a feature film script.

National Film and Television School

Beaconsfield Studios, Station Road, Beaconsfield, Buckinghamshire, HP9 1LG
☏ 01494 671234 Ⓕ 01494 674042
info@nfts.co.uk
www.nfts.co.uk
Head of Screenwriting *Corinne Cartier*
Screenwriting Co-ordinator *Chloe Davidson*

Founded in 1971, the Skillset Screen & Media Academy offers a 2-year Skillset-approved MA in Screenwriting and an 18-month part-time Diploma in Script Development. The masters course is based in a working film studio to familiarize students with the production process. All courses are 90% or more practical work. Alumni include Ashley Pharoah, Sandy Welch and Shawn Slovo.

Previous degree is not essential, but relevant experience is an advantage. Applicants should send a screenplay or another piece of writing plus a screenwriting sample. 'Dedication and a prolific output are expected'.

New York Film Academy

23 Southampton Place, London, WC1A 2BP
☏ 020 7430 2227 Ⓕ 020 7430 2772
filmuk@nyfa.com
www.nyfa.com

Screenwriting workshops presented by the US-based New York Film Academy. Students can develop an idea into a feature length screenplay in an eight-week course or enter a one-year programme.

Northern Visions Media Centre

23 Donegall Street, Belfast, BT1 2FF
☏ 028 9024 5495 Ⓕ 028 9032 6608
info@northernvisions.org
www.northernvisions.org

Self-contained two-day intensive script writing workshop, taught by filmmaker Laurence Henson.

Oxford Film and Video Makers

54 Catherine Street, Oxford, OX4 3AH
☏ 01865 792 732
office@ofvm.org
www.ofvm.org
Contact *Richard Duriez*

Founded in 1987 to offer screenwriting courses at introductory and advanced levels. Part-time 22-week programmes designed by the UK Film Council as a first step for aspiring talent. £225 per module. Applications online.

Performing Arts Labs

6 Flitcroft Street, London, WC2H 8DJ
☏ 020 7240 8040
screenwriters@pallabs.org
www.pallabs.org

PAL's screenwriters labs are among the longest-established screenwriters' programmes in the UK. Over 200 alumni since 1989.

Raindance Ltd

81 Berwick Street, London, W1F 8TW
☏ 020 7287 3833 Ⓕ 020 7439 2243
courses@raindance.co.uk
www.raindance.co.uk

Offers array of courses for aspiring filmmakers and writers. Taught by industry professionals and aims to give students an insight into the realities of the filmmaking world. Also offers script registration service.

Royal Holloway University of London

Media Arts Centre, Arts Building, Egham Hill, Egham, Surrey, TW20 0EX
☏ 01784 443734 Ⓕ 01784 443832
MediaArts@rhul.ac.uk
www.rhul.ac.uk/media-arts

Offers an MA Feature Film Screenwriting (MAFFS). Devised in 1999 with input from many people working in the film industry. Aimed at people already employed in the business, writers in other media with an interest in writing for film and those following an educational path. Part-time over two years or full time over one year.For those wishing to write for film and television, there is an MA in Screenwriting for TV and Film (Retreat Programme). Starts in January each year (two years part-time, taught in six one-week residential blocks. Two blocks held at writers' retreats in Yorkshire and Shropshire). Students work on film and television ideas, outlines, and then drafts for one of a variety of televisual or cinematic genres.

Screen Academy Scotland

Napier University, 2a Merchiston Avenue, Edinburgh, EH10 4NU
☏ 0131 455 2572 Ⓕ 0131 455 2538
info@screenacademyscotland.ac.uk
www.screenacademyscotland.ac.uk
Programme Leader *James Mavor*

Screen Academy Scotland is a collaboration between Napier University and Edinburgh College of Art (eca). Backed by Skillset, the UK Film Council Lottery Fund, the Scottish Executive, Scottish Screen, SMG and BBC Scotland.

Courses available: MA Screenwriting (full and part-time modes available, one or two years); PG Certificate in Screenwriting (online/distance learning over 30 weeks); MFA in Advanced Film Practice (one year); BA Honours in Photography and Film (undergraduate, four years); Introduction to Screenwriting (22 week, UKFC-approved, evening class); Intermediate Screenwriting (UKFC approved evening class). Various short courses. For the MA students study theoretical and craft elements of screenwriting for film, television and interactive screen industries. Links with industry and practising professionals integral to all courses.

Postgraduate courses require a previous degree though evidence of equivalent professional experience can be considered under the university's Recognition of Prior Learning (RPL) criteria.

Screenlab

17 Langland Gardens, London, NW3 6QE
Ⓣ 020 7435 1330
enquiries@screen-lab.co.uk
www.screen-lab.co.uk

Offers workshops and personal tuition to help students in all aspects of screenwriting.

The Script Factory

The Square, 61 Frith Street, London, W1D 3JL
Ⓣ 020 7851 4890
general@scriptfactory.co.uk
www.scriptfactory.co.uk

Established in 1996, The Script Factory is a screenwriter and script developer organization set up to bridge the gap between writers and the industry, and to promote excellence in screenwriting. Specializing in training, screenings, masterclasses and various services from its base in central London, The Script Factory operates throughout the UK and internationally. For the best source of information on upcoming activities, courses and latest news, and to join the free mailing list, check the website.

Signals Media Arts

Advanced Screenwriting, Victoria Chambers, St Runwald Street, Colchester, CO1 1HF
Ⓣ 01206 560255 Ⓕ 01206 369086
rachel@signals.org.uk
www.signals.org.uk

Open access course lasting ten weeks. Students work on a piece of their own work, from pitch to developed script. Some basic knowledge of screenwriting and understanding of writing for a visual medium is expected.

South Thames College

Wandsworth High Street, Wandsworth, London, SW18 2PP
Ⓣ 020 8918 7777
studentservices@south-thames.ac.uk
www.south-thames.ac.uk

Offers an introductory course teaching the skills and techniques needed to create an effective script for film and television. Focus is on feature films.

Southampton Solent University

East Park Terrace, Southampton, Hampshire, SO14 0YN
Ⓣ 023 8031 9653
fmas@solent.ac.uk
www.solent.ac.uk
Senior Lecturer *Mike Lynch*

Offers a BA in screenwriting, designed in collaboration with working screenwriters. Specializes in writing for all major screen formats, including drama documentary, soap opera, comedy and feature film.

Entry Level: 180 points, with at least 160 points from 6 or 12-unit awards. Particularly welcomes mature students, and those who may have relevant experience.

TAPS (Training and Performance Showcase)

Shepperton Studios, Studios Road, Shepperton, Middlesex, TW17 0QD
Ⓣ 01932 592 151 Ⓕ 01932 592 233
admin@tapsnet.org
www.tapsnet.org
Executive Director *Jill James*

Founded in 1993, offering workshops and masterclasses on writing for comedy, continuing drama and games, as well as script editing. Alumni have worked on series including *Coronation Street*, *EastEnders* and *The Bill*, and received commissions for their own original series ideas.

Thames Valley University

Ealing Campus, St Mary's Road, Ealing, London, W5 5RF
Ⓣ 0800 0368 888
learning.advice@tvu.ac.uk
www.tvu.ac.uk

MA in creative screenwriting. Designed to help writers find a personal voice as well as understand the requirements of the industry.

Applicants will typically have at least an upper second-class honours degree in Literature, Theatre or Media or related discipline, or an equivalent overseas qualification.

Ty Newydd Writers' Centre

Llanystumdwy, Cricieth, LL52 0LW
ⓣ 01766 522811 ⓕ 01766 523095
post@tynewydd.org
www.tynewydd.org

Residential writers' centre set up by the Taliesin Trust with the support of the Arts Council of Wales to encourage and promote writing in both English and Welsh. Most courses run from Monday evening to Saturday morning. Each course has two tutors and a maximum of 16 participants. A wide range of courses for all levels of experience. Early booking essential. Fees from £400 inclusive. People on low incomes may be eligible for a grant or bursary. Course leaflet available.

University College Falmouth

Woodlane, Falmouth, TR11 4RH
ⓣ 01326 211077
admissions@falmouth.ac.uk
www.falmouth.ac.uk

PgDip/MA professional writing programme. An intensive vocational writing programme developing skills in fiction, magazine journalism/features, and screenwriting. Students work on an extended writing project and form links with other PgDips such as television production and broadcast journalism.

University of Birmingham

Department of Drama and Theatre Arts
Edgbaston, Birmingham, B15 2TT
ⓣ 0121 414 3344 ⓕ 0121 414 3971
dramapg@contacts.bham.ac.uk

The MPhil in Playwriting Studies, established by playwright David Edgar in 1970, was the UK's first postgraduate course in playwriting. An intensive course which encourages students to think critically about dramatic writing, assisting them to put these insights into practice in their own plays.

University of Bolton

Bolton, BL3 5AB
ⓣ 01204 903237
amz1@bolton.ac.uk
www.bolton.ac.uk

Course Leader, Media, Writing and Production
Jenny Shepherd
Admissions Administrator *Nicola Dunn*

Founded in 2005, the university offers an undergraduate degree in Media, Writing and Production. Students may specialize in screenwriting, digital filmmaking or film studies strands. Creative Writing degree also offers screenwriting options. The best student scripts are produced by filmmaking students.

Applications should be made through UCAS (240 points required).

University of Central Lancashire

Preston, Lancashire, PR1 2HE
ⓣ 01772 201201
cenquiries@uclan.ac.uk
www.uclan.ac.uk

Offers a modular BA in screenwriting: Provides opportunity to study both screenwriting and film-making up until the second year.

University of Derby

Kedleston Road, Derby, DE22 1GB
ⓣ 01332 591736
adtenquiry@derby.ac.uk
www.derby.ac.uk

Offers a course in media writing, providing broad coverage. Options to specialize in scriptwriting for radio, television and film in later stages.

University of East Anglia

School of Literature and Creative Writing, Norwich, Norfolk, NR4 7TJ
ⓣ 01603 592154
j.camplin@uea.ac.uk
www.uea.ac.uk/cm/home/schools/hum/lit/Courses/Postgraduate/scriptwritingma
Admissions Assistant *Jack Camplin*

School of Literature and Creative Writing offers a masters degree in Creative Writing: Scriptwriting. Explores writing for stage, screen and radio through series of workshops and seminar modules, culminating in the writing of a full-length script. Classes are supplemented with regular masterclasses with industry professionals. Alumni include Ian McEwan, Tracy Chevalier and Bill Gallagher.

Applicants should hold a first degree or relevant experience. Courses cost £4,200 (full-time) or £2,100 (part-time) per year (2008).

University of Edinburgh

Graduate School of Literatures, Languages and Cultures, 19 George Square, Edinburgh, EH1 1JZ
ⓣ 0131 650 3068

rjamieso@staffmail.ed.ac.uk
www.ed.ac.uk
MSc Convenor *R Alan Jamieson*

Masters in Creative Writing. Aims to develop creative and reflective understanding of the chosen genre of writing via creative and literary-critical courses.

University of Exeter

Room 206, Department of English, Queen's Building, Queen's Drive, Exeter, EX4 4QH
℡ 01392 264265 🅕 01392 264361
english@ex.ac.uk
www.ex.ac.uk/english

Offers BA (Hons) in English with 2nd and 3rd year options in creative writing, poetry, short fiction, screenwriting and creative non-fiction. MA English; MA Creative Writing: poetry, novella, screenwriting, life writing and novel. PhD Creative Writing.

University of Glamorgan

Department of English, Treforest, Pontypridd, CF37 1DL
℡ 01443 668579
wmason@glam.ac.uk
www.glam.ac.uk
Contact *Wyn Mason*

MA in Scriptwriting (Theatre, Film, TV or Radio): a two-year part-time Masters degree for scriptwriters (held at the Cardiff School for Creative and Cultural Industries).

University of Hull

Scarborough Campus, Filey Road, Scarborough, YO11 3AZ
℡ 01723 362392 🅕 01723 370815
s.andrews@hull.ac.uk
www.hull.ac.uk
Director of Studies *Stuart Andrews*

BA Single Honours in Theatre and Performance Studies incorporates opportunities in writing and other media across each level of the programme. Works closely with the Stephen Joseph Theatre and its artistic director Alan Ayckbourn. The theatre sustains a policy for staging new writers. The campus hosts the annual National Student Drama Festival which includes the International Student Playscript Competition (details from The National Information Centre for Student Drama; nsdf@hull.ac.uk).

University of Leeds, Bretton Hall Campus

School of Performance and Cultural Industries, Leeds, LS2 9JT
℡ 0113 343 8710

enquiries-pci@leeds.ac.uk
www.leeds.ac.uk/paci
Course Director *Garry Lyons*

MA in Writing for Performance and Publication. This postgraduate programme (launched in September 2006) is particularly relevant to aspiring writers with professional ambitions, especially in the areas of theatre, film, television and radio drama, as well as the published novel and other culturally significant genres. Studies are offered over one-year full-time and two-years part-time and the tutors are established authors in their chosen fields. The course director is award-winning playwright and screenwriter Garry Lyons (*The Bill, The Worst Witch, Leah's Trials*). The School of Performance and Cultural Industries moved to a new £4 million theatre complex on the main Leeds campus in the autumn of 2007. Students are not only given the opportunity to work on their own writing projects but are also encouraged to collaborate with colleagues on other MAs such as Performance Studies, with a view to seeing their work staged.

University of Liverpool

Continuing Education, 126 Mount Pleasant, Liverpool, L69 3GR
℡ 0151 794 6900/6952 🅕 0151 794 2544
conted@liverpool.ac.uk
www.liv.ac.uk/conted
Course Organiser *John Redmond*

Courses include: Introduction to Creative Writing; Scriptwriting; Journalism; Writing for Children; Writing Your Life Story and a series of Saturday courses on aspects of writing. Most courses take place on one evening or daytime meeting weekly, with some Saturdays and linked days. Courses may be taken to pursue a personal interest or to gain university credit towards an award. CE offers a Certificate in Higher Education (Creative Writing) – 120 credits. For most courses no previous knowledge is required. There are fee concessions for those who are receiving certain benefits or are retired. Free prospectus on request or see the website.

University of Portsmouth

School of Creative Arts, Film and Media, Portsmouth, PO1 2EG
℡ 023 9284 5138 🅕 023 9284 5152
creative@port.ac.uk
www.port.ac.uk/departments/academic/scafm

Offers creative writing and combined creative writing degrees across a wide range of undergraduate programmes, and in all genres. BA, MA and PhD study is available. The School is a key partner in a national AHRC Research training project in Creative Writing. Staff include a number of national and/or international award-winning writers, short story writers/ novelists, playwrights, screenwriters, new-media writers and poets. The School runs regular national and international creative writing events.

University of Salford

Postgraduate Admissions, School of Media, Music & Performance, Adelphi Building, Peru Street, Salford, M3 6EQ
T 0161 295 6026 F 0161 295 6023
r.humphrey@salford.ac.uk
www.smmp.salford.ac.uk

MA in Television and Radio Scriptwriting. Two-year, part-time course taught by professional writers and producers. Also offers masterclasses with leading figures in the radio and television industry.

University of Sheffield

Institute for Lifelong Learning 196–198 West Street, Sheffield, S1 4ET
T 0114 222 7000 F 0114 222 7001
till@sheffield.ac.uk
www.shef.ac.uk/till

Certificate in Creative Writing (Degree Level 1) and a wide range of courses, from foundation level to specialist writing areas, open to all. Courses in poetry, journalism, scriptwriting, comedy, short story writing, travel writing. Brochures and information available from the address above.

University of Surrey, Roehampton

Erasmus House, Roehampton Lane, London, SW15 5PU
T 020 8392 3000
enquiries@roehampton.ac.uk
www.roehampton.ac.uk

Offers undergraduate and postgraduate degrees in creative writing, including components in screenwriting.

University of Wales, Bangor

Bangor, Gwynedd, LL57 2DG
T 01248 351151
postgraduate@bangor.ac.uk
www.bangor.ac.uk

Offers an MA in creative writing, allowing students to develop their specific interests in their preferred genre and style.

University of West of England

Faculty of Creative Arts, Bower Ashton Campus, Kennel Lodge Road, Bristol, BS3 2JT
T 0117 32 84716 F 0117 32 84745
eileen.elsey@uwe.ac.uk
www.uwe.ac.uk/amd
Programme Leader, MA Media *Eileen Elsey*

Faculty established as the Bristol School of Practical Art in 1853. Became Faculty of Creative Arts within UWE in 1992.

Offers MA in Media (Screenwriting) which seeks to teach a professional approach to screenwriting for short film, television and features, including how to pitch ideas to the industry. Full- or part-time.

Applicants should have skill and experience in creative writing and critical analysis. Will normally have an honours degree but applicants with other qualifications or professional experienced are encouraged to apply. Application forms available via website, phone or post. Applicants should also include a 10-page script. Course made up of four 30-point modules (£609 each) and one 60-credit module (£1218).

University of Westminster

Harrow Campus, Watford Road, Northwick Park, Harrow, Middlesex, HA1 3TP
T 020 7911 5903 F 020 7911 5955
harrow-admissions@wmin.ac.uk
www.wmin.ac.uk
Course Leader *Steve May*

Founded in 1838 as the Polytechnic Institution; showed the first public moving-picture show in 1896. Offering MA courses in Film and TV: Theory, Culture and Industry and Screenwriting and Producing for Film and TV. Also undergraduate courses on Contemporary Media Practice and Film and Television Production. Alumni include Michael Jackson (Channel 4 and BBC), Asif Kapadia and Seamus McGarvey.

Applications for undergraduate courses through UCAS; MA courses direct to the institution.

University of Winchester

Winchester, Hampshire, SO22 4NR
T 01962 841515 F 01962 842280
course.enquiries@winchester.ac.uk
www.winchester.ac.uk

The university was founded in 1840.

Offers BA courses in creative writing and media production and MA courses in writing for children and creative and critical writing.

Application criteria and other course-specific information available in the university prospectus.

Westminster Adult Education Service

Ebury Bridge Centre, Sutherland Street,
London, SW1V 4LH
☎ 020 7297 7297 🖷 020 7641 8140
info@waes.ac.uk
www.waes.ac.uk
Course Team Leader for Video and Make-up
Karen Fraser

Offering courses on scriptwriting for short and
feature films. Scriptwriting for Film incorporates
script analysis of short films, screenings,
group discussions and individual written
work. Scriptwriting for Feature Films offers an
introduction to screenwriting, examing structure,
character and genre.

Beginners and those with previous experience
are eligible.

The Writers Bureau

Sevendale House, 7 Dale Street, Manchester,
M1 1JB
☎ 0161 236 9440 🖷 0161 228 3533
advisory@writersbureau.com
www.writersbureau.com
Director of Studies *Diana Nadin*

Founded in 1989, offering a comprehensive
writing course covering scriptwriting for TV,
radio and theatre. Accredited by the Open and
Distance Learning Quality Council.

Admission is open to anyone. Courses cost £274.

Writers' Holiday at Caerleon

School Bungalow Church Road, Pontnewydd,
Cwmbran, NP44 1AT
☎ 01633 489438
gerry@writersholiday.net
www.writersholiday.net
Contact *Anne Hobbs*

Annual six-day comprehensive conference
including 12 courses for writers of all standards
held in the summer at the University of Wales'
Caerleon Campus. Courses, lectures, concert and
excursion all included in the fee. Private, single
and en-suite, full board accommodation. Courses
have included Writing for Publication and
Writing for the Radio.

Wye Valley Arts Centre

Hephzibah Gallery, Llandogo, NP25 4TW
☎ 01594 530214 🖷 01594 530321
wyeart@cwcom.net
www.wyeart.cwc.net

Courses (held at Hephzibah Gallery, Llandogo in
the Wye Valley and 'at our house in Cornwall')
include Creative Writing and Fiction Workshop.
All styles and abilities.

Yorkshire Art Circus

School Lane, Glasshoughton, Castleford,
Yorkshire, WF10 4QH
☎ 01977 550 401
admin@artcircus.org.uk
www.artcircus.org.uk
Manager of Core Activities *Lesley Wilkinson*
Project Officer *Beccy Stirrup*
Administrator *Angela Sibbit*

Community arts charity organizing creative
writing courses and a professional writer
development programme which offers
masterclasses, workshops and peer analysis. All
tutors are working writers and artists, and include
Bafta award winners and Orange Prize nominees.

Day courses in writing cost from £37; 63-hour
writers development programme costs £450.

Canada

The Canadian Screen Training Centre

61A York Street, Ottawa, Ontario, K1N 5T2
☎ 613 789 4720 🖷 613 789 4724
info@cstc.ca
www.cstc.ca

A non-profit organization dedicated to advancing
the development of the Canadian film, television
and new media industry. Provides intensive,
short-duration workshops to develop the skills
demanded by the screen industry Screenwriting
courses include: An Introduction to
Screenwriting; Polishing Your Craft: Developing
Your Script; The Next Big Thing: Writing for
Television; The Feature Screenplay: Writing for
the Big and Small Screen.

CFC (Canadian Film Centre)

Windfields 2489 Bayview Avenue, Toronto,
Ontario, M2L 1A8
☎ 416 445 1446 🖷 416 445 9481
www.cfccreates.com

Originally a film training centre, now offering
a wide range of programmes and funding
opportunities for practitioners of television, film
and new media. Also involved in production
across a wide range of media.

Praxis Centre for Screenwriters

School for the Contemporary Arts, Simon Fraser
University, Suite 3120, 515 W. Hastings Street,
Vancouver, British Columbia, V6B 5K3
☎ 604 268 7880 🖷 604 268 7882
praxis@sfu.ca
www.praxisfilm.com

Non-profit organization devoted to the professional development of Canadian screenwriters and filmmakers. As well as offering courses, the centre runs competitions and a scripts-for-option service.

University of Regina

Faculty of Fine Arts, Department of Media Production & Studies, Room ED243, Education Building, Regina, Saskatchewan, S4S 0A2
T 306 585 4796 F 306 585 4439
film@uregina.ca
www.uregina.ca/finearts

Offers a 4-year Bachelor of Fine Arts degree in Film and Video production, which includes course options on scriptwriting.

University of Toronto

School of Continuing Studies, 158 St George Street, Toronto, Ontario, M5S 2V8
T 416 978 2400 F 416 978 6666
learn@utoronto.ca
learn.utoronto.ca

Offers screenwriting courses which can be used towards the SCS Certificate(s) in Creative Writing. Students develop the practical skills necessary to write good screenplays, using a hands-on workshop format. Students are expected to participate in class discussions, produce work for each class and develop a good portion of a screenplay. Suitable for both beginners and more experienced screenwriters.

Vancouver Film School

2nd Floor, 198 West Hastings Street, Vancouver, British Columbia, V6B 1H2
T 604 685 5808 F 604 685 5830
www.vfs.com

The school offers a one-year comprehensive programme in screenwriting. Topics covered in the programme include writing a TV spec, comedy sketch, features, shorts and documentaries. The course also covers pitching and story editing.

York University (Graduate Program in Film)

224 Centre For Film And Theatre, 4700 Keele Street, Toronto, Ontario, M3J 1P3
T 416 736 5149
film@yorku.ca
www.yorku.ca

The Centre offers a Master of Fine Arts degree with the option to specialize in screenwriting.

Ireland

FÁS Screen Training Ireland

Adelaide Chambers, Peter Street, Dublin 8
T 01 483 0840 F 01 483 0842
film@fas.ie
www.screentrainingireland.ie

Aims to provide continuing training for professionals in film, television, animation and digital media. Runs a bursary award scheme.

Huston School of Film & Digital Media

University College Galway, Galway
T 091 495076
info@filmschool.ie
www.filmschool.ie

Based at the National University of Ireland, Galway, the Schools inaugural programme was an MA in screenwriting. Offers advanced training in screenplay writing integrated with courses in film history and analysis. Supported by regular film screenings and workshops. The School seeks to make a virtue of the West of Ireland's special position as a potential 'contact zone' between the disparate traditions of American and European cinema. Activities are intended to foster a critical awareness of both mainstream and alternative film traditions.

Bill Keating Centre

Milltown Park Sandford Road, Ranelagh, Dublin 6
T 01 219 6510
info@bilkeatingcentre.com
www.bilkeatingcentre.com

Provides comprehensive training and practical studio experience in all aspects of television, with an emphasis on presentation, production and direction. Also offers a programme development service.

National Film School

School of Creative Art, Dun Laoghaire Institute of Art Design and Technology, Kill Avenue, Dun Laoghaire, County Dublin
T 01 214 4600 F 01 214 4700
celine.blacow@iadt.ie
www.iadt.ie

The National Film School (NFS) at IADT was launched in 2003 as a centre of excellence for education and training in film, animation, broadcasting and digital media. Courses are full- or part-time. The MA in Screenwriting provides writers with the opportunity to learn the craft of screenwriting while producing a full-length (90-minute) screenplay. The programme is delivered over thirty weeks and study is supported by lectures and seminars from prominent writers, directors and producers.

Students are required to have an existing portfolio of work that is relevant to their chosen course. Applicants can either apply through the CAO (Central Application Office) system or directly to IADT. Application forms are available on the website.

National Youth Film School

St Joseph's Studios, Waterford Road, Kilkenny
T 056 776 4677 F 056 775 1504
info@yifm.com
www.yifm.com

Founded in 2001, a film school for young Irish people aged 13 to 20. A residential school for 20 film makers run over five weeks during the summer holidays.

Students applying for this programme should be able to show some previous involvement or interest in film making. Application forms available from website. Costs €800 (excluding accommodation).

UCD School of Film

Room 109, Arts Annexe, University College, Dublin, Belfield, Dublin 4
T 01 716 8301 F 01 716 8684
film.studies@ucd.ie
www.ucd.ie/film
Administrator *Rowena Kelly*

Conducts practical courses in film production and screen writing. Set up in 1992 on the initiative of actor Gregory Peck, film producer Noel Pearson, former UCD President Patrick Masterson and UCD Professor Richard Kearney. Aims to cater for the changing demands of an expanding Irish film industry. The School conducts film production and screen writing components on the MA degree programme on behalf of its academic sister body, the O'Kane Centre for Film Studies.

Early application is advised. Normal qualification for admission is a First Class Honours or Second Class Honours, Grade I Primary Degree. Applications from graduates in a wide range of disciplines and subjects encouraged. Candidates with other equivalent qualifications or experience also welcome to apply. Industry experience may be taken into account when assessing applications.

Writing Movies

T 01 239 0410
writingmovies@ireland.com
www.writingmovies.net

An intensive two-day course in screenwriting by award-winning Scottish film-maker Laurence Henson, given regularly in Dublin since 1998. Costs €195.

Australia

Australian College QED

PO Box 730, Bondi Junction, NSW 1355
T 02 9386 2500
studentservices@acq.edu.au
www.acq.edu.au

The school of Journalism, Media, and Photography offers a course on film and television that involves working on scripts, screenwriting and television production. Students learn how to write for television, film and video, develop their own screenplay and how to market it.

Australian Film, Television and Radio School

PO Box 2286, Strawberry Hills, NSW 2012
T 1300 131 461 F 02 9887 1030
info@aftrs.edu.au
www.aftrs.edu.au

Offers both full-time and short courses in screenwriting. In the graduate diploma in film and television, screenwriting students develop dramatic and visual storytelling craft skills through a series of production workshops and a three-day film shoot which develops short comedy, serious drama and genre-based scripts. Students in the Master of Arts degree write and co-produce series television drama, complete at least one industry-ready feature-length screenplay and/or a long- or short-form digital or interactive media project.

Bond University

University Drive, Robina, QLD 4226
T 07 5595 1024
information@bond.edu.au
www.bond.edu.au

Offers a 2-year full-time Bachelors degree in Film and Television with courses in screenwriting. Also offers a Master of Creative Arts in screenwriting.

Curtin University of Technology

GPO Box U1987, Perth, WA 6845
T 08 9266 9266 F 08 9266 2255
www.curtin.edu.au
Area Coordinator, Film & Television *Howard Worth*

Offers a 1-year Graduate Diploma in Film and Television with courses in screenwriting. The screenwriting courses focus on writing for documentaries and dramas. Also offers BA in

Film & Television with units in screenwriting for corporate documentary, interactive screen, independent documentary and drama.

Film and Television Institute

PO Box 579, Fremantle, WA 6959
T 08 9431 6700 F 08 9335 1283
fti@fti.asn.au
www.fti.asn.au

Offers an introductory course on the basic principles of writing short drama scripts. Over the four sessions, participants will write 3–5 minute scripts and explore character development, point of view, plot, the script editing process and other techniques.

Filmmaking Summer School at Melbourne University

605 Rathdowne Street, Carlton North, VIC 3054
T 03 9347 5035 F 03 9349 4443
info@summerfilmschool.com
www.summerfilmschool.com

Run since 1994 as part of the University's Summer programme, all courses are taught by film practitioners in their area of expertise. Specialized workshops on screenwriting, adaptation, writing for television, comedy writing and writing for children.
  Anyone can apply. Cost varies from daily rate of AUS$95 to discounted full course fees.

The International Film School Sydney

27 Rosebery Avenue, Rosebery, NSW 2018
T 02 9663 3789 F 02 9663 0816
Katrina@ifss.edu.au
www.ifss.edu.au

Offers a 2-year intensive filmmaking training course. In the first year, students attend the screenwriting programme, studying the moment-to-moment construction of cinema. Also offers a 6-day short course on screenwriting.

La Trobe University

VIC 3086
T 03 9479 2390 F 03 9479 3638
lhuss@latrobe.edu.au
www.latrobe.edu.au

Offers a Bachelor of Creative Arts degree with courses in screenwriting. Students learn various industry-accepted screenwriting formats used for fiction film and television drama, including film treatments and television 'bibles'.

Metro Screen

PO Box 299, Paddington, NSW 2021
T 02 9356 1818 F 02 9361 5320
training@metroscreen.org.au
www.metroscreen.org.au

Offers a 5-month full-time certificate IV in Screen, covering screenwriting, producing, directing, camera, lighting, sound and editing. Also offers short courses in screenwriting. The introductory course for beginners covers topics such as plot, structure, dialogue, character and industry standard formatting. The advanced course focuses on how to rewrite and progress through successive drafts of a script as well as how to ask for feedback and collaborate with other writers and filmmakers. Also offers specialized courses in structuring your screenplay to sell.

OPEN Channel

Victoria Harbour, Shed 4, North Wharf Road, Docklands, VIC 3008
T 03 8610 9300
info@openchannel.org.au
www.openchannel.org.au

Offers a short course on screenwriting where sessions cover script formatting, structure and narrative development. Also offers a scriptwriting workshop where learners develop and write a script over a four-month period under the support and guidance of industry professionals. Learners will be trained in script conventions, techniques and format.

Queensland School of Film & Television

PO Box 2378, Graceville, QLD 4075
T 07 3392 7788 F 07 3392 7511
www.qsft.qld.edu.au
Contact *Candida Paltiel*

Offers a short course in screenwriting, and covers topics such as script formatting, creating ideas, developing narrative, characters, and three-act structure.

RMIT University

GPO Box 2476V, Melbourne, VIC 3001
T 03 9925 2000 F 03 9663 2764
brendan.lee@rmit.edu.au
www.rmit.edu.au
Programme Administrator *Brendan Lee*

Offers Advanced Diploma of Arts in Professional Screenwriting, providing specialist training in writing for film, television and digital media across all genres. Aims to encourage students to develop their own creative strengths but also expose them to professional contexts and practices.

Mature applicants should have relevant experience or 'evidence of ability to meet the demands of the programme'.

Screenplay.com.au

PO Box 300, Artarmon, NSW 1570
℡ 02 8011 3533 ℻ 02 8915 1533
info@screenplay.com.au
www.screenplay.com.au
Contact *Jeff Bollow*

Established in 2000, offering a 2-day weekend course, plus Pro-Series: Down Under, a 6-month online workshop. Courses run by a production company seeking screenplays with local content to be produced locally.
    Costs AUS$350; Pro Series: Down Under AUS$1,299.

Sydney Film School

82 Cope Street, Waterloo, NSW 2017
℡ 02 9698 2244 ℻ 02 9698 2246
www.sydneyfilmschool.com

Offers a Diploma course in screenwriting, which covers topics such as the nature of a screenplay, formatting a script, creating characters and structuring a story.

University of Melbourne

Faculty of the Victorian College of the Arts, 234 St Kilda Road, Southbank, VIC 3006
℡ 03 9685 9300 ℻ 03 9682 1841
www.vca.unimelb.edu.au

Offers both undergraduate and postgraduate degrees in Film and Television with courses in Screenwriting. In the Masters degree the Screenwriting course involves the study of characterization and dramatic form at an advanced level. Each Masters student is expected to develop a script that they will take into production.

University of Technology, Sydney

PO Box 123, Broadway, Sydney, NSW 2007
℡ 02 9514 1593 ℻ 02 9514 2778
Margot.Nash@uts.edu.au
www.uts.edu.au
Senior Lecturer *Margot Nash*

Founded in 1988, offering 1-year Graduate Certificate in Screenwriting, as well as screenwriting streams in Masters writing courses. The 3-year Master of Arts: Writing course gives screenwriters the opportunity to focus on a major project under supervision.
    Application is open to anyone through the website.

New Zealand

Auckland University

Film Television and Media Studies, Private Bag 92019 Auckland Mail Centre, Auckland 1142
℡ 09 373 7599 ext. 87464 ℻ 09 373 8764
media@auckland.ac.nz
www.arts.auckland.ac.nz

Offers courses in film and TV production, including screenwriting elements.
    Must have A-levels.

Massey University

School of English & Media Studies, Private Bag 11 222, Palmerston North
℡ 06 356 9099 ext. 7311 ℻ 06 350 5672
g.l.slater@massey.ac.nz
ems.massey.ac.nz

Offers various undergraduate and graduate creative writing and performance papers including media script writing.

New Zealand Film Academy

PO Box 10-222, Dominion Road, Auckland 1030
℡ 09 374 5709 ℻ 09 309 8136
info@nzfilmacademy.com
www.nzfilmacademy.com

Offers range of 6- to 18-week courses underpinned by principles of education through experience and the importance of storytelling.

South Seas Film and Television School

PO Box 34-579, Birkenhead, Auckland 1310
℡ 09 444 3253 ℻ 09 444 7821
info@southseas.tv
www.southseas.co.nz

Leading film, television, onscreen acting and animation training institution.

Unitec New Zealand

School of Performing and Screen Arts, Private Bag 92025, Carrington Road, Mt Albert, Auckland
℡ 09 849 4180
www.unitec.ac.nz

Offers three-year full-time Bachelor of Performing and Screen Arts (Directing and Writing for Screen and Theatre) for those interested in a career as a writer or director for film, TV or theatre.

University of Waikato

Department of Screen and Media Studies, Private Bag 3105, Hamilton 3240

Ⓣ 07 838 4543 Ⓕ 07 838 4767
screenandmedia@waikato.ac.nz
www.waikato.ac.nz

Aims to provide students with the skills to work within the rapidly-evolving media environment. Includes modules on screenwriting.

Victoria University

IIML Victoria University, PO Box 600, Wellington
Ⓣ 04 463 6854 Ⓕ 04 463 6865
modernletters@vuw.ac.nz
www.victoria.ac.nz/modernletters

Taught from the International Institute of Modern Letters at Glenn Schaeffer House on the Kelburn Campus, offering internationally-respected creative writing programmes. Applications should be made directly to the International Institute of Modern Letters.

Whitireia Community Polytechnic

Private Bag 50 910, Porirua 5240
Ⓣ 0800 944 847 Ⓕ 04 237 3101
info@whitireia.ac.nz
www.whitireia.ac.nz

Yearlong online creative courses with options in scriptwriting.

Societies and Organizations

Stage scriptwriters have to remember that they are writing physically for a different space. It's a different time limit with greater spatial constraints. All in all it's actually pathetically limited, which is why writers are so important. They are the only people who can put colour into it. On-screen the writer is the last person you want putting colour into the proceedings because you've got Mr Stuntman, Mr Special Effects, Mr Music Composer et al.

James Christopher in 'From stage to screen'

Societies and Organizations

USA

Academy of Motion Picture Arts and Sciences

8949 Wilshire Boulevard, Beverly Hills, CA 90211
T 310 247 3000
www.oscars.org

Founded in 1927, the Academy is a professional honorary organization composed of over 6,500 filmmakers. The purpose of the organization is: to foster cooperation among creative leaders; recognize outstanding achievements; cooperate on technical research and improvement of methods and equipment; and provide a common forum and meeting ground for various film-related crafts. Hosts of the Oscars.

Academy of Television Arts & Sciences (ATAS)

5220 Lankershim Boulevard, North Hollywood, CA 91601
T 818 754 2800 F 818 761 2827
www.emmys.org

Founded in 1946, ATAS awards the Primetime Emmys and is the creator of the Television Hall of Fame, whilst its East Coast counterpart, the National Academy of Television Arts & Sciences (NATAS), distributes Emmy Awards in various categories including Daytime, Sports, News and Documentary and Public Service.

Membership can be either active status or associate status subject to requirements (see website for further details).

Alliance of Motion Picture & Television Producers

15301 Ventura Boulevard, Sherman Oaks, CA 91403
T 818 995 3600
www.amptp.org

Founded in 1982, the primary trade association with respect to labour issues in the motion picture and television industry. The association negotiates 80 industry-wide collective bargaining agreements on behalf of over 350 motion picture and television producers.

American Film Institute (AFI)

2021 N. Western Avenue, Los Angeles, CA 90027
T 323 856 7600 F 323 467 4578
www.afi.com

Founded in 1967, a non-profit organization established to train the next generation of filmmakers and to preserve the American film heritage. The AFI Conservatory trains filmmakers by focusing on hands-on experience with established figures, and maintains America's film heritage through the AFI Catalogue of Feature Films.

American Screenwriters Association

269 S. Beverly Drive, Suite 2600, Beverly Hills, CA 90212-3807
T 866 265 9091
asa@goasa.com
www.asascreenwriters.com

This non-profit organization provides writers with practical resources, networking opportunities, and services and programmes. It has an international membership of more than 40,000 members from 40 countries located throughout North America, Europe, the Pacific and the Middle East. The association works with many leading screenwriters such as Gary Ross (*Seabiscuit, Pleasantville, Big*), Aaron Sorkin (*The West Wing, An American President*) and Shane Black (*Lethal Weapon, The Last Boy Scout*). Also offers script registration service.

In order to become a member, a membership application must be completed and annual fees paid (there is no set criteria on becoming a member).

Atlanta Screenwriters Group

info@atlscript.org
www.atlscript.org/group.shtml

Founded in 1998 to help local screenwriters develop their ideas, swap industry stories and give constructive feedback.

Membership is open to all. An email should be sent to notify attendance prior to the meeting and an idea for discussion is mandatory (a complete script is not necessary).

The Authors Guild

31 East 32nd Street, 7th Floor, NY 10016
T 212 563 5904 F 212 564 5363
staff@authorsguild.org
www.authorsguild.org

Founded in 1912 by a group of writers, the Guild now represents more than 8,000 authors. Provides web services, seminars on subjects of professional interest and legal advice on contracts and publishing disputes.

Writers can qualify for 'Regular Membership' if they have been published by an established American publisher or had three works published in a periodical of general circulation in the last 18 months. 'Associate Membership' is available to those with a contract for a work not yet published.

CineStory

427 N. Canon Drive, Suite 202, Beverly Hills, CA 90210
T 310 968 0086
info@cinestory.org
www.cinestory.org

Founded in 1995, it has developed a 3-day Retreat where aspiring writers have the opportunity to have one-on-one meetings, informal hook-ups, and movie clip discussions with industry leaders. Every writer meets with three Retreat mentors each to discuss their script submissions and careers.

Click & Copyright

212 N. 3rd Avenue, Suite 570, Minneapolis, MN 55401
T 612 455 2290
customerservice@clickandcopyright.com
www.clickandcopyright.com

Founded in 1999, assisting independent artists to protect their work, by drafting and preparing the correct US Copyright Office forms. Copyright registration of a screenplay with the US Copyright Office protects the screenplay for the author's entire life and for a further 70 years, and it is also the only way to protect a screenplay from infringement in Federal Court.

Dramatists Guild of America

1501 Broadway, Suite 701, New York, NY 10036
T 212 398 9366 F 212 944 0420
jlevine@dramatistsguild.com
www.dramatistsguild.com

Established over 80 years ago, a professional organization for playwrights, composers and lyricists working in the American theatre market. Its business affairs division provides model contracts and agreements, and advice in negotiating contracts with theatres, collaborators, publishers and others.

There are three categories of membership. Membership as a student is available for those who are enrolled in a dramatic writing course. Membership as an Associate is open to those who have written at least one stage play and Active membership is available to those playwrights who have had at least one play produced in a Broadway, Off-Broadway, or LORT theatre.

Film Independent

9911 West Pico Boulevard, 11th Floor, Los Angeles, CA 90035
T 310 432 1200 F 310 432 1203
membership@filmindependent.org
www.filmindependent.org

Established over 25 years ago, a non-profit organization dedicated to independent film and independent filmmakers. Members have access to discounted equipment, editing suites, casting rooms, a valuable resource library, over 120 free screenings and 150 educational events every year. Film Independent also produces the Los Angeles Film Festival and the Independent Spirit Awards.

Membership costs a nominal fee and is open to all.

Independent Film & Television Alliance

10850 Wilshire Boulevard, 9th Floor, Los Angeles, CA 90024-4321
T 310 446 1000 F 310 446 1600
info@ifta-online.org
www.ifta-online.org

Founded in 1980 as the American Film Marketing Association, the trade association for the independent film and television industry worldwide. Its members include 160 companies from 22 countries, spanning production, distribution and financing of independent film and television programming.

International Game Developers Association (IGDA)

19 Mantua Road, Mt. Royal, NJ 08061
T 856 423 2990 F 856 423 3420
dawn@igda.org
www.igda.org
Administrative Assistant *Dawn Durkee*

A non-profit membership organization that aims to advance the careers and lives of game developers, including writers. Offers networking opportunities, professional development and advocates on issues affecting the developer community.

Northwest Screenwriters' Guild

4616 25th Avenue, NE PMB #439, Seattle, WA 98105
T 206 760 0462
illovox@comcast.net
www.nwsg.org

Dedicated to helping screenwriters in the Northwest further their craft and careers. Provides workshops and opportunities for direct interface with industry professionals.

Membership open to anyone. Those who have completed screenplays can apply for professional 'compendium' standing (see website for details).

Organization of Black Screenwriters

PO Box 347, North Hollywood, CA 91603
T 323 735 2050 F 323 735 2051
rjones@obswriter.com
www.obswriter.com

Founded in 1986 to address the lack of black writers represented in the entertainment industry. Provides members with intense training, hands-on experience, expansive resources and extensive industry exposure.

There are various levels of membership, including student, general (emerging writers) and professional (writers who have been accepted into a writing fellowship).

ProtectRite

Nat'l Creative Registry, 1106 Second Street, Encinitas, CA 92024
T 800 368 9748
info@protectrite.com
www.protectrite.com

Founded in 1994, a leader in online intellectual property registration. Establishes the completion date of registered materials and offers long-term storage if original is lost or destroyed.

Screenwriters Federation of America

4337 Marina City Drive, Suite 1141, Marina del Rey, CA 90292
www.screenwritersfederation.org

Formerly known as Screenwriters Guild of America, a federation of industry professionals including screenwriters, directors, producers, literary agents, managers, educators, authors and studio executives. Aims to assist screenwriters in personal and professional development, build a networking community for professional screenwriters, and to administer a Federation-sanctioned, industry-standard, free marketplace for greenlit scripts. Also offers a script registration service.

Sundance Institute

8530 Wilshire Boulevard, 3rd Floor, Beverly Hills, CA 90211-3114
T 310 360 1981 F 310 360 1969
Institute@sundance.org
www2.sundance.org

Founded in 1981 by Robert Redford, a non-profit organization which runs highly competitive programmes for emerging and aspiring producers, directors, writers and film composers. Runs a Screenwriters Lab focusing entirely on script and story development every January. Also owns and runs the annual Sundance Film Festival.

US Copyright Office

101 Independence Avenue, S.E., Washington, D.C. 205
T 202 707 3000
www.copyright.gov

Allows for copyright protection of manuscripts.

Writers Guild of America

East: 555 West 57th Street, Suite 1230, New York, NY 10019
T 212 767 7800 F 212 582 1909
info@wgaeast.org
www.wgaeast.org
West: 7000 Third Street, Los Angeles, CA 90048
T 323 951 4000 F 323 782 4800
www.wga.org

In 1951, the Writers Guild of America, East and the Writers Guild of America, West were formed to represent writers in motion pictures, television and radio. Since 1954 they have negotiated and administered minimum basic agreements with major film producers, networks and stations. Offers range of other services and support to writers. The total membership of the Guild (East and West) is approximately 11,000. It also offers script registration which protects the writers work for a period of 10 years.

Membership of the Guild works on a unit system basis. In order to be eligible for current membership a writer must acquire a minimum of 24 units in the three years preceding application. A writer may be eligible for Associate membership if he or she has had writing employment and/or sales within the Guild's jurisdiction and with a signatory company but has acquired less than 24 units in the three years preceding application. For full details regarding the unit system see organization's website.

UK

BECTU Script Registration Service

373-377 Clapham Road, London, SW9 9BT
☎ 020 7346 0900 🄵 020 7346 0901
info@bectu.org.uk
www.bectu.org.uk

Enables members to register scripts to provide proof that a script was in existence by a given date. Members send their script to BECTU Head Office. The union logs the script, seals it and returns it to the member concerned for safekeeping. Does not archive such scripts or keep copies. Submissions information available on the website.

Bristol Screenwriters

☎ 0782 515 7496
bristolscreenwriters@hotmail.co.uk
www.bristolscreenwriters.org

Membership organization that gives writers of film screenplays the opportunity to have their work read to an audience for feedback and comment. Also runs events that give writers the opportunity to explore different aspects of the film industry. Supported by Watershed Media Centre.

The British Academy of Film & Television Arts (BAFTA)

195 Piccadilly, London, W1J 9LN
☎ 020 77340022 🄵 020 77341792
info@bafta.org
www.bafta.org

Supports, develops and promotes the moving image industry, including film, TV and video games. Membership of 6,500 individuals professionals around the world. Hosts high-profile annual industry awards.

British Film Institute

Stephen Street Office/BFI National Library, 21 Stephen Street, London, W1T 1LN
☎ 020 7255 1444
www.bfi.org.uk

Founded in 1933 to promote understanding and appreciation of Britain's film and television heritage and culture. BFI Southbank (formerly the National Film Theatre) hosts over 1,000 screenings per year as well as many special events.

The BFI also runs the IMAX cinema on London's South Bank as well as publishing extensively on the industry, playing a key role in the Times BFI London Film Festival and the London Lesbian and Gay Film Festival, administering the national archive of the moving image and the national film library. It oversees a programme of cinema, video and DVD release and publishes *Sight & Sound*, a monthly film magazine.

Channel 4 British Documentary Film Foundation

PO Box 60415, London, E2 6WQ
help@britdoc.org
www.britdoc.org

Aims to promote and develop the British documentary by looking beyond television to develop, fund and distribute the work of the next generation of UK documentary filmmakers. Offers grants for short films, particularly by new filmmakers, feature-length projects with the potential to break through, experimental films, passion projects by established filmmakers, documentaries by artists from other mediums such as photography or art and ambitious development projects. Main fund is worth over £500,000 per annum to fund one-off documentary films. Also administers a competition for four emerging filmmakers to make totally unique three minute films for Channel 4, Studio Artois and the BRITDOC festival.

Easiest way to make contact is to sign up and explain how you want to work with the organization. Potential filmmakers can then be emailed when the Foundation is accepting submissions, with details of how to apply.

Club Panico@The London Film Academy

club@londonfilmacademy.com
www.londonfilmacademy.com/about/
ClubPanico_LFA.asp

A club for emerging filmmakers and an aftercare service financially subsidized by the LFA. Supports graduates by helping them find work, expanding their contacts and supporting their independent productions. Principally for LFA graduates but also open to those with at least one year's relevant professional experience.

To be considered for membership, email CV and two industry references.

Euroscript

64 Hemingford Road, London, N1 1DB
☎ 07958 244656
ask@euroscript.co.uk

www.euroscript.co.uk
Director *Anne Woods*

Founded in 1996 as part of the EU Media programme, the organization has been an independent development company since 2000. Aims to develop high quality screenplays with writers and producers from around the world. Offers weekend and residential workshops, individual consultancy and an annual screenplay competition.

No membership required, courses and services available at a variety of prices.

Federation of Screenwriters in Europe

C/O UNI-Europa, Box 9 31 rue de l'Hopital, Brussels, B-1000
manager@scenaristes.org
www.scenaristes.org
Chair *Christina Kallas*
Manager *Pyrrhus Mercouris*

A non-profit organization which aims to defend freedom of artistic expression and to promote the work and rights of scriptwriters. Represents 21 national guilds and 9,000 writers. Membership is open to any organization representing screenwriters in a European country. A yearly contribution is payable.

The European Screenwriters Manifesto

Spearheaded by the Federation of Screenwriters in Europe, the following manifesto was initially signed by 125 writers from 22 European countries at the European Conference on Screenwriting in November 2006. It was officially presented at the Berlin International Film Festival in February 2007, where the number of signatories passed 1,000. To add your name to the list, visit scenaristes.org.

Stories are at the heart of humanity and are the repository of our diverse cultural heritage. They are told, retold and reinterpreted for new times by storytellers. Screenwriters are the storytellers of our time. European writing talent should be trusted, encouraged and supported. The European film industries need to find ways to attract and keep its screenwriters in the cinema and in their craft.

We assert that:

1. The screenwriter is an author of the film, a primary creator of the audiovisual work.
2. The indiscriminate use of the possessory credit is unacceptable.
3. The moral rights of the screenwriter, especially the right to maintain the integrity of a work and to protect it from any distortion or misuse should be inalienable and should be fully honored in practice.
4. The screenwriter should receive fair payment for every form of exploitation of his work.
5. As author the screenwriter should be entitled to an involvement in the production process as well as in the promotion of the film and to be compensated for such work. As author he should be named in any publication accordingly, including festival catalogues, TV listing magazines and reviews.

We call on:

6. National governments and funding agencies to support screenwriters by focusing more energy and resources, whether in form of subsidy, tax breaks or investment schemes, on the development stage of film and television production and by funding writers directly.
7. Scholars and film critics to acknowledge the role of screenwriters, and universities, academies and training programmes to educate the next generations in accordance to the collaborative art of the medium and with respect towards the art and craft of screenwriting.
8. Festivals, film museums and other institutions to name the screenwriters in their programs and plan and screen film tributes to screenwriters just as they do to directors, actors and countries.
9. National and European law should acknowledge that the writer is an authorof the film.
10. National and European law should ensure that screenwriters can organise, negotiate and contract collectively, in order to encourage and maintain the distinct cultural identities of each country and to seek means to facilitate the free movement of writers in and between all nations.

We will: Distribute this manifesto to industry members and the press in our respective countries. Campaign for the implementation of the agenda defined by this manifesto. Seek the transition into national and European law of the legal changes demanded by this manifesto.

Film Agency for Wales

Suite 7, 33-35, West Bute Street, Cardiff, CF10 5LH
T 029 2046 7480 F 029 2046 7481
enquiries@filmagencywales.com
www.filmagencywales.com

The national development agency for the film industry in Wales, established in July 2006. Aims to increase the range and number of films being produced by new, emerging and established Welsh talent. A particular interest in supporting Welsh writers, writer-directors, directors and producers. Funding available.

Film Education

21-22 Poland Street, London, W1F 8QQ
T 020 7851 9450 F 020 7439 3218
mattp@filmeducation.org
www.filmeducation.org

Founded in 1985, a registered charity funded by the UK film industry and the BFI. Aims to meet the increasing demand by teachers for current educational material on film and film making. Organizes training courses, conferences, workshops, seminars and events including National Schools Film Week.

Film London

Suite 6, 10 The Tea Building, 56 Shoreditch High Street, London, E1 6JJ
T 020 7613 7676 F 020 7613 7677
info@filmlondon.org.uk
www.filmlondon.org.uk
Contact *Hilary Dugdale*

Promotes and develops London as an international filmmaking capital. Includes all film, television, commercials and new interactive media based in London. The organization has a long history of supporting film productions, from experimental work to narrative fiction, animation and documentaries. Funding awards include the Pulse Digital Shorts scheme, London Production Fund, Microwave Micro-Budget Feature Film Fund, Borough Production Funds and London Artists' Film and Video Awards.

First Light Movies

Unit 6, Third Floor, The Bond 180–182 Fazeley Street, Birmingham, B5 5SE
T 0121 753 4866
info@firstlightmovies.com
www.firstlightmovies.com
CEO *Pip Eldridge*
Marketing and Communications Manager *Clare Lewis*

Launched in 2001, First Light provides funding and expertise for 5- to 18-year-olds from all backgrounds to make short digital films with professional filmmakers. Distributes National Lottery money through UK Film Council funding. Enabled more than 10,000 young people to make more than 750 films.

Funding is given through an application process 3 times a year. There are 3 funding awards available worth a total of £25,000.

Isle of Man Film

Hamilton House, Peel Road, Douglas, Isle of Man, IM1 5EP
T 01624 687173 F 01624 687171
iomfilm@dti.gov.im
www.isleofmanfilm.com

Founded in 1995, the organization (formerly known as The Isle of Man Film Commission) promotes the activities of the island's film industry and actively markets the Isle of Man as a film location. Helped produce and finance over 80 productions to date. The Isle of Man Media Development Fund offers to assist films with up to 25% of their budget as direct equity investment if the project is largely filmed on the Isle of Man and benefits local service providers.

Full application details can be found on the organization's website.

New Producers Alliance

NPA Film Centre, 7.03 Tea Building, 56 Shoreditch, High Street, London, E1 6JJ
T 020 7613 0440 F 020 7729 1852
queries@npa.org.uk
www.npa.org.uk

Founded in 1993, a national membership organization offering training and networking opportunities to filmmakers.

New Writing North

Culture Lab Grand Assembly Rooms, Newcastle University, King's Walk, Newcastle-upon-Tyne, NE1 7RU
T 0191 222 1332 F 0191 222 1372
www.newwritingnorth.com

The writing development agency for the north-east of England. Works with writers across genres to develop career opportunities, new commissions, projects, residencies, publications and live events.

New Writing South

9 Jew Street, Brighton, BN1 1UT
T 01273 735353
chris@newwritingsouth.com
www.newwritingsouth.com
Administrator *Kata Gyongyosi*

New Writing South works to create an environment in the south east of England in which new creative writing can flourish. We offer practical resources to writers, build partnerships between writers and those who may produce their work, and encourage new writing opportunities.

Membership is open to all creative writers in the region. Annual subscription fee of £35 (concessions £20).

Northern Ireland Screen

Alfred House, 21 Alfred Street, Belfast, BT2 8ED
℡ 028 9023 2444 📠 028 9023 9918
info@northernirelandscreen.co.uk
www.northernirelandscreen.co.uk

Incorporated in 1997 as the Northern Ireland Film and Television Commission. Involved in many aspects of promotion and development of screen-based industries. Allocates various funds for the development and production of the moving image, including feature film, shorts, animation, documentaries, television drama series and new media.

Northern Screenwriters

info@northernscreenwriters.co.uk
www.northernscreenwriters.co.uk

A non-profit organization based in Manchester. Membership of film-makers and scriptwriters based in the North West. Core activities are a script development programme, script workshops and groups and a searchable script database.

PACT (Producers Alliance for Cinema and Television)

Procter House, 1 Procter Street, Holborn, London, WC1V 6DW
℡ 020 7067 4367 📠 020 7067 4377
www.pact.co.uk

Trade association representing the commercial interests of independent feature film, television, animation and interactive media companies. Also has an office in Glasgow.

Player–Playwrights

The Secretary, 9 Hillfield Park, London, N10 3QT
playerplaywrights@groups.msn.com
groups.msn.com/playerplaywrights

Established in 1948 as a writers' co-operative. Meets upstairs at the Horse and Groom in Great Portland Street, W1. Membership of 100–150, mostly writers or aspiring writers but also actors (professional and amateur) who perform members' works. Current presidents are Laurence Marks and Maurice Gran.

Regional Screen Agencies

Nine self-governing regional development agencies to support and develop the media sectors throughout England and to promote public access to film culture. They are:

EM Media – East Midlands (35-37 St Mary's Gate, Nottingham, NG1 1PU. T: 0115 934 9090. F: 0115 950 0988. Web: www.em-media.org.uk. Email: info@em-media.org.uk)

Film London (Suite 6.10, The Tea Building, 56 Shoreditch High Street, London, E1 6JJ. T: 020 7613 7676. F: 020 7613 7677. Web: www.filmlondon.org. uk. Email: info@filmlondon.org.uk)

North West Vision (Manchester Film Office, Ground Floor, BBC New Broadcasting House, Oxford Road, Manchester, M60 1SJ. T: 0870 609 4481. Web: www.northwestvision.co.uk. Email: info@visionandmedia.co.uk)

Northern Film and Media (Central Square, Forth Street, Newcastle-upon-Tyne, NE1 3PJ. T: 0191 269 9200. F: 0191 269 9213. Web: www. northernmedia.org. Email: info@northernmedia. org)

Screen East (2 Millennium Plain, Norwich, NR2 1TF. T: 01603 776 920. F: 01603 767191. Web: www.screeneast.co.uk. Email: info@screeneast. co.uk)

Screen South (The Wedge, 75-81 Tontine Street, Folkestone, Kent, CT20 1JR. T: 01303 259777. F: 01303 259786. Web: www.screensouth.org. Email: info@screensouth.org)

Screen West Midlands (9 Regent Place, Birmingham, B1 3NJ. T: 0121 265 7120. F: 0121 265 7180. Web: www.screenwm.co.uk. Email: info@ screenwm.co.uk)

Screen Yorkshire (Studio 22, 46 The Calls, Leeds, LS2 7EY. T: 0113 294 4410. F: 0113 294 4989. Web: www.screenyorkshire. co.uk. Email: info@screenyorkshire.co.uk)

South West Screen (St Bartholomews Court, Lewins Mead, Bristol, BS1 5BT. T: 0117 952 9977. F: 0117 952 9988. Web: www.swscreen.co.uk. Email: info@swscreen.co.uk)

Rocliffe Forum

Rocliffe PO Box 37344, London, N1 8YB
scripts@rocliffe.com
www.rocliffe.com

A production company that runs New Film Forums and monthly New Writing Forums. Established by Farah Abushwesha in 2000 and run by Pippa Mitchell, Kerry Appleyard and Abushwesha. Now has over 6,000 members. The Writing Forum is a monthly platform for new writing and a networking event. Three 7–8 minute script extracts (any genre) are cast in-house and rehearsed by professional actors and directors on the evening. A narrator sets the piece in context and the extracts are then performed to an audience of producers, development executives, directors, actors and literary agents.

The writer then receives feedback and answers questions from the audience. Attendance is by reservation only.

Royal Television Society

5th Floor, Kildare House, 3 Dorset Rise, London, EC4Y 8EN
℡ 020 7822 2810 🅕 020 7822 2811
info@rts.org.uk
www.rts.org.uk
President *Sir Robert Phillis*

Founded in 1927. A leading forum for discussion and debate on all aspects of the television community, providing opportunities for making contact with people at all levels across the industry. Hosts a high-profile annual industry awards ceremony, as well as talks, lectures and conferences. Society archive (going back to its foundation) open to members and academic researchers by appointment.

Scottish Screen

249 West George Street, Glasgow, G2 4QE
℡ 0845 300 7300
info@scottishscreen.com
www.scottishscreen.com

Formed in 1997, the national screen agency for Scotland with responsibility for developing all aspects of screen culture and industry across the country. Acts as a government advisor as well as an advocate for the industry and a development agency.

The Script Factory

The Square, 61 Frith Street, London, W1D 3JL
℡ 020 7851 4890 🅕 020 7851 4858
general@scriptfactory.co.uk
www.scriptfactory.co.uk

Founded in 1996 and now one of Europe's leading development organizations for screenwriters. Aims to identify and develop new talent, to support the people who work with screenwriters and to present live screenwriting events. Has an in-house staff of ten plus a large pool of freelancers to deliver an annual training programme catering for in excess of 1,000 people per year. The board includes Daniel Battsek (president of Miramax Films), Andrew Cripps (president of United International Pictures), Mike Figgis, Anthony Minghella and Meera Syal. Also offers a script registration service.

The Script Vault

www.thescriptvault.com

Offers a script registration and deposit service that safeguards copyright by registering the author of a piece of work and also establishing the date it was written. Costs £10 per script for ten years.

SKILLSET

Focus Point, 21 Caledonian Road, London, N1 9GB
℡ 020 7713 9800
info@skillset.org
www.skillset.org

The sector skills council for the audio visual industries (broadcast, film, video, interactive media and photo imaging). Jointly funded by industry and government. Aims to ensure that the UK's audio-visual industries have people with the right skills, in the right place, at the right time. Conducts consultation work with industry, publishes research and strategic documents, runs funding schemes and project work, and provides information. Also provides careers advice for aspiring new entrants and established industry professionals.

The Society of Authors

84 Drayton Gardens, London, SW10 9SB
℡ 020 7373 6642 🅕 020 7373 5768
info@societyofauthors.org
www.societyofauthors.org
General Secretary *Mark Le Fanu*
Secretary, Broadcasting Group *Jo Hodder*

An independent trade union with more than 8,000 members. As well as campaigning for the profession, the society advises members on negotiations with publishers, agents and other organizations, and offers guidance on business aspects of their writing. Also assists with complaints and legal disputes. Among the society's publications are *The Author*, published quarterly, and the *Quick Guide* series on various aspects of writing (free to members). Groups within the organization for different genres and types of writing.

Membership is open to all published authors. Annual subscription costs £85, with some concessions available.

Southwest Scriptwriters

15 Quarry Steps, Clifton, Bristol, BS8 2UD
℡ 0117 909 5522
info@southwest-scriptwriters.co.uk
www.southwest-scriptwriters.co.uk
Secretary *John Colborn*
Artistic Director *Tim Massey*

Organization founded in 1994 for keen writers of stage, screen and radio. Members must be over 18. Workshop meetings take place weekly at Bristol Old Vic theatre. Discussion events with established professionals, rehearsed readings and occasional productions of new work also take place. Meetings cost £1 plus a £6 annual subscription fee.

Talent Circle

www.talentcircle.co.uk

Online network founded in 2003. Caters for the independent film industry including screenwriters, filmmakers or production staff. Provides access to jobs, industry news and resources online. Other activites include seminars, screenings and networking events. Free registration gives full access to the site.

UK Film Council

10 Little Portland Street, London, W1W 7JG
Ⓣ 020 7861 7861 Ⓕ 020 7861 7862
info@ukfilmcouncil.org.uk
www.ukfilmcouncil.org.uk

The government-backed strategic agency for film in the UK, aiming to promote the UK industry and increase enjoyment and understanding of it. Runs a development fund with £12 million over three years to support the development of high quality and commercially-viable screenplays. There is also a premiere fund with £24 million over three years to facilitate the production of more mainstream films. The new cinema fund has £15 million over three years to back innovative filmmakers, especially new talent, and to explore new production technologies. Its short film schemes have produced over 450 films.

UK MEDIA Desk

c/o UK Film Council, 10 Little Portland Street, London W1W 7JG
Ⓣ 020 7861 7511 Ⓕ 020 7861 7950
england@mediadesk.co.uk
www.mediadesk.co.uk
Director of UK MEDIA Desk *Agnieszka Moody*

MEDIA Programme Development Funding

The MEDIA Programme of the European Union offers support to cover the costs of development of European films, TV programmes and multimedia in the following categories:

- Fiction with minimum duration of 50 minutes
- Creative documentary with minimum duration of 25 minutes (one-off or series)
- Animation with minimum duration of 24 minutes (one-off or series)

Applications are accepted from film and television production companies with a track record. Individual writers cannot apply but their interests are safeguarded by the requirement that applicant companies must demonstrate that they have acquired the rights from writers by submitting a copy of relevant rights transfer agreement.

Eligible development costs:

- The acquisition of rights
- Research (including archive research)
- All scriptwriting up to and including the final draft (provided that it is paid before the first day of principal photography or pre-production)
- Storyboards
- Identification of the key cast and crew
- Preparation of the production budget and schedule
- Preparation of financing plan
- Identification of industry partners, co-producers and financiers
- Initial marketing and sales plans (attending markets and attracting buyers)

The schemes are published on an annual basis as the so-called "Calls for Proposals". One call usually has two deadlines. A typical budget for a Call for Proposals is around €20 million (for all 32 participating countries).

Funding is offered as a non-repayable grant of up to €60,000 per project. A producer must raise an equivalent amount of money to comply with a principle of 50–50% matching funding. Costs are eligible from the submission date, with the exception of those related to rights acquisition which can be backdated up to 12 months to the submission date.

There are two different schemes aimed at companies of different experience and financial capacity: single project and slate funding.

Single Project is aimed at companies with limited financial capacity registered for at least 12 months and wishing to invest in the development of just one project.

The company must be able to provide evidence that they have completed, as the majority producer, a previous work similar to the eligible projects described above. It must also show that this work has been distributed during the period between 1 January 2005 and the date of submission of the application.

Slate Funding is aimed as a company development grant for medium-sized companies with previous experience at an international level and the financial capacity to develop several projects simultaneously. The company must have been registered for at least 36 months before the submission date to be eligible for slate funding.

Within five years leading to the submission date, the company must have produced as a majority producer two projects in a similar eligible category and these projects must have had international distribution within the same period.

If the company is not eligible for funding because if the lack of the company's track record, then the individual track record of professionals who are part of the company can be used even if they produced the previous work for other companies or worked on a freelance basis. In this case the individual must at the time of submission be a shareholder or CEO/MD of the company or have been on the company's payroll for the last 12 months. On a previous (qualifying) production they must have a producer's, executive producer's or delegate producer's credit.

The UK has been using MEDIA development support with great success. Between 2001–07 over €14 million has been invested by the MEDIA Programme in UK-based companies. Examples of UK projects developed with MEDIA support include *The Other Boleyn Girl*, *It's a Free World* and *Love+Hate*.

The local UK MEDIA Desk is a local information office for the MEDIA Programme in the UK. In addition to the Desk, there are regional Antenna offices in Glasgow, Cardiff and Belfast. If you are based in the UK, these offices are your first port of call for advice on the application process as well as other opportunities offered by the MEDIA Programme (for example, screenwriting training courses or international industry events). As well as its own website, the UK MEDIA Desk also publishes a monthly e-bulletin to subscribers.

MEDIA Antenna Scotland (Emma Valentine, 249 West George Street, Glasgow, G2 4QE. T 0141 302 1776; F 0141 302 1778; E scotland@mediadesk.co.uk)

MEDIA Service Northern Ireland (Cian Smyth, c/o Northern Ireland Screen, 3rd Floor, Alfred House, 21 Alfred Street, Belfast BT2 8ED. T 028 9023 2444; F 028 9023 9918; E northernireland@mediadesk.co.uk)

MEDIA Antenna Cymru/Wales (Ruth Appleby, c/o Creative Industries, 33-35 Stryd Gorllewin, Bute/West Bute Street, Cardiff CF10 5LH. T 02920 436112; F 02920 495598; E wales@mediadesk.co.uk)

Women in Film and Television UK (WFTV)

2 Wedgewood Mews, 12-13 Greek Street, London, W1D 4BB
T 020 7287 1400
info@wftv.org.uk
www.wftv.org.uk

Founded in 1990. A membership association open to women with at least one year's professional experience in the television, film and/or digital media industries. Currently over 800 members. Seeks to protect and enhance the status, interests and diversity of women working across the media.

Writernet

Cabin V, Clarendon Buildings, 25 Horsell Road, London, N5 1XL
T 020 7609 7474
info@writernet.org.uk
www.writernet.org.uk
Director *Jonathan Meth*

Founded in 1985, the organization works with new writing in all performance contexts. Formerly known as the New Playwrights Trust. Provides industry networking opportunities, advice, guidance and career development training.

Writers Copyright Association

Ealing Film Studios, Ealing, London, W5 5EP
mail@wcauk.com
www.wcauk.com
Managing Director *Jon Newman*
Office Manager *Valerie Harte*

Founded in 2000, offering copyright protection of literary material and intellectual property.

Registration fees: £25 for 5 years (UK), £99 for 10 years (worldwide).

Writers' Guild of Great Britain

15 Britannia Street, London, WC1X 9JN
T 020 7833 0777 F 020 7833 4777
erik@writersguild.org.uk
www.writersguild.org.uk
General Secretary *Bernie Corbett*
Deputy General Secretary *Anne Hogben*
Assistant General Secretary *Naomi MacDonald*
Events and Communications Manager *Moe Owoborode*

The TUC-affiliated union for professional writers. Established in 1958, the Guild represents writers working in TV, film, radio, theatre, books, poetry and videogames. Negotiates minimum fees, royalties and other terms with all major UK production groups. The organization offers its members: legal advice and contract-vetting; a pension scheme for screen and radio writers;

weekly email bulletin and quarterly *UK Writer* magazine; and regular events.

Full membership is available to anyone who has received payment for a piece of written work under a contract with terms not less than those negotiated by the Guild. Writers who do not qualify can join as Candidate Members and those on accredited writing courses or theatre attachments can become Student Members.

Canada

Academy of Canadian Cinema and Television

172 King Street East, Toronto, Ontario, M5A 1J3
T 416 366 2227 F 416 366 8454
info@academy.ca
www.academy.ca
Managing Director *Jeanette Slinger*

Non-profit body to promote Canadian cinema and television mostly through the Genie Awards, Gemini Awards and Les Prix Gemeaux. Other activities include the Academy Speaker Series, Kodak Soirees, Screening Series and industry-partnered events. Created in 1979. Now over 4,000 members. Four types of membership with varying eligibility requirements: voting membership (annual fee CA$150), friend of the Academy (CA$100), alumni membership (CA$75), student membership (CA$25).

Alliance for Children and Television (ACT)

1400 René-Lévesque Boulevard East, Office 713, Montreal, Quebec, H2L 2M2
T 514 597 5417
alliance@act-aet.tv
www.act-aet.tv

Seeks to enrich the screen-based media children experience by providing ongoing professional training, rewarding high-quality programmes and ensuring that children and youth have regular access to high-quality programming.

Canadian Film and Television Production Association (CFTPA)

151 Slater Street, Suite 902, Ottawa, Ontario, K1P 5H3
T 613 233 1444 F 613 233 0073
ottawa@cftpa.ca
www.cftpa.ca

Aims to promote and stimulate the Canadian production industry. A non-profit trade association representing almost 400 film, television and interactive media companies across the country. Runs a national mentorship programme to train young people in the demands of the industry. Offices in Ottawa, Toronto and Vancouver.

Canadian Film Institute

2 Daly Avenue, Suite 120, Ottawa, Ontario, K1N 6E2
T 613 232 6727 F 613 232 6315
cfi-icf@magma.ca
www.cfi-icf.ca

Incorporated in 1935 as a federally-chartered, non-governmental, non-profit cultural organization. Merged with the Conservatory of Cinematographic Art (Montreal) to form Cinematheque Canada in 1988 but retains its autonomous programming, budgetary, and administrative activities in Ottawa. Aims to encourage and promote the production, diffusion, study, appreciation, and use of moving images for cultural and educational purposes in Canada and abroad. Runs an ongoing public film programme, distributes a small collection of films, and publishes books and monographs on Canadian cinema.

Canadian Screenwriter Collection Society (CSCS)

366 Adelaide Street, West Suite 401, Toronto, Ontario, M5V 1R9
T 416 979 7907 F 416 979 9273
info@wgc.ca
www.writersguildofcanada.com/cscs/

Created by the Writers Guild of Canada to claim, collect, administer and distribute secondary authors' levies that film and television writers are entitled to under the national copyright legislation of several European countries and other jurisdictions.

Membership open to writers with a writing credit on a produced Canadian feature film or television programme.

Canadian Women in Communications

67 Yonge Street, Suite 804, Toronto, Ontario, M5E 1J8
T 416 363 1880 F 416 363 1882
cwcafc@cwc-afc.com
www.cwc-afc.com

Founded in 1991, a national, bilingual organization dedicated to the advancement of women in the communications sector through strategic networking and targeted professional development.

Membership open to women and men currently working in the communications industry, a student in the field or any person who supports CWC's mandate.

Documentary Organization of Canada

215 Spadina Avenue, Suite 126, Toronto, Ontario, M5T 2C7

Ⓣ 416 599 3844 x3 Ⓕ 416 979 3936
execdir@docorg.ca
www.docorg.ca
Executive Director *Samantha Hodder*

A national non-profit professional and advocacy organization dedicated to supporting the art of independent documentary filmmaking and filmmakers in Canada. Represent over 500 directors, producers, and craftspeople in six regional chapters (Atlantic, Quebec, Toronto, British Columbia, Newfoundland and Ottawa-Gatineau).

National Film Board of Canada

PO Box 6100, Station Centre-ville, Saint-Laurent, Quebec, H3C 3H5

Ⓣ 1 800 267 7710 Ⓕ 514 283 7564
www.nfb.ca

A federal cultural agency created in 1939. Its mandate is to produce and distribute distinctive, culturally diverse, challenging and relevant audiovisual works that 'provide Canada and the world with a unique Canadian perspective'. There is a French Program to produce films of importance to Francophones in Quebec and the rest of Canada and an English Program that works with documentary directors in every region of Canada.

Société des Auteurs de Radio, Télévision et Cinéma

1229 rue Panet, Montréal, Québec, H2L 2Y6

Ⓣ 514 526 9196
information@sartec.qc.ca
www.sartec.qc.ca

Founded in 1949 to represent writers of radio, television and cinema works in the French language. Around 800 members.

Women in Film and Television Toronto

110 Eglinton Avenue East, Suite 601, Toronto, Ontario, M4P 2Y1

Ⓣ 416 322 3430 ext. 221 Ⓕ 416 322 3703
wift@wift.com
www.wift.com

Dedicated to the advancement of women and under-represented groups in screen-based media. Offers year-round training and business skills development through its Centre for Media Professionals, networking events and industry awards. Represents over 3,000 women and men in Canadian screen-based media. Offers much useful advice on filmmaking on the affiliated canadianfilmmaker.com

Women in Film and Television Vancouver

Suite #462, 411 Dunsmuir Street, Vancouver, British Columbia, V6B 1X4

Ⓣ 604 685 1152 Ⓕ 604 685 1124
info@womeninfilm.ca
www.womeninfilm.ca
Managing Director *Yvette Dudley-Neuman*

An internationally affiliated, non-profit society committed to advancing and celebrating women in screen based media. Offers networking opportunities, professional development and advocacy.

See website for membership information.

Writers Guild of Canada

366 Adelaide Street, West Suite 401, Toronto, Ontario, M5V 1R9

Ⓣ 419 979 7907 Ⓕ 416 979 9273
info@wgc.ca
www.wgc.ca
Executive Director *Maureen Parker*
President *Rebecca Schechter*

Originally part of the Association of Canadian Television and Radio Artists (ACTRA). Independent as of 1995. Represents 1,800 professional screenwriters across Canada. Activities include: negotiating, administering and enforcing collective agreements with producers and broadcasters; working with government agencies to improve support; and increasing the profile of screenwriters. Produces Canadian Screenwriter Awards and the *Canadian Screenwriter Magazine*. Also offers script registration services.

Qualification is one writing contract in WGC jurisdiction. Occasionally offers membership incentives. Total membership fees CA$500 (CA$350 initiation fee and CA$150 annual basic dues).

Ireland

Arts Council Ireland

70 Merrion Square, Dublin 2

Ⓣ 01 618 0200 Ⓕ 01 676 1302
www.artscouncil.ie

Established in 1951, the state's development agency for the arts aims to promote the appreciation and practice of the arts. Ran a budget of €80m. in 2007.

Cork Film Centre

Civic Trust House, 50 Popes Quay, Cork
☎ 021 421 5160
info@corkfilmcentre.com
www.corkfilmcentre.com

Resource-based organization providing for people working in the medium of the moving image.

FÁS Screen Training Ireland

Adelaide Chambers, Peter Street, Dublin 8
☎ 01 483 0840 ℻ 01 483 0842
film@fas.ie
www.screentrainingireland.ie

Aims to provide continuing training for professionals in film, television, animation and digital media. Runs a bursary award scheme.

Film Base

Curved Street Building, Temple Bar, Dublin 2
☎ 01 679 6716 ℻ 01 679 6717
info@filmbase.ie
www.filmbase.ie

Founded in 1986 as a support organization for Ireland's independent film and video sector. Runs training and development programmes, hires equipment and editing suites, provides an information service and lobbies for the low budget film sector. Publishes *Film Ireland* magazine and the *Ireland on Screen* directory and administers a short film award scheme.

Galway Film Centre

Cluain Mhuire, Monivea Road, Galway
☎ 091 770 748 ℻ 091 770 746
info@galwayfilmcentre.ie
www.galwayfilmcentre.ie

Offers range of services and facilities to independent filmmakers, artists and community groups.

Irish Film and Television Academy

First Floor, Palmerstown Centre, Kennelsfort Road, Palmerstown, Dublin 20
☎ 01 620 0811 ℻ 01 620 0810
info@ifta.ie
www.ifta.ie

A not-for-profit all-Ireland organization to promote, encourage and reward creative excellence in film and television. Does this through screenings, educational events, awards, archiving and the production of a quarterly magazine.

Irish Film Board/Bord Scannán na héireann

Queensgate, 23 Dock Road, Galway
☎ 091 561 398 ℻ 091 561 405
info@filmboard.ie
www.irishfilmboard.ie
Marketing and Communications Executive *Louise Ryan*

Ireland's national film agency, founded in 1993. Funds the development and production of Irish films aimed at domestic and international audiences, and promotes Ireland as a location for international production.

Irish Film Institute

Irish Film Centre, 6 Eustace Street, Temple Bar, Dublin 2
☎ 01 679 5744
info@irishfilm.ie
www.irishfilm.ie

Founded as the National Film Institute of Ireland in 1945 with mission to promote and preserve film culture in Ireland. Maintains the Irish Film Archive, the Irish Film Centre and a travelling cinema, Cinemobile.

Irish Playwrights and Screenwriters Guild

Art House, Curved Street, Temple Bar, Dublin 2
☎ 01 670 9970
info@script.ie
www.script.ie
Chief Executive *David Kavanagh*
Chair of the Board *Audrey O'Reilly*

Formerly the Society of Irish Playwrights, which was founded in 1969, the Irish Playwrights' and Screenwriters' Guild is the representative body in Ireland for writers for the stage, screen, radio and new media.

Full membership of the Guild is available to any writer resident in Ireland who has been contracted to write for payment in Ireland for the stage or screen (including radio and the new media). Also has an associate membership open to aspiring writers.

Irish Writers Centre

19 Parnell Square, Dublin 1
☎ 01 872 1302 ℻ 01 872 6282
info@writerscentre.ie
www.writerscentre.ie

Founded in 1991 to foster writing and an audience for literature in Ireland. Runs a year-round programme of readings, workshops, lectures and seminars as well as a range of support and information services. Grant-aided by the Arts Councils of Ireland.

Screen Producers Ireland

77 Merrion Square, Dublin 2
Ⓣ 01 662 1114 Ⓕ 01 661 9949
info@screenproducersireland.com
www.screenproducersireland.com

Representative body for independent film, television and animation producers in Ireland.

Australia

Australian Film Commission

Level 4, 150 William Street, Woolloomooloo, NSW 2011
Ⓣ 02 9321 6444 Ⓕ 02 9357 3737
info@afc.gov.au
www.afc.gov.au

The commission ensures the creation, preservation and availability of Australian audiovisual content. Through the National Film and Sound Archive, the AFC collects, documents, preserves and provides access to Australia's screen and sound heritage. Also has offices in Melbourne, Brisbane and Canberra.

Australian Film Institute

236 Dorcas Street, South Melbourne, VIC 3205
Ⓣ 03 9696 1844 Ⓕ 03 9696 7972
www.afi.org.au
Operations Manager *Justine Beltrame*

Established in 1958, not-for-profit organization for industry professionals and general members. Activities include running annual AFI awards and film festival.

Professional membership or general membership: 1 year costs AUS$55, 2 years AUS$95.

Australian Script Centre

77 Salamanca Place, Hobart, Tasmania
Ⓣ 03 6223 4675
admin@ozscript.org
www.ozscript.org

Offers a script registration service. Prices and submission guidelines available on the website.

The Australian Writers' Guild

8/50 Reservoir Street, Surry Hills, NSW 2010
Ⓣ 02 9281 1554 ext. 109
comms@awg.com.au
www.awg.com.au
Admin and Project Officer *Joel Naoum*
Communications and Membership Manager *Stephen Asher*
Executive Director *Jacqueline Woodman*

Professional body representing film, theatre, television, radio and new media writers throughout Australia. Founded in 1962, the organization provides script assessment and registration services, along with free industrial and legal advice for members. Aims to help authors protect their creative rights and improve professional standards.

Membership is open to writers or 'any person in sympathy with the AWG's aims and objectives'.

Film Australia

101 Eton Road, Lindfield, NSW 2070
Ⓣ 02 9413 8777 Ⓕ 02 9416 5672
production@filmaust.com.au
www.filmaust.com.au

Founded in 1946, a government-owned company producing television documentaries and educational programmes. Provides support to the Australian documentary sector through a range of services and facilities.

Film Finance Corporation

Level 12, 130 Elizabeth Street, Sydney, NSW 2000
Ⓣ 02 9268 2555 Ⓕ 02 9264 8551
ffc@ffc.gov.au
www.ffc.gov.au

Wholly-owned government company funding only projects with high levels of creative and technical contribution by Australians, or projects certified under Australia's Official Co-Production Programme. Funds expensive programme formats such as feature films, mini-series, telemovies and documentaries, and does not normally fund cheaper formats such as current affairs, serial drama and 'infotainment'. Since its establishment, it has invested in 1,056 projects including *Strictly Ballroom*, *The Adventures of Priscilla: Queen of the Desert* and *Muriel's Wedding*.

New Zealand

Film New Zealand

23 Frederick Street, PO Box 24142, Wellington 6142
Ⓣ 04 385 0766 Ⓕ 04 384 5840
info@filmnz.org.nz
www.filmnz.com
Chief Executive Officer *Judith McCann*
Projects Manager *Susan Ord*
Systems Manager *Natalie Crane*
Office Manager *Michael Daly*

Founded in 1994 by the screen production industry. Constituted as a charitable trust with governance by a board of trustees. Provides information, introductions and support to national and international filmmakers interested in using New Zealand as a filming location.

The New Zealand Film Archive

84 Taranaki Street, Wellington 6011
Ⓣ 04 384 7647 Ⓕ 04 382 9595
information@nzfa.org.nz
www.filmarchive.org.nz
Registrar *Virginia Callanan*
Client Services Manager *Bronwyn Taylor*

Independent charitable trust established in 1981. Strives to collect, protect and project New Zealand's moving image history. Regular events include weekly screenings, community programmes, education initiatives and a research library.

New Zealand Film Commission

Level 3, 119 Ghuznee Street, PO Box 11 546, Wellington
Ⓣ 04 382 7680 Ⓕ 04 384 9719
info@nzfilm.co.nz
www.nzfilm.co.nz

Established in 1978 by act of parliament. Has the statutory responsibility 'to encourage and participate and assist in the making, promotion, distribution and exhibition of films' made in New Zealand by New Zealanders on New Zealand subjects. Provides loans and equity financing to film-makers to assist in the development and production of feature films and short films. Also active in the sales and marketing of New Zealand films, and with training and professional development.

New Zealand Writers Guild

PO Box 47 886, Ponsonby, Auckland 1144
Ⓣ 09 360 1408 Ⓕ 09 360 1409
info@nzwg.org.nz
www.nzwritersguild.org.nz

A professional association of script writers and a registered trade union, founded in 1975. Represents the interests of writers in the fields of film, television, theatre, radio, comedy and new media. Membership encompasses most of the professional script writers working in New Zealand. Activities include: publishing and providing information and advice; lobbying government; representing writers to funding bodies and other industry organizations; providing template contracts and agreements; providing opportunities for skill development; registering intellectual property; arbitrating on script credits.

Nga Aho Whakaari, Māori in Film, Video and Television Inc.

PO Box 68 626, Newton, Auckland 1145
Ⓣ 09 368 4430 Ⓕ 09 368 4431
admin@ngaahowhakaari.co.nz
www.ngaahowhakaari.com
Executive Officer *Kelvin McDonald*
Chairman *Ngamaru Raerino*
Administrative Assistance *Tui Walters*

National representative body for Māori working in film, video and television in New Zealand. Established in 1996. Focuses on the advancement of Māori moving images, culture and language, and supports the development of Māori film and television production.

Screen Directors Guild of New Zealand

PO Box 47-294 Ponsonby, Auckland 1011
Ⓣ 09 360 2102
office@sdgnz.co.nz
www.sdgnz.co.nz
Executive Director *Anna Cahill*

Has its genesis in 1995 when 18 directors met for an annual conference in Wellington and discussed the need for a representative organization. Promotes the New Zealand film and television industries as well as lobbying for better working conditions for members. Also administers a royalty payment scheme.

Full membership open to any screen director, editor or independent producer with a minimum of twenty minutes of credited screen time. Also offers secondary, associate, student and temporary memberships.

Screen Production and Development Association of New Zealand (SPADA)

Level 2 170, Cuba Street, PO Box 9567, Te Aro, Wellington
Ⓣ 04 939 6934 Ⓕ 04 939 6935
info@spada.co.nz
www.spada.co.nz

Membership-based organization, established in the 1980s and now with over 350 members. Represents the collective interests of independent producers and production companies on all issues that affect the business and creative aspects of independent screen production in New Zealand. Activities include professional development, lobbying, an e-newsletter and hosting of industry networking events.

Script to Screen

PO Box 147 263, Ponsonby, Auckland
Ⓣ 09 360 5400 Ⓕ 09 360 1409
simon@script-to-screen.co.nz

www.script-to-screen.co.nz
Executive Director *Simon Marler*

An independent, industry-wide initiative, established to develop the culture of screenwriting in New Zealand. Formerly called The New Zealand Writers Foundation, a joint initiative of the New Zealand Writers Guild and the NZ Film Commission, conceived in 2001. Works in partnership with other sectors of the film and television industries. Each month the Writer's Room hosts leading screenwriters to discuss their craft. Also runs a Script Read-Through Workshop series, where scripts are performed by professional actors, and writing workshops. Administers an international scholarship, under the terms of which a New Zealand screenwriter spends several months developing a script with a major production company.

Women in Film & Television

PO Box 6652, Wellington
Ⓣ 04 389 3862
info@wiftwellington.org.nz
www.wgtn.wift.org.nz
Contact *Zelda Edwards*

An organization offering support to women in the film and television industries via an informal network. Part of WIFT International. Men are welcome to belong as Friends of WIFT. Also runs a chapter in Auckland (PO Box 90-415, Auckland Mail Center; T 09 378 7271; W www.wiftauckland.org.nz; E office@wiftauckland.org.nz)

Festivals

Film is our youngest and most vibrant art form. Its energy and popularity comes from the constant influx of young filmmakers with their new ideas and overpowering enthusiasm.

Jake Eberts in the 'Foreword'

The screenplay is never the finished article. It is a blueprint, vitally important but not yet a fully fledged film.

Jon Croker in 'Helping the baby grow up'

Festivals

USA

AFI Festival

2021 N. Western Avenue, Los Angeles, CA 90027-1657
Ⓣ 323 856 7600 Ⓕ 323 467 4578
www.afi.com

Established in 1987, the festival features a rich slate of films from emerging filmmakers. The American Film Market (largest motion picture trade event in the world) runs concurrently with the festival. Held over 10 days in November.

African Diaspora Film Festival

535 Cathedral Parkway, Suite 14B, New York, NY 10025
Ⓣ 212 864 1760
info@africanfilm.com
www.nyadff.org
Contact *Mike Sharland*

Founded in 1993 to offer a platform for conveying African Diaspora artistic styles and craft in film. Held annually for about two weeks during November/December. There is no submission fee but, owing to volume, tapes and DVDs will not be sent back.

American Black Film Festival

abff@thefilmlife.com
www.abff.com

Established in 1997, the aim of this 5-day festival is to strengthen the black filmmaking community through resource sharing, education, artistic collaboration and development. Held annually.

There are four categories of film for consideration at the Festival, each with its own submission criteria (see website for further information).

Angelus Student Film Festival

7201 Sunset Boulevard, Los Angeles, CA 90046
Ⓣ 800 874 0999
info@angelus.org
www.angelus.org

Director *Monika Moreno*
Associate Director *Kale Zelden*

Founded in 1996 for student films up to 90 minutes. Screenings/awards ceremony held at the Directors' Guild of America, Hollywood. Also runs workshops and hosts a lunch for finalists. Held annually. Past honorary chairs include Gary Oldman, Sean Astin, Jim Caveziel and Lynn Redgrave.

Submissions welcome via website. Free admission to festival events.

Antelope Valley Independent Film Festival

3041 West Avenue K, Lancaster, CA 93536
Ⓣ 661 722 6478
info@aviff.com
www.aviff.com

Established in 1997 and has developed a reputation as a filmmaker-friendly festival with standing-room-only crowds and lively post-screening discussions. While the festival welcomes the work of first-time filmmakers, it also consistently features highlights from a variety of international festivals including Cannes, Sundance, and the Academy of Motion Picture Arts & Sciences Student Academy Awards. Held annually.

Submission of material is subject to an application fee.

Asian American International Film Festival

145 Ninth Street, STE 350, San Francisco, CA 94103
Ⓣ 415 863 0814 Ⓕ 415 863 7428
info@asianamericanmedia.org
www.asianamericanmedia.org

Established in 1982, an important launching point for Asian American independent filmmakers as well as a vital source for new Asian cinema. The festival is the largest presenter of Asian American film in the world, screening over 130 films and attracting an audience of almost 30,000 people. Held annually in March.

Submission of material is subject to an application fee and films and videos must be made by or about

Asian Americans and Asians of any nationality. All lengths and genres will be considered.

Asian CineVision

133 West 19th Street, Suite 300, New York, NY 10011
☏ 212 989 1422 🖷 212 727 3584
info@asiancinevision.org
www.asiancinevision.org

Established in 1978, the country's first festival dedicated to screening works by media artists of Asian descent. The festival has many categories that showcase the best of Asian and Asian American cinema, including features, shorts competition, screenplay competition, music video competition, works-in-progress and 'for youth by youth'. Held annually.

Only accepts material submitted by people of Asian descent.

Aspen FilmFest

Aspen Film 110 E. Hallam Street, Suite 102, Aspen, CO 81611
☏ 970 925 6882 🖷 970 925 1967
filmfest@aspenfilm.org
www.aspenfilm.org
Executive Director *Laura Thielen*
Program Director *George Eldred*
Communications Manager *Jennifer Slaughter*

Established in 1979, festival for full-length features and documentaries, with events including screenings and the annual 'Independent by Nature' award. Recent films featured included *Capote* and *Shut Up and Sing*.

Held annually in September/October, typically lasting 5 days.

Does not welcome unsolicited applications.

Aspen Shortsfest

Aspen Film 110 E. Hallam Street, Aspen, CO 81611
☏ 970 925 6882 🖷 970 925 1967
shortsfest@aspenfilm.org
www.aspenfilm.org
Executive Director *Laura Thielen*
Competition Manager *George Eldred*
Communications Manager *Jennifer Slaughter*

Events include competition screenings, free roundtable discussions, special presentations and an awards ceremony. Live action, documentary and animated shorts. Previous films entered include *The Danish Poet* and *Eramos Pocos*. Aspen Shortsfest is an Oscars qualifying event.

Held annually, typically lasting 5 days in April.

Welcomes unsolicited applications. Website has details of entry guidelines and procedures.

Atlanta Film Festival

IMAGE Film & Video Center, 535 Means Street, NW Suite C, Atlanta, GA 30318
☏ 404 352 4225 🖷 404 352 0173
aff@imagefv.org
www.atlantafilmfestival.com

Established in 1976, an annual festival featuring narrative, documentary, animation, and student films. Over ten days, showing more than 150 films. The festival provides filmmakers with the opportunity to attend panels and workshops.

Austin Film Festival

1145 West 5th Street, Suite 210, Austin, TX 78703
☏ 512 478 4795
alex@austinfilmfestival.com
www.austinfilmfestival.com
Screenplay/Teleplay Competition Director *Mary P. Hansell*

Founded in 1994, festival and conference focusing on screenwriters' contribution to the film and television industries. Includes screenplay awards for television and film across genres such as drama, comedy and science-fiction, with prizes of between $2,500 and $5,000. Previous recipients of awards include Shane Black, Harold Ramis and Barry Levinson.

Held annually in October.

Screenplay and teleplay awards are open only to writers who do not earn a living from television or film. Scripts must not have been optioned or sold.

Austin Gay and Lesbian International Film Festival

1216 East 51st Street, Austin, TX 78723
☏ 512 302 9889 🖷 512 302 1088
film@agliff.org
www.agliff.org

Established in 1987, the oldest and largest gay and lesbian film festival in the southwest. Showcases features and shorts, mainstream and avant-garde. Held annually for 9 days during the months of September and October.

Films and videos should be of interest to lesbians, gay men, bisexual, intersex and/or transgendered people, and is subject to a submission fee.

Avignon/New York Film Festival

☏ 212 650 3391
avignonfest2008@aol.com
www.avignonfilmfest.com

Established in 1994, this 5-day festival is the American version of the Avignon Film Festival.

The festival screens a selection of features and documentaries. Held annually.

Submission of material is subject to an administrative fee.

Babelgum Online Film Festival

filmfanatics@babelgum.com
www.babelgum.com

Online film festival to promote independent filmmaking and offer a showcase for aspiring filmmakers around the world. Chaired by Spike Lee.

Competition categories include: Short Film; Documentary; Animation; Social/Environment; Spot/Advertising; Music Video; Looking for Genius Award. Winners in each section receive €20,000. Entries should be of a professional standard but competition welco\s students.

Big Apple Film Festival

☎ 646 708 5910
info@bigapplefilmfestival.com
www.bigapplefilmfestival.com

Held annually in November. Screens a variety of films from the New York City independent film community, as well as additional selections from across the country and around the world. A closing night awards ceremony recognizes achievement in filmmaking, screenwriting and acting.

Big Bear Lake International Film Festival

PO Box 1981, Big Bear Lake, CA 92315
☎ 909 866 3433
bigbearfilmfest@aol.com
www.bigbearlakefilmfestival.com
Programming Director/Vice President *Sandy Steers*
Screewriting Director/Vice President *Monika Skerbelis*

Established in 2000, activities include a screenwriting competition, as well as independent film screenings, networking events, seminars and a film awards ceremony.

The festival is a 3-day event held annually in September.

Unsolicited entries of films or screenplays are welcomed by email via a call for entry form.

Boston International Film Festival

PO Box 240023, Boston, MA 02124
☎ 617 482 3900
info@bifilmfestival.com
www.bifilmfestival.com

Shows local, national and international filmmakers.

The Broad Humor Film Festival

9854 National Boulevard, Suite 166, Los Angeles, CA 90034
info@broadhumor.com
www.broadhumor.com

Showcases humourous scripts and films by women.

Bronx Independent Film Festival

The Lovinger Theatre at Lehman College, The City University of New York, 250 Bedford Park Boulevard West, Bronx, NY 10468
film@bronxstage.com
www.bronxstage.com

Founded in 2003 to celebrate the history of film in The Bronx by showcasing emerging filmmakers while promoting The Bronx as a cultural destination. The 2-day festival is held annually.

Accepts narrative, documentary, animation and experimental films. Short Films should be 29 minutes or less and feature films should run for 30 minutes or more.

Brooklyn International Film Festival

BiFF 180 South 4th Street, Suite 2 S, Brooklyn, NY 11211
☎ 718 486 8181 📠 718 599 5039
2008@wbff.org
www.wbff.org

Held annually since 2002, the successor to the Williamsburg Brooklyn Film Festival. For and by independent film makers.

Carolina Film and Video Festival

☎ 336 334 4197
cfvf@uncg.edu
www.carolinafilmandvideofestival.org

Founded in 1978, this festival focuses on works of student and/or independent filmmakers. It is a competitive festival and offers awards in the following categories: documentary, narrative, and experimental or animated film.

Works of any length from students and independent filmmakers.

Chicago City Limits Comedy Film Festival

☎ 212 888 5233
film@chicagocitylimits.com
www.firstsundays.com

The festival takes place on the first Sunday of every month and features short comedy films.

Submission of material is subject to an administrative fee. Accepts films that are under 1 minute and up to 30 minutes.

Chicago International Film Festival

30 East Adams Street, Suite 800, Chicago, IL 60603

℡ 312 683 0121 🖷 312 683 0122
info@chicagofilmfestival.com
www.chicagofilmfestival.org

Founded in 1964 by award-winning filmmaker and graphic designer Michael Kutza. The festival is a competitive event with categories for feature films, documentaries and short films (including animation and student productions). Held annually.

Submission of material is subject to an application fee, and entries that have been screened publicly in Chicago or aired on US television prior to the festival are not eligible for screening. Entries must be either in English or contain English subtitles and must be accompanied by a 25–30 word summary of the production's content.

Chicago Underground Film Festival

info@cuff.org
www.cuff.org

Running since 1994 and aims to showcase independent, experimental and documentary films from around the world that dissent radically in form, technique, or content from the indie mainstream. Held annually for one week.

Submission of material is subject to an administrative fee. Foreign language work must be dubbed or subtitled in English and a written description or synopsis must be enclosed with the entry form.

Cinequest San Jose Film Festival

PO Box 720040, San Jose, CA 95172-0040
℡ 408 295 3378 🖷 408 995 5713
info@cinequest.org
www.cinequest.org
Contact *Mike Rabehl*

Founded in 1991, showcasing independent films. Also hosts the Maverick film competition. Held annually for about 12 days. Includes a competition for screenwriters.

Submission of material is subject to an administrative fee and films of all running lengths are accepted.

Cleveland International Film Festival

2510 Market Avenue, Cleveland, OH 44113-3434
℡ 216 623 3456
mgoodman@clevelandfilm.org
www.clevelandfilm.org

Founded in 1977, the festival screens films in a variety of categories including American Independents, Gay, Lesbian, Bisexual and Transgender interest films, and Spanish language films.

Coney Island Short Film Festival

532 La Guardia Place, Suite 638, New York, NY 10012
info@coneyislandfilmfestival.com
www.coneyislandfilmfestival.com

Founded in 2001. Jury presents awards for best feature, short, documentary feature, documentary short, experimental, animation, music video and "Made in Coney Island". Runs for 3 days.

Crested Butte Reel Fest

PO Box 1733, Crested Butte, CO 81224
℡ 970 349 2600 🖷 970 349 1384
www.crestedbuttereelfest.com

Founded in 1998, a competitive festival for films under 40 minutes in the categories of animation, experimental, live-action narrative (comedy, drama, science fiction, action or musical) and documentaries under 60 minutes.

Cucalorus Film Festival

℡ 910 343 5995 🖷 910 343 5227
mail@cucalorus.org
www.cucalorus.org

Founded in 1994, a 4-day festival in Wilminton, North Carolina, showing documentaries, feature films and shorts.

Damah Film Festival

8033 Sunset Boulevard, Suite 1041, Hollywood, CA 90046
℡ 310 237 5436
www.damah.com

Founded in 2001, an annual festival of short films with spiritual topics.

Films must be under 30 minutes.

Durango Independent Film Festival

PO Box 1587, Durango, CO 81302-1587
℡ 970 375 7779 🖷 970 375 1437
info@durangofilm.org
www.durangofilm.org
Contact *Joanie Fraughton*

Founded in 2001, a competitive film festival for independent films. Accepts entries for shorts, documentaries, and feature films on all subject matter. Runs over 5 days.

Feature length films should be over 50 minutes and shorts should be 49 minutes or under in duration.

East Lansing Film Festival

☎ 517 980 5802 ⓕ 517 336 2750
www.elff.com

Founded in 1997, the festival focuses on presenting a diverse selection of independent and foreign features, documentaries, shorts and student films. The festival also hosts the Lake Michigan Film Competition which showcases films from Wisconsin, Illinois, Indiana and Michigan. Held annually over a period of 5 days.

Fiery Film Fest

☎ 505 693 0906
director@fieryfilmfest.org
www.fieryfilmfest.org

Established in 2001 in Clovis, New Mexico, showing animation, comedy, documentary, drama, action, sci-fi, horror and experimental genres.

Film Columbia

☎ 518 392 1162
filmcolumbia@aol.com
www.filmcolumbia.com

Founded in 2000. Presents work by independent filmmakers in the Hudson River region. The programme includes features, documentaries, short selections and student films.

In order to qualify, a principal of the film, (director, producer, screenwriter, actor, production company or locale) must be connected to the Hudson River region, from New York City to Canada.

A Flickering Image Festival

3233 Grand Avenue, Suite N-110, Chino Hills, CA 91709
shortsfest@actorsbone.com
www.actorsbone.com/Shorts/index.html
Festival Director *N. Barry Carver*
Festival President *Paul J. Molinaro*
Festival Co-ordinator *Marianne Friend*
Marketing Director *Tracy Eliott*

Competitive festival which screens ten winners' short films in Hollywood. All genres are welcomed but films must last a maximum of thirty minutes.

The event is held once a year, with screenings of winning films in January.

Unsolicited applications are welcomed. All entries, whether in the top ten or not, receive written feedback from judges.

Florida Film Festival

1300 South Orlando Avenue, Maitland, FL 32751
☎ 407 629 1088 ⓕ 407 629 6870
www.floridafilmfestival.com

Founded in 1992 and produced by the Enzian Theater, screening narrative and documentary features and shorts, animation and midnight movies. Hosts several educational forums. The Festival is a qualifying festival for the Oscars in the category of live action short films.

Fort Lauderdale International Film Festival

1314 East Las Olas Boulevard, Suite 007, Fort Lauderdale, FL 33301
☎ 954 760 9898
info@fliff.com
www.fliff.com

Founded in 1986, screening over 200 films from around the world. Held annually for over a month.

Full Frame Documentary Festival

324 Blackwell Street, Suite 500, Washington Building Bay 5, Durham, NC 27701
☎ 919 687 4100 ⓕ 919 687 4200
info@fullframefest.org
www.fullframefest.org

Founded in 1998 in Durham, North Carolina, a documentary film festival showcasing more than 100 films over 4 days. The festival also has panel discussions, seminars, and question and answer sessions.

Great Lakes Independent Film Festival

Great Lakes Film Association, PO Box 346, Erie, PA 16512
☎ 814 873 5069
Fest@greatlakesfilmfest.com
www.greatlakesfilmfest.com

Founded in 2002, dedicated to showcasing independent films in the digital medium. Accepts submissions of independently produced feature length and short length films and scripts in the genres of documentary, horror, experimental, religious/spiritual, animation, and all genres of music videos and gay/lesbian.

The Hamptons International Film Festival

3 Newtown Mews, East Hampton, NY 11937
☎ 631 747 7978 ⓕ 631 324 5116
info@hamptonsfilmfest.org
www.hamptonsfilmfest.org

Established in 1992, the festival highlights narrative features, documentaries and shorts, and offers diverse programming with breakthrough films by new directors, premieres by established filmmakers, panel discussions and special events with guests from the industry.

Takes place annually.

Submission of material is subject to an application fee. It is a requirement that films may not have been released theatrically or shown on television in the US prior to the festival.

Hawaii International Film Festival

680 Iwilei Road, Suite 100, Honolulu, HI 96817
T 808 528 3456 F 808 536 2707
info@hiff.org
www.hiff.org
Contact *Simeon Paskell*

Founded in 1981, showing feature films, documentaries, short films, Hawaii-themed films, music videos and student films. Runs for 2 weeks.

Hazel Wolf Environmental Film Festival

PO Box 22695, Seattle, WA 98122
T 206 624 9725
info@hazelfilm.org
www.hazelfilm.org
President *Grace Stahre*

Founded in 1998, specializing in documentaries, fiction, animation and web media with an environmental theme. Includes filmmaking and environmental workshops and panel discussions, and 'meet the filmmaker coffees'. Recent films shown include *McLibel* and *Buyer Be Fair*.

The event is held annually, lasting 3–4 days in March/April.

Unsolicited applications are welcomed by email.

Heartland Film Festival

200 S. Meridian Street, Suite 220, Indianapolis, IN 46225-1076
T 317 464 9405 F 317 464 9409
info@heartlandfilmfestival.org
www.heartlandfilmfestival.org

Founded in 1991, aiming to recognize filmmakers whose work explores the human journey and the positive values of life.

Hollywood Black Film Festival

8306 Wilshire Boulevard, Suite 2057, Beverly Hills, CA 90211
T 310 407 3596 F 310 943 2326
info@hbff.org
www.hbff.org
Founder/Executive Director *Tanya Kersey*
Director of Programming *Jacqueline Blaylock*
Marketing Director *Gerald Haynes*

An annual celebration of black cinema drawing together established filmmakers and emerging artists. Established in 1999, activities include public screenings, competitions for filmmakers and scriptwriters (with a staged reading of the winning script) and a 'Pitchathon' for those trying to sell ideas and screenplays.

Held in June, lasting 6 days.

Welcomes unsolicited applications by email.

Hollywood International Film Festival

433 North Camden Drive, Suite 600, Beverly Hills, CA 90210
T 310 288 1882 F 310 288 0060
info@hollywoodawards.com
www.hollywoodawards.com
Contact *Brian Dunnigan*

Founded in 1997, a competitive festival with a mandate to highlight international films made by the global creative community. Runs for 5 days.

Submissions are accepted monthly. Features, documentaries, short subjects and music videos are eligible for the competition.

Hope and Dreams Film Festival

PO Box 131, Hope, NJ 07844
T 908 459 5797
hopeanddreams@earthlink.net
www.hopeanddreams.com
Festival Director *C. L. Rusin*

Founded in 1998, events include workshops, social functions, talks, an awards ceremony and a DVD sales market. Screenings of features, shorts, documentaries and animation. All subjects are considered, but 'themes of hope and dreams' will be given additional consideration.

Held annually, the festival takes place over a weekend in October.

Applications through the website are welcomed.

Hot Springs Documentary Film Institute

T 501 321 4747 F 501 321 0211
hsdfi@hsdfi.org
www.hsdfi.org

Founded in 1992, showing quality documentaries on a variety of topics and themes. Held annually.

Non-fiction documentaries in all forms (features, shorts, and student films). The duration for shorts is 20 minutes or less.

Humboldt International Film Festival

T 707 826 4113
filmfest@humboldt.edu
www.humboldt.edu/~filmfest/

Founded in 1967, among the oldest student-run film festivals in the world. Celebrates independent, student and international films that test cinematic boundaries. Films must be under 30 minutes and have been created within the last 5 years.

Ivy Film Festival

info@ivyfilmfestival.com
www.ivyfilmfestival.com

Established in 2001 to garner recognition for
student filmmakers by means of a panel of
celebrity judges. Held annually on the campus
of Brown University in association with several
other Ivy League schools.

Kansas City Filmmakers Jubilee

4741 Central Suite 306, Kansas City, MO 64112
T 913 649 0244 F 913 649 0244
kcjub@kcjubilee.org
www.kcjubilee.org

Founded in 1997, a juried film festival featuring
more than 100 local, regional, national and
international films. The festival also hosts
panels and workshops for filmmakers. Held in
April.

Long Island International Film Expo

c/o Bellmore Movies, 222 Pettit Avenue (Side
entrance), Bellmore, New York, NY 11710
T 516 783 3199
debfilm@aol.com
www.liifilmexpo.org
Director *Debra Markowitz*

Founded in 1998, a week-long showcase of short
and feature-length independent films of all
genres. Some of the films featured in recent years
include *Strike the Tent, Dorian Blues, Push, My
Date with Drew* and *Falling Sky*.
    Held annually.
    Welcomes unsolicited applications which are to
be made either by email or via www.withoutabox.
com.

Los Angeles Film Festival

Film Independent, 9911 W. Pico Boulevard,
11th Floor, Los Angeles, CA 90035
T 310 432 1240
LaFilmFest@filmindependent.org
www.lafilmfest.com

Founded in 1995, showcasing upwards of 175
narrative features, documentaries, shorts,
and music videos. A qualifying festival in
all categories for the Independent Spirit
Awards and for the Short Films categories
of the Academy Awards. Runs for 10 days in
June.
    Feature films and shorts must be in English or
have English subtitles, and must not have had any
commercial theatrical or television play in the
US. Feature-length films should be 50 minutes or
more and short films should be under 50 minutes
running time.

Los Angeles International Short Film Festival

1610 Argyle Avenue, Suite 113, Hollywood,
CA 90028
T 323 461 4400
info@lashortsfest.com
www.lashortsfest.com

Founded in 1997, showcasing short films in
the following categories: drama, comedy,
documentary, animation, and experimental.

Malibu Film Festival

Malibu Film Foundation, PO Box 1133, Malibu,
CA 90265
info@malibufilmfestival.org
www.malibufilmfestival.org

Founded in 2000. Shows over 80 international
features, documentaries and short films from
around the world. Runs for 4 days in April. Films
that have not yet screened in Los Angeles are
given priority in jury selection.

Margaret Mead Film and Video Festival

American Museum of Natural History, Central
Park West at 79th Street, NY 10024
T 212 769 5305 F 212 769 5329
meadfest@amnh.org
www.amnh.org/mead
Artistic Co-director *Elaine Charnor*
Managing Co-director *Kathy Brew*
Festival Co-ordinator *Tamar Goelman*
Festival Co-ordinator *Natalie Tschechaniuk*

An annual festival of experimental non-fiction
and documentary films. Founded in 1977,
events include screenings, Q&A sessions
with filmmakers, panel discussions and
performances.
    The festival lasts 5 days and is held in
November each year.
    Unsolicited applications are welcomed via the
website.

Media That Matters Film Festival

Arts Engine 104 W. 14th Street, 4th Floor,
New York, NY 10011
T 646 230 6368 F 646 230 6388
info@artsengine.net
www.mediathatmattersfest.org
Contact *Emily Hayward*

Founded in 2001, showcasing short films on
topical themes. The jury selects work on a
variety of topics including gay rights and global
warming, made by independent filmmakers many
of whom are under the age of 21. Films include
documentaries, music videos, animations and
experimental work. All films run for 8 minutes
or less.

Submission of material is subject to an entry fee (free for students). Films must run for 8 minutes maximum. Films of all genres are accepted. Interested in films on social and environmental issues.

Miami International Film Festival

c/o Miami Dade College, 25 NE 2nd Street, Room 5501, Miami, FL 33132
℡ 305 237 3456
info@miamifilmfestival.com
www.miamifilmfestival.com

Founded in 1984, showing films from over 50 countries. Cash awards are given in four competition categories: World Features, Ibero-American Features, World & Ibero-American Documentaries, and Short Films. Grand Jury prizes of $25,000 awarded in dramatic and documentary competitions.

Minneapolis-St Paul International Film Festival

309 Oak Street, SE, Minneapolis, MN 55414
℡ 612 331 7563 Ⓕ 612 378 7750
info@mnfilmarts.org
www.mnfilmarts.org
Festival Director *Al Milgrom*
Office Director *Jim Brunzell III*
Public Relations Manager *Vince Muzik*

Activities include a children's festival and an emerging filmmakers award, as well as screenings, premieres and an awards ceremony. Running since 1983, the festival shows around 150 films a year, incorporating both local and international films.
Held annually for 2 weeks in April.
Welcomes unsolicited applications but prefers phone call first.

Moondance International Film Festival

℡ 303 545 0202
director@moondancefilmfestival.com
www.moondancefilmfestival.com
Executive Director *Elizabeth English*

Held in Boulder, Colorado. Aims to promote and encourage screenwriters and independent filmmakers. When emailing type 'MOONDANCE' in the subject line.

Napa Sonoma Wine Country Film Festival

12000 Henno Road, PO Box 303, Glen Ellen, CA 95442
℡ 707 935 FILM
wcfilmfest@aol.com
www.winecountryfilmfest.com

Founded in 1986, the festival presents features, documentaries, shorts and animation in six categories: world cinema, Latin cinema, US cinema, arts in film, cinema of conscience and eco cinema.

The New Fest

139 Fulton Street, Suite PH-3, New York, NY 10038
℡ 212 571 2170 Ⓕ 212 571 2179
info@newfest.org
www.newfest.org

Founded in 1989, showing local, national and international pictures to foster a greater sense of awareness and community among gay, bisexual and transgender audiences. Runs for 11 days.
Submission of material is subject to an administrative fee and films/videos should be by, about or of interest to lesbians, gay men, bisexuals or transgendered people.

New Hampshire Film Festival

155 Fleet Street, Portsmouth, NH 03801
℡ 603 647 6439
info@nhfilmfestival.com
www.nhfilmfestival.com

Aims to present the best in recent independent cinema from the USA and around the world. Running since 2001. Award for best screenplay.

New Orleans Film Festival

843 Carondelet Street, New Orleans, LA 70130
℡ 504 309 6633 Ⓕ 504 309 0923
info@neworleansfilmfest.com
www.neworleansfilmfest.com
Executive Director *Ali Duffey*
Artistic Director *John Desplas*

Founded in 1989, the festival includes screenings, workshops, talks and an awards ceremony.
An 8 day event held annually in early October.
Welcomes unsolicited applications. A call for entries is listed on the website. Films can be entered in 5 competitive categories.

The New York City Horror Film Festival

PO Box 8582, Woodcliff Lake, NJ 07677
NYCHorrorfest@aol.com
www.NYCHorrorfest.com
Festival Director *Michael J. Hein*

Focusing solely on horror, science-fiction and thrillers, the festival 'celebrates genre films and those who make them'. Founded in 2001, events include special screenings, informative panel discussions and 'frightening fun parties'. Films include short and feature narratives, documentaries, animation and music videos.

Held annually in October, the festival runs for a week.

Unsolicited applications are welcomed by email.

New York Film Festival

The Film Society of Lincoln Center, 70 Lincoln Center Plaza, New York, NY 10023-6595
⊤ 212 875 5610
festival@filmlinc.com
www.filmlinc.com

Founded in 1963. Shows 30 features and 12 short films each year. Runs for 17 days.

All filmmakers, regardless of experience, are invited to submit material. There is no entry fee and films of all lengths and genres are considered.

The New York International Documentary Festival

159 Maiden Lane, New York, NY 10038
⊤ 212 668 1100 Ⓕ 212 943 6396
mail@docfest.org
www.docfest.org

Founded in 1998, a 5-day event focussing on new and classic international documentaries. Every screening and seminar is followed by a panel discussion with the filmmaker, film subject(s) and other guests.

Selection of material to the festival is by invitation only. If interested in submitting material, send a short synopsis, technical information, press material and director's biography to the festival and they will review it and decide on whether to extend an invitation. US films considered for the festival must have been completed within twelve months prior to the festival, and films made outside the US must have been completed within 24 months prior to the festival.

New York International Independent Film and Video Festival

505 East Windmill Lane, Suite 1B-102, Las Vegas, NV 89123
⊤ 702 361 1430 Ⓕ 702 361 6309
filmmfest@aol.com
www.nyfilmvideo.com

A competitive event founded in 1993. Hosts film, music and art events in New York and Los Angeles and has attracted major figures from around the world. Hosted four times per year.

See website for submission procedure.

New York International Latino Film Festival

419 Lafayette Street, 6th Floor, New York, NY 10003

⊤ 646 723 1428 Ⓕ 212 228 6149
info@nylatinofilm.com
www.nylatinofilm.com

Founded in 1999, showing movies from North, Central and South America and the Caribbean made by, about and/or for the Latino community.

Films of all genres are considered and can be of any form including feature narrative, short films, documentary and experimental shorts. Films must be in English or have English subtitles, and cannot have been broadcast or distributed commercially in the US prior to the festival (exceptions are made for short films and documentaries).

Newport International Film Festival

PO Box 146, Newport, RI 02840
⊤ 401 846 9100 Ⓕ 401 846 6665
info@newportfilmfestival.com
www.newportfilmfestival.com

Founded in 1998, with over 100 screenings of feature length narrative films, documentaries and shorts. Hosts international and US premieres. Also runs daily panel discussions with well-known industry professionals. Held annually.

All entries must either be in English or subtitled in English. Feature films and documentaries should be 60 minutes or longer, and short films should be 30 minutes or less in duration.

Nihilist Film Festival

Box 36422, Los Angeles, CA 90036
nihilist01@aol.com
www.nihilists.net/film.html
Director *Elisha Shapiro*

Specializes in 'amusing and challenging films', including documentaries, animation, drama and comedy. Founded in 1999, the festival includes 'the blessing of the TVs'. 'If you've made a video that has appalled and offended other film festivals, we're looking for you.'

Held annually in December.

Welcomes unsolicited applications by mail.

Omaha Film Festival

⊤ 402 203 8173
omahafilmfestival@gmail.com
www.omahafilmfestival.org
Program Director *Marc Longbrake*

Established in 2005, exhibiting independent film and running an extensive film education programme. Awards prizes for short length and feature length screenplays, as well as the 'Nebraska Screenplay' prize.

Palm Beach Jewish Film Festival

3151 N. Military Trail, West Palm Beach,
FL 33409
ⓣ 561 689 7700 ext. 158 Ⓕ 561 478 3060
pbjhh@jcconline.com
www.palmbeachjewishfilm.org
Director *Karen Davis*

Founded in 1989, screening feature films,
documentaries and shorts with a Jewish focus.
Recent films featured include *More than 1000
Words* and *A Love to Hide*.
  10-day event held annually in
November/December.
  Unsolicited applications are welcomed by mail.

Palm Springs International Film Festival

1700 E. Tahquitz Canyon Way, Suite 3,
Palm Springs, CA 92262
ⓣ 760 322 2930 Ⓕ 760 322 4087
info@psfilmfest.org
www.psfilmfest.org
Marketing and Publicity Manager *David Lee*

Founded in 1990, the festival includes screenings,
panel discussions, international gala screenings
and an awards ceremony. Covers a wide range
of genres including live action, animation,
documentaries and short films. Films entered in
previous years include *Black Book* and *The Tiger's
Tail*.
  The event is held annually for 12 days in
January.
  On-line applications are welcomed.

Pan African Film Festival

3775 Santa Rosalia Drive, Los Angeles, CA 90008
ⓣ 323 295 1706 Ⓕ 323 295 1952
info@PAFF.org
www.paff.org

Founded in 1992 in Los Angeles, run by
a non-profit corporation dedicated to the
promotion of cultural and racial tolerance
and understanding. Each year over 100 films
shown.
  The festival accepts films and videos made by
and/or about people of African descent. Films
should preferably depict positive and realistic
images and can be of any genre.

Portland International Film Festival

1219 SW Park Avenue, Portland, OR 97205
ⓣ 503 221 1156 Ⓕ 503 294 0874
info@nwfilm.org
www.nwfilm.org

Founded in 1978. Focuses mainly on foreign
films but does include American features,
documentaries and shorts. Each year the festival
screens nearly 100 films from over 30 countries.
Runs for two weeks each year.
  Unsolicited preview tapes with entry may
be sent. Entries must include a synopsis of
the film and a one-page bio/filmography of
the filmmaker's background and previous
works.

Real to Reel Film and Video Festival

111 S. Washington Street, Shelby, NC 28150
ⓣ 704 484 2787 Ⓕ 704 481 1822
violet.arth@ccartscouncil.org
www.ccartscouncil.org/realtoreel

Founded in 2000, providing a platform for
independent film, video and multimedia artists
from around the world. Runs for 4 days.
  Four categories: documentary, feature length
(over 61 minutes in length), shorts (under 60
minutes in length) and animation. Each entry
must include all promotional material for the film
such as credits, film stills, and synopsis.

Red Bank International Film Festival

Freedom Film Society Inc., PO Box 447,
Red Bank, NJ 07701
ⓣ 732 741 8089 Ⓕ 732 741 8093
mleckstein@monmouth.com
www.rbiff.org

Founded in 2001, showing all genres made by
both industry professionals and students from
all over the world. Cash awards for all major
category winners. Runs for 3 days.
  Entries must include a synopsis of 200 words
or less.

Reel Affirmations Film Festival

One in Ten PO Box 73587, Washington, D.C. 200
ⓣ 202 986 1119 Ⓕ 202 939 0981
info@oneinten.org
www.reelaffirmations.org
Programs Manager *Joe Bilancio*

Founded in 1991, one of the largest gay and
lesbian festivals in the world. Held annually.
  The festival seeks features, shorts,
documentaries and animated works in 35mm,
DVD, Beta SP and VHS formats.

Rhode Island International Film Festival

PO Box 162, 96 Second Street, Newport,
RI 02840
ⓣ 401 861 4445 Ⓕ 401 490 6735
info@film-festival.org
www.film-festival.org

Founded in 1997, a juried competition and
showcase for independent filmmakers from all
over the world. The festival's short film winner

qualifies for Academy Award consideration. Runs for 6 days.

No category restrictions. Films must be presented in their original language with English language subtitles. Feature films should run for 46 minutes or longer, and short films should be 45 minutes or less.

Rochester International Film Festival

Movies On A Shoestring Inc., PO Box 17746, Rochester, NY 14617

🕾 585 234 7411
president@rochesterfilmfest.org
www.rochesterfilmfest.org

Founded in 1959. The world's oldest continuously-held short film festival. Programme includes narrative films, documentaries and animations submitted by independent filmmakers from all parts of the world. Includes films by students and professionals.

Open to films and videos in all genres. Maximum length of entries is 30 minutes. No more than two entries per filmmaker.

Sacramento Film and Music Festival

10374 Ambassador Drive, Rancho Cordova, CA 95670

🕾 916 600 7029
questions@sactofilmfest.com
www.sacfilm.com

Founded in 2000. Submission-based and highly selective. Invites films in all genres and of all lengths. Since its inception, has screened over 400 films and videos. Annual.

Entries should include a brief synopsis. All entries must either be in English or subtitled in English.

San Diego Film Festival

7974 Mission Bonita Drive, San Diego, CA 92120

🕾 619 582 2368 🖷 619 286 8324
info@sdff.org
www.sdff.org
Executive Director *Robin Laatz-Kozak*
Programming Director *Karl Kozak*

Established in 2002 to support up-and-coming independent filmmakers. Incorporates screenings, panel discussions, workshops for screenwriters and filmmakers and awards. Features full-length films, documentaries, shorts and music videos.

The festival is held annually in September, lasting 5 days.

Unsolicited applications are welcomed, via application form on website.

San Francisco International Film Festival

39 Mesa Street, Suite 110, The Presidio, San Francisco, CA 94129

🕾 415 561 5000 🖷 415 561 5099
gga@sffs.org
www.sffs.org

Founded in 1957, the longest-running film festival in the Americas and the first to showcase international productions. Screens more than 200 films and hosts more than 100 filmmakers each year. Runs for 2 weeks.

San Francisco International Gay and Lesbian Film Festival

145 Ninth Street, #300, San Francisco, CA 94103

🕾 415 703 8650 🖷 415 861 1404
info@frameline.org
www.frameline.org

Founded in 1977, the world's largest and oldest LGBT film festival, screening films from around the world. The festival awards a $10,000 First Feature Award, the $10,000 Best Documentary Award and a $7,500 Audience Award.

Santa Barbara International Film Festival

1528 Chapala Street, Suite 203, Santa Barbara, CA 93101

🕾 805 963 0023
www.sbfilmfestival.org

Running since 1986. A juried festival showing over 200 feature films and 50 short films. Also panel discussions and a student screenwriting and filmmaking competition. Annual.

Santa Fe Film Festival

PO Box 2167, Santa Fe, NM 87504

🕾 505 988 7414
info@santafefilmfestival.com
www.santafefilmfestival.com

Founded in 2000, the festival premiers local New Mexican film, new American and foreign film including revivals, retrospectives, independent productions and mini-festivals. Also hosts panel discussions, demonstrations, talks and workshops.

Savannah Film Festival

c/o Savannah College of Art and Design, PO Box 3146, Savannah, GA 31402-3146

🕾 912 525 5051
filmfest@scad.edu
www.scad.edu/filmfest/

Founded in 1998, showcasing feature-length films and shorts from both award-winning professionals and emerging student filmmakers.

The festival also hosts panel discussions and presentations by visiting artists.

Feature films and videos must be no less than 40 minutes in length. Short works must not exceed 40 minutes. All entries must either be in English or subtitled in English. The entry must include a brief synopsis not longer than 100 words.

Screamfest Horror Film Festival

Ⓣ 310 358 3273
screamfestla@aol.com
www.screamfestla.com

Founded in 2001 to give filmmakers and writers in the horror/sci-fi genres a venue to have their work showcased to the industry. Los Angeles-based.

Seattle International Film Festival

400 9th Avenue, N., Seattle, WA 98109
Ⓣ 206 464 5830 Ⓕ 206 264 7919
info@seattlefilm.org
www.seattlefilm.org
Artistic Director *Carl Spence*
Managing Director *Deborah Person*

Established in 1975, incorporating screenings, tributes, workshops and demonstrations, panel discussions and an awards ceremony. Over 400 feature films and documentaries and shorts from more than 60 countries are presented to an audience of 160,000 each year. Films featured in recent years include *The Illusionist, The Science of Sleep, Half Nelson* and *Me and You and Everyone We Know*.

Held annually in May/June and lasting 25 days.

Unsolicited applications are welcomed via the website.

Sedona International Film Festival & Workshop

PO Box 162, Sedona, AZ 86339
Ⓣ 928 282 1177 Ⓕ 928 282 5912
info@sedonafilmfestival.com
www.sedonafilmfestival.com

Founded in 1997. Features more than 125 films, including features, documentaries, foreign films, shorts, animation and student films. Many of the films screened at the festival have gone on to garner Academy Award nominations including *Genghis Blues, Spellbound* and *Why Can't We Be A Family Again*. Lasts for 5 days.

ShockerFest International Film Festival

PO Box 580450, Modesto, CA 95358
Ⓣ 209 537 5221 Ⓕ 209 531 0233
director@shockerfest.com
www.shockerfest.com

Founded in 2002, the festival showcases films in the science fiction, fantasy and horror genres. Runs for 2 days every year.

Shriekfest Film Festival & Screenplay Competition

PO Box 920-444, Sylmar, CA 91392
Ⓣ 818 367 9161 Ⓕ 818 367 9861
shriekfest@aol.com
www.shriekfest.com
Founder/Co-director *Denise Gossett*
Co-director *Todd Beeson*

Established in 2001, dedicated to horror, thrillers, science fiction and fantasy films and screenplays. Includes adult and junior competitions for screenplays and films, as well as panel discussions and screenings. Recent films include *Dark Remains, The Other Side*, and *The Cellar Door*.

Held annually in October, lasting 4 days.

Welcomes unsolicited applications by mail or email.

Silver Lake Film Festival

2658 Griffith Park Boulevard, Suite 389, Los Angeles, CA 90039
info@silverlakefilmfestival.org
www.silverlakefilmfestival.org

Founded in 2000 to provide a showcase for independent film in Los Angeles. It screens over 200 narrative features, documentaries and short films. In addition to its annual festival, the organization presents a monthly screening of short films throughout the year. Held annually for a period of 10 days.

Silverdocs

info@silverdocs.com
www.silverdocs.com

International film festival in Silver Spring, Maryland, celebrating independent film and documentary. Running since 2003. Provides networking opportunities and talks by leading industry figures.

Slamdance Film Festival

5634 Melrose Avenue, Los Angeles, CA 90038
Ⓣ 323 466 1786 Ⓕ 323 466 1784
mail@slamdance.com
www.slamdance.com

Founded in 1995, the festival takes place each year in Utah at the same time as the Sundance Film Festival. The festival was launched to showcase undistributed films by emerging filmmakers. Festival discoveries have included directors Marc Forster (*Monster's Ball*) and Jared Hess (*Napoleon Dynamite*). Held annually in January.

Films of all genres and from any country accepted.

The Sonoma Valley Film Society Festival

589 First Street West, Sonoma, CA 95476
T 707 933 2600 F 707 933 2612
info@SonomaFilmFest.org
www.sonomafilmfest.org

Founded in 1998, screening about 75 independent films annually, including feature length narratives, world cinema, documentaries, shorts and student films. Most screenings are followed by a Q&A with the stars and filmmakers.

See website for submission details.

Starz Denver Film Festival

Denver Film Society @ The Starz Film Center 900 Auraria Parkway, Denver, CO 80204
T 303 595 3456 F 303 595 0956
info@denverfilm.org
www.denverfilm.org

Founded in 1978, the festival screens films from around the world including new international releases, cutting-edge features, documentaries, and short subjects. Hosts panels and seminars, as well as Q&A sessions with the director, cast and crew of the films. Annually for 10 days.

Sundance Film Festival

Sundance Institute, PO Box 684429, Park City, Utah, UT 84068
T 310 360 1981 F 310 360 1969
Institute@sundance.org
www.sundance.org/festival

Founded in 1978 as the Utah/US Film Festival, it was renamed the Sundance Film Festival in 1991. It is the largest independent cinema festival in the US. The festival comprises competitive sections for American and international dramatic and documentary films, and a group of non-competitive showcase sections, including the Sundance Online Film Festival.

Held annually in January.

Submission of material is subject to an application fee.

Tampa International Gay and Lesbian Film Festival

PO Box 18445, Tampa, FL 33679-8445
T 813 879 4220 F 813 514 6428
amy@cliptampabay.com
www.tiglff.com

Founded in 1990, focussing on films with homosexual themes as well as transgender issues and gender roles. Held annually.

Films and videos submitted should be of interest to lesbian, gay, bisexual, intersex and/or transgendered people. The festival accepts narrative, documentary, experimental and animated films and videos (both features and shorts).

Telluride Film Festival

800 Jones Street, Berkeley, CA 94710
T 510 665 9494 F 510 665 9589
Mail@telluridefilmfestival.org
www.telluridefilmfestival.org

Founded in 1974, a non-competitive festival for independent film. The festival screens features and student films. Films such as *Slingblade* have premiered at the festival. Annual.

Unsolicited works may be submitted for consideration from May 1st to July 15th each year. Movies of any length, in any format, and in all genres and disciplines are eligible for consideration. Features (60 minutes or longer) which have had any public exposure in North America prior to the Labor Day event are not eligible for consideration.

Temecula Valley International Film & Music Festival

27740 Jefferson Avenue, Suite 100, Temecula, CA 92590
T 951 699 5514 F 951 506 4193
festival@tviff.com
www.tviff.com

Founded in 1995. Has since screened over 700 films from more than 20 countries. Annual. Open to US and foreign films in five categories: full length features, shorts, documentaries, animation and student films. Entries must include a film synopsis, complete cast/crew credits list, stills, and complete technical information (running time). Non-English language produced films must have subtitles or be dubbed in English.

Terror Film Festival

PO Box 823, Frazer, PA 19355
terrorfilmfest@aol.com
www.terrorfilmfestival.net

Independent international film festival held each October in Philadelphia. Offers over $10,000 in cash and prizes, including awards for excellence in screenwriting.

Tiburon International Film Festival

1680 Tiburon Boulevard, Tiburon, CA 94920
T 415 381 4123 F 415 388 4123
info@TiburonFilmFestival.com
www.TiburonFilmFestival.com

Founded in 2002, the aim of the festival is to strengthen cultural awareness and to create a platform for independent filmmakers.

Tribeca Film Festival

375 Greenwich Street, New York, NY 10013
⊤ 212 941 2400 Ⓕ 212 941 3939
festival@tribecafilmfestival.org
www.tribecafilmfestival.org

The Tribeca Film Festival was founded in 2002 by Robert De Niro, Jane Rosenthal and Craig Hatkoff as a response to the attacks on the World Trade Centre. Conceived to foster the economic and cultural revitalization of Lower Manhattan through an annual celebration of film, music and culture, the Festival's mission is to promote New York City as a major filmmaking centre and allow its filmmakers to reach the broadest possible audience.

Held annually.

United Nations Association Film Festival

PO Box 19369, Stanford, CA 94309
⊤ 650 724 5544
info@unaff.org
www.unaff.org
Festival Director *Jasmina Bojic*

Focusing on international short and feature-length documentaries, includes screenings, talks and an awards ceremony.

Lasting 5 days, the festival is held in October. Call for entries is released on website.

US International Film and Video Festival

713 South Pacific Coast Highway, Suite A, Redondo Beach, CA 90277-4233
⊤ 310 540 0959 Ⓕ 310 316 8905
filmfestinfo@filmfestawards.com
www.filmfestawards.com

Founded in 1967, devoted exclusively to recognition of outstanding business, television, documentary, educational, entertainment, industrial and informational productions.

Vermont International Film Festival

1 Main Street, Suite 307, Burlington, VT 05401
⊤ 802 660 2600 Ⓕ 802 860 9555
info@vtiff.org
www.vtiff.org

Founded in 1985 by peace and social justice activists, George and Sonia Cullinen, the festival has provided a forum for films dealing with issues of war and peace, justice and human rights, and the environment. Most years, the festival accepts independent films from around the world in any length and genres that fit into one of the three categories: Justice and Human Rights, War and Peace, or the Environment.

Vistas Film Festival

1000 West Page Avenue, Dallas, TX 75208
⊤ 214 887 9048 Ⓕ 214 887 1734
www.vistasfilmfestival.org

Founded in 1999, a 5-day international festival of movies by or about Latinos or Latino culture. Annual.

Submissions are accepted throughout the year and all entries are considered for the main festival in September, as well as for the monthly screening series that occurs throughout the year. Non-competitive. Films must fall within the festival's focus of works by or about Latinos or Latino culture. Feature length and short films are considered for presentation.

Washington, D.C. Independent Film Festival

2950 Van Ness Street, NW Suite 728, Washington, D.C. 200
⊤ 202 333 6615
info@dciff.org
www.dciff.org
Founder/Executive Director *Carol Bidault de l'Isle*

Established in 1999, screening over 100 independent feature films, shorts, documentaries and animated films from around the world. Other events include seminars, discussions with filmmakers and a music festival.

Held in March annually, the festival lasts 11 days.

Welcomes applications from around the world. 'All films (except opening and closing night) are chosen from an open call'.

Waterfront Film Festival

PO Box 387, Saugatuck, MI 49453
⊤ 269 857 8351
info@waterfrontfilm.org
www.waterfrontfilm.org

Founded in 1999, this non-competitive festival is open to independent films of any genre including features, shorts, documentaries and animation.

Non-English language-produced films must have English subtitles or be dubbed in English.

Westchester Film Festival

148 Martine Avenue, Room 107, White Plains, NY 10601
⊤ 914 995 2917 Ⓕ 914 995 2948
www.westchestergov.com

Founded in 2000, this juried festival has the following categories: feature, documentary feature, narrative short, documentary short, international, animation, screenplay, student resident and student non-resident.

Applicants must be 18 years or older.

Williamstown Film Festival

PO Box 81, Williamstown, MA 01267
☎ 413 458 9700 ℻ 413 458 2702
contactus@williamstownfilmfest.com
www.williamstownfilmfest.com

Running since 1998, aiming to celebrate the industry—past, present and future. As well as classic screenings, there are panels, seminars and discussions, and explorations of new technologies.

Withoutabox

10920 Ventura Boulevard, Studio City, CA 91604
☎ 818 980 8161 ℻ 818 980 8163
www.withoutabox.com

Online application submission service for film festivals, connecting festivals with filmmakers and screenwriters. Also helps independent filmmakers to distribute and profit from their work.

The Woods Hole Film Festival

PO Box 624, 87B Water Street, Woods Hole, MA 02543
☎ 508 495 3456 ℻ 508 495 3456
info@woodsholefilmfestival.org
www.woodsholefilmfestival.org
Executive Director *Judy Laster*
Managing Director *J.C. Bouvier*

Founded in 1989, incorporating screenings, screenplay competition with staged readings of winners, workshops and social events. Programme features around 40 features and 50 shorts including comedy, drama, documentary, animation and experimental films. All films are in competition.

Held annually from last Saturday in July to first Saturday in August.

Welcomes unsolicited applications online.

Woodstock Film Festival

86 Mill Hill Road, PO Box 1406, Woodstock, NY 12498
☎ 845 679 4265 ℻ 509 479 5414
info@woodstockfilmfestival.com
www.woodstockfilmfestival.com

Founded in 2000, the festival showcases films, concerts, workshops, celebrity-led panels, and an awards ceremony. Maverick Awards are presented for Best Feature, Best Documentary, Best Short Documentary, Best Short Film, Best Student

Film, Best Cinematography, Best Editing and Best Animation. Annual.

Worldfest Houston International Film Festival

PO Box 56566, Houston, TX 77256
☎ 713 965 9955 ℻ 713 965 9960
mail@worldfest.org
www.worldfest.org

Founded in 1961 by producer/director Hunter Todd. Dedicated to the independent feature and short film and does not accept films from major studios. Presents around 55 new indie feature films and 100 short films each year. Runs for 10 days in April. Attended by over 500 filmmakers each year, films can compete in more than 200 awards categories.

A 50 to 100 word synopsis is required for all entries.

UK

10 Sec Film Fest

Mobile Media Entertainment Ltd, 2 Percy Street, London, W1T 1DD
contact@10secfilmfest.com
www.tensec.com

Described as the 'shortest film festival in the world, instant entertainment for an attention deficient universe.' Best submissions are shown on the website.

Must be a member to submit films. No fee for entering. Download and fill in a film submission form from the website for each film entered.

2 Days Later Short Film Competition

The Community Pharmacy Gallery, 16 Market Place, Margate Old Town, Kent, CT9 1ES
☎ 01843 223 800 ℻ 01843 223 800
info@2dayslater.co.uk
www.2dayslater.co.uk
Competition Director *Mick Etherton*

Founded in 2002. Runs a competition for micro budget short horror films made in 48 hours to be shown at a screening event. Also runs weekend filmmaking workshops for novice film makers.

A competition brief is available in July on website. Deadline is 10 days before the Haloween screening. Festival is free. To become involved, approach by email in first instance.

Africa In Motion

14/9 Viewforth Square, Edinburgh, EH10 4LW
☎ 07807 485 058
info@africa-in-motion.org.uk
www.africa-in-motion.org.uk

Edinburgh-based festival showing a broad range of African films (old and new) and complementary events. Particularly interested in attracting young, emerging African filmmakers to submit shorts and documentaries and to talk to audiences about their work. The festival then tours other parts of the UK.

The Angel Film Festival

138 Upper Street, Islington, London, N1 1QP
T 020 7288 2233
chris@angelfilmfestival.org
www.angelfilmfestival.org
Director *Chirs Timms*

Competition with screenings, talks and award ceremonies for new short films. All genres welcomed. Established in 2005. Takes place annually over four weeks.
Submissions welcomed via email.

Animate the World

Barbican Centre, Silk Street, London, EC2Y 8DS
T 020 7638 4141
info@barbican.org.uk
www.barbican.org.uk/animate

The Barbican's annual children's film festival, founded in 2001. Shows children's animated feature films and shorts from around the world, together with family animation workshops and special events.

Animex International Festival of Animation and Computer Games

School of Computing, University of Teesside, Middlesbrough, Tees Valley, TS1 3BA
T 01642 342631 F 01642 342691
chris@animex.net
animex.tees.ac.uk
Festival Director *Chris Williams*

Annual festival since 2000 with its roots in the creative side of animation and computer games. Brings together animators, directors, students, artists, designers, writers and educators for talks, presentations, workshops, screenings and parties.

Aspects Literature Festival

North Down Borough Council Town Hall, The Castle, Bangor, BT20 4BT
T 028 9127 0371 F 028 9127 1370
www.northdown.gov.uk

Annual festival attracting leading Irish writing talent. See website for details.

Aurora (Norwich International Animation Festival)

Francis House, Redwell Street, Norwich, NR2 4SN
T 01603 756280
info@aurora.org.uk
www.niaf.org.uk

Festival includes screenings, seminars, debates, live performance and installation works. Firmly committed to showing cross-disciplinary work and aims to present animation within the context of the wider moving image.
Submission forms are available via the website. Not only accepts animation but also manipulated live action, artists' films and installation work.

Bath Film Festival

2nd Floor, Abbey Chambers Kingston Parade, Bath, BA1 1LY
T 01225 401149 F 01225 401149
www.bathfilmfestival.org

Established in 1991 by members of Bath Film Society, screening invited previews, arthouse films and documentaries. Also includes a film-makers discussion and film-related workshops. Annual.

Belfast Festival at Queen's

Culture & Arts Division, 8 Fitzwilliam Street, Belfast, BT9 6AW
g.farrow@qub.ac.uk
www.belfastfestival.com
Festival Director *Graham Farrow*

Ireland's biggest international arts festival, established in 1962. Covers all art forms including theatre, dance, classical music, literature, jazz, comedy, visual arts, folk music and popular music, attracting over 50,000 visitors.

Belfast Film Festival

3rd Floor, 23 Donegall Street, Belfast, BT1 2FF
T 028 90 325 913 F 028 90 329 397
info@belfastfilmfestival.org
www.belfastfilmfestival.org
Programmer *Stephen Hackett*
Programmer *Michael Besnard-Scott*

Established in 2000. Runs annually in late March for 11 days. Includes screenings, talks, workshops, masterclasses, special events and short film and documentary competitions. Particular interest in international and European cinema, Japanese anime, music documentaries, documentaries with an international focus and Irish shorts.
Submissions accepted via a form on the website.

Berwick Film & Media Arts Festival

56-58 Castlegate, Berwick-upon-Tweed,
Northumberland, TD15 1JT
☎ 01289 303355
info@berwickfilm-artsfest.com
www.berwickfilm-artsfest.com

A contemporary film and media arts festival, first
held in 2005. Shows mainstream and art house
films as well as artists' videos and films in a range
of venues including the Town Hall Prison Cells,
The Berwick Gymnasium, The Barrel's Ale House
pub, the Ice House and the Black Hole. Other
events include an artists' symposium, talks by
industry professionals, an educational outreach
programme run by ISIS Arts and a mobile cinema.

Betting on Shorts

London Consortium, Institute of Contemporary
Arts, 12 Charlton House, Terrace, London,
SW1Y 5AH
☎ 020 7839 8669 ☏ 020 7930 9896
submissions@bettingonshorts.com
www.bettingonshorts.com
Director *Ricarda Vidal*
Director *Louise Höjer*

Founded in January 2005. Run annually at the end
of November. Calls for short films of all genres
between two and ten minutes. Features screenings,
awards and audience feedback. Audiences are
invited to bet on which film will win.

Submissions encouraged via email if relevant to
an announced festival theme. Past themes include
"Playtime", "Vacancy" and "Mad or Bad". See
website for more details. Deadline end of August.

Beyond TV International Video Festival

Undercurrents, Old Exchange, Pier Street,
Swansea, SA1 1RY
☎ 01792 455900
beyondtv@undercurrents.org
www.undercurrents.org

Shows a broad range of short movies,
documentaries, music videos and animation
themed around social or environmental activism.

Check website for information on submitting
films.

BFM International Film Festival

Suite 13, 5 Blackhorse Lane, Walthamstow
Stratford, London, E17 6DS
☎ 020 8531 9199 ☏ 020 8531 9199
festival@bfmmedia.com
www.bfmmedia.com

Established in 1998, week-long festival screening
the work of black filmmakers from the UK and
the rest of the world. Other events include Q&A
sessions and the BFM Short Awards which
recognize new and emerging black talent. Films
featured in recent years include *Four Brothers* and
Life and Lyrics, as well as films from the African
continent.

The festival is held annually in September.

Birds Eye View Film Festival

Birds Eye View, Unit 306, Aberdeen Centre, 22-24
Highbury Grove, London, N5 2EA
☎ 020 7704 9435
info@birds-eye-view.co.uk
www.birds-eye-view.co.uk
Festival Director *Rachel Millward*
Project Manager *Rosie Strang*

Annual festival founded in 2005 running for
six days. Next held in summer/autumn 2008.
Screens international features, documentaries
and short films by female directors. Also features
masterclasses and workshops on all aspects of
filmmaking.

Unsolicited applications welcomed. Submission
guidelines and forms appear on the website.

Bite the Mango Film Festival

National Media Museum, Bradford,
West Yorkshire, BD1 1NQ
btm@nationalmediamuseum.org.uk
www.bitethemango.org.uk
Director *Addy Rutter*
Festival Programmer *Tom Vincent*
Festival Co-ordinator *Ben Eagle*
Festival Administrator *Jennifer Hall*

Hosted at the National Media Museum since
1995, bringing audiences an impressive collection
of the best cinema from all four corners of
the world, drawing influences from Africa,
South Asia, Central America and the Far East.
Showcases an eclectic mix of features, shorts and
documentaries and a selection of master classes,
seminars and workshops delivered by industry
professionals providing a culturally vibrant
atmosphere as well as providing an international
platform for new and aspiring filmmakers to
showcase their work. Held in September.

Unsolicited scripts invited via email.

Bradford Animation Festival

National Media Museum, Bradford, W. Yorkshire,
BD1 1NQ
☎ 01274 203364
www.nationalmediamuseum.org.uk/baf
Director *Deb Singleton*
Festival Co-ordinator *Ben Eagle*
Booking Co-ordinator *Jennifer Hall*

Founded in 1994, the UK's longest running and largest animation festival. Host to screen talks, workshops and special events with some of the industry's top names, the Festival's culminates in the annual BAF Awards, which celebrate the best in new animation from around the world. Past guests include Nick Park, Dave Burgess, Joanna Quinn, Marc Craste, Andreas Hykade and Georges Schwizgebel. Held in November.

Information on how to submit entries available via the website.

Bradford International Film Festival

National Media Museum, Bradford, West Yorkshire, BD1 1NQ
☏ 0870 70 10 200
BAFtickets@nationalmediamuseum.org.uk
www.bradfordfilmfestival.org.uk
Artistic Director *Tony Earnshaw*
☏ 01274 203 320 ℱ 01274 203 387
Festival Co-ordinator *Ben Eagle*
Festival Programmer *Tom Vincent*
Industry Weekend Co-ordinator *Addy Rutter*
Festival Administrator *Jennifer Hall*

Founded in 1995, showcasing new US independents, new European cinema and CineFile, a strand dedicated to documentaries on films and filmmakers. Also includes the Widescreen Weekend, focusing on classic titles screening in CinemaScope, 70mm and Cinerama. The festival features screenings, 'Screentalk' interviews, retrospectives, masterclasses, industry events and silent films with live musical accompaniment. Held annually at the UK's National Media Museum for 16 days in March.

Britdoc

☏ 020 7033 2565
festival@britdoc.org
www.britdoc.org

One of the UK's most important festivals of documentary production, run by the Channel 4 British Documentary Film Foundation. Held annually over three days, attracting international film producers, distributors and financiers.

Cambridge Film Festival

Arts Picturehouse, 38-39 St Andrews Street, Cambridge, CB2 3AR
☏ 01223 500082 ℱ 01223 462555
info@cambridgefilmtrust.org.uk
www.cambridgefilmtrust.org.uk

Major film festival held each July, established in 1977 and relaunched in 2001. Shows selection of films alongside an eclectic range of specialist interest programmes at the three-screen Arts Picturehouse and assorted other arts venues in the town. Also runs several touring events.

Those wishing to submit a film can either download a form from the website or use the submission page on the withoutabox.com service. Particularly keen to receive locally- or regionally-produced work.

Can Leicester International Short Film Festival

☏ 0116 2621265
info@lineout.org
www.lineout.org

Annual festival which began in 1996 to serve the East Midlands but has since gained an international profile.

Seeks films from all over the world, of any genre and by filmmakers of any age. Submission is free and there is no limit on number of times an individual can enter. Download the submission form from website.

Celtic Media Festival

249 West George Street, Glasgow, G2 4QE
☏ 0141 302 1737 ℱ 0141 302 1738
jude@celticmediafestival.co.uk
www.celticfilm.co.uk

First held in 1982. Aims to promote the languages and cultures of the Celtic countries on screen and in broadcasting during an annual three-day event.

Chichester Film Festival

Chichester Cinema, New Park Road, Chichester, West Sussex, PO19 7XY
☏ 01243 786650 ℱ 08715 227780
info@chichestercinema.org
www.chichestercinema.org

Running since 1992. Includes audience-selected awards for best feature and best film short.

Cinem@tic

Technology Innovation Centre, Millennium Point, Curzon Street, Brighton, B4 7XG
☏ 0121 331 5400
cinem@tic.ac.uk
Centre Manager *Steve Smith*
Festival Director *Stephen Gordon*
Festival Administrator *Nina Parmer*

Short film festival founded in 2002 featuring free screenings, talks and a prize competition. Runs annually for two days.

Unsolicited applications accepted via email.

Cinemagic International Film Festival for Young People

49 Botanic Avenue, Belfast, BT7 1JL
⊤ 028 9031 1900 Ⓕ 028 9031 9709
info@cinemagic.org.uk
www.cinemagic.org.uk

Established in 1989 and now the largest children's film festival in Great Britain and Ireland. Runs for 17 days and includes screenings, premieres, special guests, discussions, competitions and masterclasses in all aspects of filmmaking. There are prizes for best feature and best short for a teenage audience (judged by a jury aged 15–18) and best feature and best short for a children's audience (judged by a jury aged 8–14).

Entry form and regulations available on the website.

Co-operative Young Film-Makers Film Festival

The Co-operative Group, 5th Floor, New Century House, Manchester, M60 4ES
⊤ 0161 246 2216
info@youngfilm-makers.coop
www.youngfilm-makers.coop
Festival Organiser *Phaedra Patrick*

Young Film Makers is organised and funded by The Co-operative Group. The festival is held at the National Media Museum in Bradford. Over two days there are screenings of more than 100 short films as well as 20 workshops and masterclasses. Screenings are free. Some events have a small charge.

Films must be under 6 minutes long and you must be under 21. For all other criteria and guidelines see the website www.youngfilm-makers.coop

Cornwall Film Festival

Krowji West Park, Redruth, Cornwall, TR15 3AJ
⊤ 01209 204655
info@cornwallfilmfestival.com
www.cornwall-film-festival.co.uk

Annual festival with programme of screenings, skills development and networking. Submissions to the festival are restricted to films where a member of the creative team (writer, director, producer) are resident in Cornwall. Selection is made in three different categories: competition; screening; videotheque. Submission forms available on the website.

Dead by Dawn

www.deadbydawn.co.uk

Horror film festival run in association with Filmhouse and held in Scotland. Instructions on how to submit a film available on the website (for both features and shorts).

Deaffest, the Deaf Film and Television Festival

Light House, The Chubb Buildings, Fryer Street, Wolverhampton, WV1 1HT
⊤ 01902 716055 Ⓕ 01902 717143
www.deaffest.co.uk
www.light-house.co.uk

Established in 1998, catering for deaf film-lovers. Its programme includes films, panel discussions, networking opportunities and social events, all of which are free.

Discovering Latin American Film Festival

info@discoveringlatinamerica.org
www.discoveringlatinamerica.com

Established in 2002, showing Latin American feature films, shorts and documentaries. Also hosts debates, talks and Q&A sections. Held annually.

Document International Human Rights Documentary Film Festival

c/o Mona Rai, 268 Albert Drive, 2/1 Pollokshields, Glasgow, G41 2RJ
⊤ 0141 429 0185
doc6subs@googlemail.com
www.docfilmfest.org.uk

Established in 2003, providing a platform for both established and emerging documentary filmmakers to screen their work related to international human rights issues.

East End Film Festival

⊤ 020 7613 7676
east.filmlondon.org.uk

Founded in 2001, exploring the potential of cinema to cross cultural, political and artistic boundaries. Films in the programme come from around the world. Screenings take place in venues across East London. Nitin Sawhney is the festival's patron.

Edinburgh International Film Festival

88 Lothian Road, Edinburgh, EH3 9BZ
⊤ 0131 228 4051 Ⓕ 0131 229 5501
submissions@edfilmfest.org.uk
www.edfilmfest.org.uk
Submissions Coordinator *Katie Crook*

Running since 1947, one of the UK's most important film festivals. Notable films screened in the last few years include *Mrs Brown*, *The Full Monty*, *La Vie Revée des Anges*, *East is East*, *Run Lola Run*, *Billy Elliot*, *American Splendor*, *Motorcycle Diaries* and *Tsotsi*.

Accepts submissions including shorts, features, animation, documentary and music video that are no older than 18 months at the time of the festival. Download the relevant forms and regulations from the website or submit via the online submissions service withoutabox.com. Deadline for all features and shorts is in February and for animation and Mirrorball films in March.

Edinburgh Mountain Film Festival

stevie@edinburghmountainff.com
www.edinburghmountainff.com

Established in 2003 and held each October in Edinburgh. An independent festival geared towards outdoor enthusiasts. Seeks 'exciting, unusual and intriguing new films' to show.

Emergeandsee

☏ 0781 7288 797
nathalie@emergeandsee.org
www.emergeandsee.org

Voluntary organization founded in 2000, showing student work at regular cinema events in London (Curzon Cinema, Soho), Berlin and Budapest. Also distributes show reels to international festivals and cinemas. Submissions can be made via the website.

Encounters International Short Film Festival

Watershed Media Centre, 1 Canon's Road, Harbourside, Bristol, BS1 5TX
☏ 0117 929 9188 🖷 0117 952 9988
info@encounters-festival.org.uk
www.encounters-festival.org.uk

Major short film festival.

Filmmakers submitting work are requested to complete an online submission form and to send a DVD copy of the film for consideration. Deadline for submissions is normally end of May but check on website.

End of the Pier International Film Festival

PO Box 213, Bognor Regis, West Sussex, PO21 2ZL
☏ 01243 841775
info@eotpfilmfestival.com
www.eotpfilmfestival.com

Held each May. Showcases high quality independent film from around the globe. Aims to gain recognition for new, low budget and independent filmmakers via a competition. Runs a wide-ranging education programme plus talks, seminars and master classes. Entry guidelines available on website.

Fantastic Films Weekend

National Media Museum, Bradford, West Yorkshire, BD1 1NQ
tony.earnshaw@nmsi.ac.uk
www.nationalmediamuseum.org.uk/fantastic
Artistic Director *Tony Earnshaw*
Festival Co-ordinator *Ben Eagle*
Booking Co-ordinator *Jennifer Hall*

The UK's fastest growing festival dedicated to horror, fantasy and sci-fi cinema and television. Founded in 2002 as a weekend event focusing on classic ghost stories and the supernatural. Over the last five years it has evolved into a must-see event showcasing a disparate line-up of classic chillers, sci-fi shockers, fantasy epics, vintage TV shows and rarely seen gems from the vaults. FFW runs the gamut from monochrome classics of the silent era through to the latest digital epics. Held in June.

Filmstock International Film Festival

c/o 24 Guildford Street, Luton, LU1 2NR
☏ 01582 402 200
contact@filmstock.co.uk
www.filmstock.co.uk

Running since 2000, devoted to independent shorts, features and documentaries. Shows around 150 films, selected from an open submission scheme. Paper entries can be downloaded from the website or via withoutabox. (£15 for short film/short documentary up to 25 minutes; £20 for features/feature documentary). Runs Filmschlock, 'an evening of how film should not be done'.

Firecracker Showcase: London's East Asian Film Festival

Firecracker Media Ltd, 27 Old Gloucester Street, London, WC1N 3XX
☏ 020 7870 3427
info@firecracker-media.com
www.firecracker-media.com

Organizes festival-style events in London to bring exposure to the cinema of East Asia.

FLIP International Animation Festival

Light House, Media Centre, The Chubb Buildings, Fryer Street, Wolverhampton, WV1 1HT
☏ 01902 716055
flip@light-house.co.uk
www.flipfestival.co.uk
Festival Co-ordinator *Peter McLuskie*

Founded in 2004, a festival of animated films including screening, talks, workshops and awards.

Films featured have included *Toy Story*, *Howls Moving Castle* and *Spirited Away*.

Held over three days each Oct.

For tickets call the box office. To become involved contact by email.

Frightfest

10 Wiltshire Gardens, Twickenham, TW2 6ND
℡ 020 8296 0555
info-frightfest@blueyonder.co.uk
www.frightfest.co.uk
Co-Director *Ian Rattray*

UK's premiere horror film festival. Runs annually, featuring screenings and presentations of fantasy, thriller and horror films. Established in 2000.

Unsolicited material welcomed. Submissions process outlined on website.

Future Shorts

34-35 Berwick Street, London, W1F 8RP
℡ 020 7734 3883 ℻ 020 7439 3255
info@futureshorts.com
www.futureshorts.com

A worldwide monthly film festival since 2003. Filmmakers have the chance to get their work seen all over the world from a single submission. Films that are selected join a monthly programme shown at upwards of 25 UK venues (theatrical and non-theatrical).

Glasgow Film Festival

Glasgow Film Theatre, 12 Rose Street, Glasgow, G3 6RB
℡ 0141 332 6535
www.glasgowfilmfestival.org.uk
www.gft.org.uk

Informal celebration of international cinema.

Hull International Short Film Festival

Suite 12, The Danish Buildings, 44/46 High Street, Hull, HU1 1PS
℡ 01482 381512 ℻ 01482 381517
office@hullfilm.co.uk
www.hullfilm.co.uk
Festival Director *Laurence Boyle*

Established in 2000, dedicated to the exhibition and creation of short films. Activities include screenings, lectures, masterclasses and educational events. Committed to promoting new work and developing critical discussion.

The festival is held annually and lasts 4–5 days.

Unsolicited applications are welcomed by mail.

i-blink Film Festival

ceMAP Coventry University, Priory Street, Coventry, CV1 5FB
info@i-blink.org
www.i-blink.org

A film festival and international scriptwriting competition for cutting edge short films. Prizes for the ten winners include loan of cameras, edit suite and accessories, training at Pinewood Studios and a copy of professional scriptwriting software FinalDraft 7. The three top films will also win: 1st Prize - €5000; 2nd Prize - €2000; 3rd Prize - €1000. All entrants receive professional feedback on their scripts.

Images of Black Women Film Festival

PO Box 54145, London, W5 9DA
℡ 07876 155 228
info@imagesofblackwomen.com
www.imagesofblackwomen.com
Festival Director *Sylviane Rano*

An annual celebration of women of African descent working both in front of and behind the camera. Held at the Tricycle Cinema, in Kilburn, London in 2008.

International Festival of Fantastic Films

gil@glaneyoung.freeserve.co.uk
www.fantastic-films.com

Weekend-long festival of science fiction, fantasy and horror movies, held in Manchester. Includes guest interviews, discussions, panels, special events, presentations, auctions, artshow and poster exhibition, dealer room, themed dinner, parties and screenings. Also runs competitions for independent and amateur films.

International Manga and Anime Festival

First Floor, London County Hall, Belvedere Road, London, SE1 7BP
www.imaf.co.uk

First held in 2004 to meet the demand for Manga- and Anime-related activities. Administers a drawing/animation competition, which had a prize pot of $75,000 in 2006. Also runs screenings, art exhibitions, workshops, talks and opportunities to meet the professionals.

International Screenwriter's Festival

c/o Arturi Films Ltd, Unit 24C Daniels Ind. Estate Bath Road, Stroud, Gloucestershire, GL5 3TL
℡ 01453 753 440 ℻ 01453 753 990
info@screenwritersfestival.com
www.screenwritersfestival.com

Festival Director *David Pearson*
Festival Co-ordinator *Kenny MacDonald*

Founded in 2006. Annual three day event exploring the art, craft, education and business of screenwriting. Advisory board members and regular contributors include Julian Fellowes (*Gosford Park*), Olivia Hetreed (*Girl With a Pearl Earring*) and Bill Nicholson (*Gladiator*). Takes place in and around Cheltenham Film Studios. Sessions include masterclasses, keynote speeches, networking events, screenings, seminars, workshops and educational events for schools. Caters for all skill levels. Scheduled for the first week of July 2009. A ticket for the duration of the festival cost £330 in 2008.

Manuscripts only accepted for specific festival initiatives. Application instructions available on the website.

Italian Film Festival

30 Lauder Road, Edinburgh, EH9 2JF
℡ 0131 667 2979 🖷 0131 667 2979
richard.mowe@which.net
www.italianfilmfestival.org.uk
Director *Richard Mowe*

Founded in 1994, showing films (features, shorts and documentaries) by established Italian directors at venues throughout the UK. Held annually.

Kendal Mountain Festival

Riverside Centre Yard 39 - Highgate, Kendal, Cumbria, LA9 4ED
℡ 01539 738669 🖷 01539 738669
filmentry@mountainfilm.co.uk
www.mountainfilm.co.uk
Director *John Porter*
Director Extreme Film School *Brian Hall*
Development Director/Book Festival Director *Julie Tait*
Festival Co-ordinator *Abigail Willis*

Founded in 1979. Runs annually in the third week of November for 10 days. Includes screenings, competitions, five day film school, 48-hour filmmaking marathon, book festival, workshops, seminars and exhibitions. All film types encouraged but especially those with a link to adventure, the culture of mountain communities and the environmental/economic issues affecting them. 2003 featured the gala premiere of *Touching the Void*.

Unsolicited applications welcomed. Entry details and regulations on the website.

Kinofilm - Manchester International Short Film Festival

johnw@kinofilm.org.uk
www.kinofilm.org.uk

Kino aims to promote the short film format through exhibition, distribution, promotion and education. For information on entering films for the festival, see the website. Currently seeking new funding for future events.

Landcrab Film Festival

information@landcrabfilmfestival.co.uk
www.landcrabfilmfestival.co.uk

Short film festival held annually in Dorset, featuring films from recognized filmmakers and amateurs alike.

Latin American Film Festival

436A New Cross Road, London, SE14 6TY
℡ 020 8692 6925
info@latinamericanfilmfestival.com
www.latinamericanfilmfestival.com

Founded in 1990. Dedicated to promoting the distribution and appreciation of Latin American films in the UK, featuring a mix of film screenings and associated events.

Leeds International Film Festival

Town Hall, The Headrow, Leeds, LS1 3AD
℡ 0113 247 8398 🖷 0113 247 8397
filmfestival@leeds.gov.uk
www.leedsfilm.com

Reputation as one of the UK's most progressive and important film festivals and the UK's largest regional film festival. Running since the 1980s.

Looks for striking and innovative features and shorts of any genre from across the world. To download an entry form see the website or email for more information.

Liverpool Film Festival

Toxteth TV, 37–45 Windsor Street, Liverpool, L8 1XE
℡ 07763 486 141
info@liverpoolff.com
www.liverpoolff.com

Annual festival, screening a mixture of local and international films. Aims to unite filmmakers, commercial organizations, academic institutions, enthusiasts and community groups. Has a partnership with the Future Shorts network to host the Liverpool Future Shorts monthly screenings event for short films.

Llanberis Mountain Film Festival

℡ 01286 871534
info2008@llamff.co.uk
www.llamff.co.uk

Eclectic programme of films, talks, poetry, photography and art, from Wales and beyond. Aims to capture the spirit of adventure and celebrate mountain art, sport and culture worldwide.

London Australian Film Festival

film@barbican.org.uk
www.barbican.org.uk/australianfilm/home

Annual since 1994, showing all major Australian feature film releases. Based at the Barbican Centre in London. Also offers a shorts programme in association with Flickerfest, Australia's leading international short film festival.

London Children's Film Festival

☏ 020 7382 7368
film@barbican.org.uk
www.londonchildrenfilm.org.uk

Launched in 2005 and held over 9 or 10 days at London's Barbican Centre. Film programme includes previews and premieres, new world cinema, documentaries, archive titles, shorts, films made by young people and sing-a-long film. Also runs an extensive education programme before and during the festival.

London International Animation Festival

c/o The Film and Video Workshop, Hungerford Primary School, Hungerford Road, London, N7 9LF
☏ 020 7607 8660 🅕 020 7607 8660
info@liaf.org.uk
www.liaf.org.uk
Director *Nag Vladermersky*

Annual festival founded in 2003. Lasts one week, usually the first week of September. Strives to screen the best and most recent animated films from Britain and abroad. Also features retrospective and themed sessions, workshops and international guests.
    Submissions via email.

London Lesbian and Gay Film Festival

www.llgff.org.uk

Established in 1986, the third largest film festival in the UK. Usually held in March/April and then tours from May to September around 40 towns and cities across the UK and Ireland. Presents an eclectic programme of feature films, shorts, artists' film and video and experimental work, as well as discussions and special events.

London Short Film Festival

Curzon Soho Cinema, 99 Shaftesbury Avenue, London, W1D 5YD
info@shortfilms.org.uk
www.shortfilms.org.uk
Co-Director *Philip Ilson*
Co-Director *Katie Taylor*

Annual film festival lasting about 10 days and founded in 2004. Screens all types of short film: drama, animation, documentary, experimental, low-budget. Also features talks and multi-media events.
    Unsolicited applications welcomed via email.

Lovebytes Digital Arts Festival

The Workstation, 15 Paternoster Row, Sheffield, S1 2BX
☏ 0114 268 2080
info@lovebytes.org.uk
www.lovebytes.org.uk

Held annually during March/April, bringing together film-makers, computer programmers, graphic designers and musicians from all over the world.

MediaGuardian Edinburgh International Television Festival (MGEITF)

117 Farringdon Road, London, EC1R 3BX
☏ 020 7278 9515 🅕 020 7278 9495
info@mgeitf.co.uk
www.mgeitf.co.uk

Major annual event for the television industry, founded in 1976 and hosted at the Edinburgh International Conference Centre. Explores key industry issues and offers excellent networking opportunities.

Meniscus Film Festival

158 Welholme Road, Great Grimsby, North East Lincolnshire, DN32 9LP
☏ 01472 507367
info@meniscusfilms.com
www.meniscusfilms.com

Held annually at the Whitgift Cinema in Grimsby, screening a range of independent cinema releases and hosting several other events throughout the year.

Mid Ulster Film Festival

☏ 075 1500 5595
info@midulsterfilmfestival.com
www.midulsterfilmfestival.com

Held in An Creagan in Omagh. Now in its 5th year. Programme includes feature films, short films and documentaries.
    Entry applications available via the website.

Midlands BFM International Film Festival

T 0121 678 6039
midlands@bfmmedia.com
www.bfmmedia.com

Festival with stated aim of being a vehicle for black British talent. The closing night includes the Short Film Awards, a highlight for new writers, directors, actors and cinematographers. Linked with Time Out's 'London on Screen' season.

National Association of Writers' Groups (NAWG) Open Festival of Writing

The Arts Centre Biddick Lane, Washington, Tyne & Wear, NE38 2AB
www.nawg.co.uk

The Association launched in 1995 and encompasses 150 affiliated groups and over 100 individual members. Aims to bring cohesion and fellowship to isolated writers' groups and individuals. Has held several festivals, originally in Washington, Tyne & Wear and latterly at St Aidan's College, University of Durham.

Northern Lights Film Festival

Tyneside Cinema Old Town Hall, West Street, Gateshead, NE8 1HE
T 0191 261 7674 F 0191 221 0535
info@nlff.co.uk
www.nlff.co.uk
Festival Director *Stephanie Little*

One week festival held annually in November. Founded in 2003. Features screenings, talks, educational workshops and the largest short film award in the UK. Specializes in features, documentaries, shorts and animation from the UK and northern Europe.

Script submissions welcomed via the website.

Onedotzero

Unit 212c Curtain House, 134-146 Curtain Road, London, EC2A 3AR
F 020 7729 0057
info@onedotzero.com
www.onedotzero.com

Founded in 1996, since when it has collated and commissioned over 200 hours of original programming for its annual digital film festival and related projects. Now tours to 60 major cities around the world.

Optronica

mail@optronica.org
www.optronica.org

Festival of visual music presented by the British Film Institute, Addictive TV and Cinefeel.

Features live audiovisual performances, screenings, illustrated talks, workshops and related special presentations.

Oska Bright Film Festival

Carousel Community Base, 113 Queens Road, Brighton, East Sussex, BN1 3XG
T 01273 234 734 F 01273 234 735
enquiries@carousel.org.uk
www.oskabright.co.uk
Artistic Director *Mark Richardson*

First film festival anywhere in the world run by, and for, people with a learning disability. Established in 2004. On odd years the Brighton-based festival lasts two days and offers screenings, workshops, bursaries and an awards ceremony. On even years the previous year's festival tours the UK.

Submissions must be no longer than 10 minutes. Further details via email or phone.

Outsiders: Liverpool Lesbian and Gay Film Festival

15 Sandon Street, Liverpool, L8 7NS
T 0151 703 0548
mattfox@outsidersfilmfestival.com
www.outsidersfilmfestival.com
Festival Director *Matthew Fox*
Co-programmer *Joan Burnett*
Co-programmer *Michelle Mangan*

Screens current and classic shorts and feature films with gay, lesbian, bisexual or transgender themes. Other activities include masterclasses, exhibitions, workshops, social events and monthly screenings throughout the year. Past films featured include *Shortbus*, *Unveiled* and *Pink Narcissus*.

The festival is held annually over a period of two weeks.

Welcomes any film submissions with gay, lesbian, bisexual or transgender content or makers.

Oxdox International Documentary Film Festival

The Jam Factory, 27 Park End Street, Oxford, OX1 1HU
info@oxdox.com
www.oxdox.com

Festival with a theme of 'Documenting the Real World'. Subjects in 2007 included China, the Mafia, women in Islamic countries and city life. Administers an audience award for the Best Feature Documentary and Best Short film while a panel selects Best Student Film and Best Dance Film.

Purbeck Film Festival

8 Salisbury Road, Swanage, BH19 2DY
℡ 07939 968238
info@purbeckfilm.com
www.purbeckfilm.org.uk

Long-running rural film festival. Aims to show wide range of themed films and to serve areas without easy access to public cinema. Presents over eighty films along with their associated notes, personal introductions and discussions.

QuickFlick World London

℡ 0772 944 9107
saint@qfworld.tv
www.qfworld.tv

A global digital film festival held simultaneously once a month in cities across the globe. Each festival has a theme and a set technical parameter. The finished films are then screened and critiqued.

RAI International Festival of Ethnographic Film

Royal Anthropological Institute, 50 Fitzroy Street, London, W1T 5BT
℡ 020 7387 0455 ℻ 020 7388 8817
film@therai.org.uk
www.raifilmfest.org.uk
RAI Film Officer, Festival Manager *Susanne Hammacher*

Biennial festival specializing in international ethnographic documentary and anthropological filmmaking. Features screenings, workshops, talks and exhibitions.

Unsolicited applications welcomed via email/website.

Raindance Film Festival

81 Berwick Street, London, W1F 8TW
℡ 0207 287 3833 ℻ 0207 439 2243
info@raindance.co.uk
www.raindance.co.uk
Founder *Elliot Grove*

International festival screening around 90 feature films and documentaries and 150 shorts. Founded in 1993, the event has previously hosted UK premieres of *Pulp Fiction*, *Blair Witch Project* and *Memento*.

Held annually for 12 days in September/October.

Rushes Soho Shorts Festival

66 Old Compton Street, London, W1D 4UH
℡ 020 7437 8676
info@sohoshorts.com
www.sohoshorts.com
Festival Director *Joe Bateman*

Annual event for both established filmmakers and newcomers. Founded in 1999 and now receives over 2,000 entries from around the globe. Awards festival held in Leicester Square and the main venue is the Curzon Cinema in Soho.

Films must be no longer than 12mins without credits and should be submitted on both DVD and Beta Tape. (SAE required for return). Entries to the festival must have been completed between beginning of the previous year up to and including the month of April in the year the festival is being held. The festival competition categories are: Documentary, Short Film, Music Video, Broadcast Design, Animation and Newcomer.

Sand-Swansea Animation Day

SAND Office, Swansea Institute of Higher Education, Mount Pleasant, Swansea, West Glamorgan, SA1 6ED
℡ 01792 481194 ℻ 01792 481158
sand@sihe.ac.uk
www.sand.org.uk

Founded in 2000, an international CGI (computer generated imagery) event which includes a conference, workshops, seminars and film screenings. Tailored primarily for the Creative IP Industries and individuals interested in animation and digital media. Held annually over 5 days.

Sheffield DocFest

The Workstation, 16 Paternoster Row, Sheffield, S1 2BX
℡ 0114 272 5141 ℻ 0114 276 1849
info@sidf.co.uk
www.sheffdocfest.com
Director *Heather Croall*
Festival Programmer *David Teigeler*
Festival Manager *Emma Ryan*

Established in 1994, the programme includes public screenings, pitching competitions and networking events. Among the workshop activities are a 'Student DocDay Afternoon' and a Newcomers Day for aspiring filmmakers. Also hosts 'MeetMarket', an event matching documentary makers with UK and international buyers. Recent films featured include *Black Gold* and *A Crude Awakening*.

Held annually, the festival lasts between 4 and 7 days.

Unsolicited applications are welcomed through an online submission process.

Showcomotion Young People's Film Festival

Showroom Cinema, Paternoster Row, Sheffield, S1 2BX
℡ 0114 276 3534 ℻ 0114 279 6522

info@showcomotion.org.uk
www.showcomotion.org.uk
Festival Director *Kathy Loizou*
Festival Administrator *Kate Allen*

Founded in 1998, the festival screens short and
feature films for young people from across
the world. Incorporates talks and discussions,
workshops, premieres and award ceremonies.
Past films featured include *Hoodwinked, Offside,
Eat Dog Cat Mouse* and *Midsummer Dream*. The
festival is also host to the Showcomotion Children's
Media Conference, bringing together producers,
broadcasters, filmmakers and commissioners.

Held annually in July, lasting for two weeks.

A call for submissions is made in January but
entries are accepted all year round.

Signals International Short Film Festival

Victoria Chambers, St Runwald Street,
Colchester, CO1 1HF
℡ 01206 560 255
info@signals.org.uk
www.signals.org.uk
Director *Andy Roshay*

Short film festival running yearly for six days.
Established in 2005. Includes screenings, talks,
exhibitions, workshops and an awards ceremony.

Unsolicited material accepted via email.

Super Shorts

c/o Portobello Post, 274d Portobello Road,
London, W10 5TE
℡ 020 8969 3236
admin@supershorts.org.uk
www.supershorts.org.uk

Over 700 short films shown across London
venues. Films then go on a national tour.

Sutton Film Festival

10 Endale Close, Carshalton, Surrey, SM5 2HB
℡ 020 8669 0855
info@suttonfilmfestival.co.uk
www.suttonfilmfestival.co.uk
Organiser *Marq English*
Organiser/Technical *David Wells*

One-day festival founded in 2002 to showcase
new film talent. Includes screenings and
networking events. Encourages all genres of short
films alongside music promos, documentaries
and trailers.

Submissions via email/letter. No phone calls.

The Times BFI London Film Festival

www.lff.org.uk

London showcase of the best new films from
around the world. Shows mix of films that would

not otherwise get a UK screening and those that
will get a release in the autumn or spring.

UK Brasilian Film Festival

5 Durham Yard, Unit 10, Teesdale Street, London,
E2 6QF
℡ 020 7729 1332
info@somethingfrombrasil.org
www.somethingfrombrasil.org

Festival celebrating the diversity of Brazilian films
and paying homage to Black Brazilian cinema.
Programme includes features, medium-length
films and shorts covering a broad panorama of
documentary, fiction and experimental formats.

UK Jewish Film Festival

4.07 Clerkenwell Workshops, 27-31 Clerkenwell
Close, London, EC1R 0AT
℡ 020 3176 0048
info@ukjff.org.uk
www.ukjewishfilmfestival.org.uk

Founded in 1997 in Brighton, home of the
founder, Judy Ironside. In 2003 the festival went
to eight venues including Bristol, Birmingham,
Manchester and Glasgow and has subsequently
become a national event. Screenings are often
accompanied by visiting directors and speakers.

Viva! Spanish Film Fest

70 Oxford Street, Manchester, M1 5NH
viva@cornerhouse.org
www.vivafilmfestival.com

Founded in 1995, showcasing Spanish and Latin
American features, shorts and documentaries. All
films are in Spanish with English subtitles.

Watershed

1 Canon's Road, Harbourside, Bristol, BS1 5TX
℡ 0117 927 6444
info@watershed.co.uk
www.watershed.co.uk
Communications Coordinator *Anja Datton*

Founded in 1982, a dedicated media centre that
hosts a variety of events such as the Encounters
Short Film Festival, the Depict! Awards for Short
Films and a Japan Film Season. Also home to the
Bristol Screenwriters' Group.

Wildscreen Festival

Ground Floor, The Rackhay Queen Charlotte
Street, Bristol, BS1 4HJ
℡ 0117 328 5950 Ⅎ 0117 328 5955
info@wildscreen.org.uk
www.wildscreenfestival.org

The Wildscreen Festival searches for the world's
best wildlife and environmental films. Founded

by Sir Peter Scott in 1982, the festival has been organised every other year for the past 25 years. Held in Bristol, the world's centre for wildlife filmmaking, it attracts hundreds of delegates from around the globe who work in film, television and the press, as well as those actively involved in working to conserve the environment.

Refer to website for submission information.

Winchester Writers' Conference

Faculty of Arts, University of Winchester, West Hill, Winchester, Hampshire, SO22 4NR
℡ 01962 827238
Barbara.Large@winchester.ac.uk
www.writersconference.co.uk
Founder-Director *Barbara Large*

Running for 29 years, held in late June and followed by a week of writing workshops. Offers workshops, mini courses, lectures, seminars and one-to-one appointments to help writers harness their creative ideas and to develop their writing, editing and marketing skills. Many lectures cater for those wishing to write for the screen. Honorary Patron: Dame Beryl Bainbridge.

Wood Green International Short Film Festival

Film and Video Workshop, Hungerford School, Hungerford Road, London, N7 9LF
℡ 020 8365 8631
wgtcm@aol.com
www.woodgreenfilmfest.com

Founded in 2003, showing a selection of films and specialist programmes over three days. Focuses on short films of 5–10 minutes.

Wow! Wales One World Festival

Taliesin Arts Centre, University of Wales, Swansea
℡ 01239 615066 ℻ 01239 615066
dave@oneworldfilm.com
www.oneworldfilm.com

Founded in 2002, with over 30 different films from around the world. Held at various venues throughout Wales.

Canada

The 3 Americas Film Festival

www.fc3a.com

Five-day festival held in Quebec City, showcasing films from across the Americas with a focus particularly on independent films.

Submissions welcomed through the website.

Antimatter Underground Film Festival

636 Yates Street, Victoria, British Columbia, V8W 1L3
℡ 250 385 3327 ℻ 250 385 3327
info@antimatter.ws
www.antimatter.ws
Curator *Deborah de Boer*

Founded in 1998. Dedicated to the exhibition and nurturing of film and video as art. Showcases all genres of experimental film (primarily shorts), through screenings, installations, performances and media hybrids. Strives to be completely free from commercial and industry agendas. Runs annually for 9 days.

Unsolicited applications welcomed via email.

Atlantic Film Festival, Halifax

PO Box 36139, Halifax, Nova Scotia, B3J 3S9
℡ 902 422 3456 ℻ 902 422 4006
festival@atlanticfilm.com
www.atlanticfilm.com

Founded 1980 and held annually in September. Screenings run alongside a 'Strategic Partners' co-production conference, the 'alfresco filmFesto' outdoor summer film series and several industry panel, discussion, master class and lecture initiatives. Sister project for young people, 'ViewFinders: International Film Festival for Youth', takes place annually in April.

Script submissions encouraged. Application via website.

Banff Mountain Film Festival

The Banff Centre Box 1020, Banff, Alberta, T1L 1H5
℡ 403 762 6369 ℻ 403 762 6277
mountainculture@banffcentre.ca
www.banffcentre.ca

One-week festival held annually, founded in 1975. Screens around 50 short films with a focus on high-adrenaline mountain stories. Also presents talks, shows and discussion opportunities with top filmmakers and adventurers. Prize-money for competitions up to CA$4,000. Later tours 30 countries with a total audience of more than 170,000 people.

Submissions welcomed. Details and deadlines on the website.

Banff World Television Festival

Achilles Media Ltd, 102 Boulder Crescent, Suite 202, Canmore, Alberta, T1W 1L2
℡ 403 678 1216 ℻ 403 678 3357
info@achillesmedia.com
www.bwtvf.com

Television festival established in 1980. Focused on the development of business and creative opportunities for those in television programming. Features seminars, master classes, pitching opportunities and the high-profile Banff World Television Awards (in Partnership with Alberta Film).

Calgary International Film Festival

Suite 210, 308-11 Avenue, S.E., Calgary, Alberta, T2G 0Y2
℡ 403 283 1490 ℻ 403 283 1498
www.calgaryfilm.com

Established in 1999 as a non-profit organization. Aims to provide an annual world-class cinema event for the citizens of Calgary and the surrounding area.

Canadian Comedy Festival and Awards

PO Box 39, 72924 Airport Line, Hensall, Ontario, N0M 1X0
cass@bayleygroup.com
www.canadiancomedyawards.ca

Celebration of the best Canadian comedic talent in stand-up, television and film. Held annually in London, Ontario by The Comedy Network. Features four days of stand-up shows, workshops and awards.

Canadian Film Centre's Worldwide Short Film Festival

c/o CFC, 2489 Bayview Avenue, Toronto, Ontario, M2L 1A8
℡ 416 445 1446 ext. 815
shortfilmfest@cfccreates.com
www.worldwideshortfilmfest.com

Held annually in Toronto. Activities include screenings of international short films, a symposium on short filmmaking and an awards ceremony. Features Screenplay Giveaway Prize with winner receiving prizes worth $100,000. Held in June, lasting six days.

Canadian Filmmaker's Festival

℡ 416 846 3378
berneuler@canfilmfest.ca
www.canfilmfest.ca

Non-profit organization devoted to the celebration, promotion and advancement of Canadian filmmakers. Exclusively features Canadian films and provides valuable showcasing and networking opportunities for homegrown talent.

Cinefest – Sudbury International Film Festival

49 Durham Street, Sudbury, Ontario, P3E 3M2
℡ 705 688 1234 ℻ 705 688 1351
cinefest@cinefest.com
www.cinefest.com

Established in 1989, screening more than 100 films each year including domestic and international productions, films from Northern Ontario and aboriginal productions.
Submissions are welcomed via the website.

Dawson City International Short Film Festival

Klondike Institue of Art and Culture, Bag 8000, Dawson City, Yukon, Y0B 1G0
℡ 867 993 5005 ℻ 867 993 5838
dawsonarts@yknet.ca
www.dawsonfilmfest.com

Founded in 1999, screening over 60 films each year. Other events include workshops, guest speakers and panel discussions.

Edmonton International Film Festival

Suite 201, 10816A – 82 Avenue, Edmonton, Alberta, T6E 2B3
℡ 780 423 0844 ℻ 780 447 5242
info@edmontonfilmfest.com
www.edmontonfilmfest.com

Held annually for nine days in October. Emphasis placed on 'discovery'. Features a mix of feature-length films, light documentaries and short films from around the world.

Fantasy Worldwide Film Festival

37 Langford Avenue, Upper Level, Toronto, Ontario, M4J 3E4
℡ 416 406 2224 ℻ 416 406 2224
info@fantasyworldwide.com
www.fantasyworldwide.com
Executive Director *Johanna Kern*
Director of Operations *Deborah Pagliuca*

Festival launched in February 2005 to promote fantasy filmmaking in Canada and around the world. Includes screenings, seminars and award ceremonies. Interested in feature films, shorts, documentaries and animation focusing on world mythology, fantasy (no horror), mysticism, magical realism, science fiction, historical fiction, legend and artetype. Runs annually in October for 3–5 days.
Welcomes submissions via email. All filmmakers from beginner to veteran encouraged to apply.

Freeze Frame International Festival of Films for Kids of All Ages

465-70 Rue Arthur Street, Winnipeg, Manitoba, R3B 1G7
℡ 204 943 5341 ℻ 204 957 5437
info@freezeframeonline.org
www.freezeframeonline.org

Provides film screenings and video production workshops for kids and teens. Goes on tour and runs a family film day, showing an hour of international animated shorts followed by a 3-hour animation workshop.

Giggleshorts

⊤ 416 924 7201
giggles@giggleshorts.com
www.giggleshorts.com

Festival of international comedy shorts, held at the Brunswick Theatre in Toronto.

Hot Docs Canadian International Documentary Festival

110 Spadina Avenue, Suite 333, Toronto, Ontario, M5V 2K4
⊤ 416 203 2155 Ⓕ 416 203 0446
www.hotdocs.ca

North America's largest documentary festival, founded in 1993. Annually shows over 100 cutting-edge domestic and international documentaries. Runs industry programmes to promote professional development, and market and networking opportunities for documentary professionals. Also hosts the Toronto Documentary Forum, a limited-seating international market event for buyers and sellers.

Image and Nation: Montréal International LGBT Film Festival

#404, 4067 Boulevard St. Laurent, Montréal, Québec, H2W 1Y7
info@image-nation.org
www.image-nation.org
Director of Programming *Katharine Setzer*
Executive Director *Charlie Boudreau*

Lesbian, gay, bisexual and transgender festival founded in 1987. Screens feature length, short and documentary films along with filmmaker talks and special presentations. Interested in film works from all genres made by and for an LGBT audience. Does accept television productions upon occasion. Past screening examples include *20 Centimetres* by Ramon Salazar (2006) and *Reinas* by Manuel Gomez Pereira (2006). Takes place annually over 11 days.

Unsolicited applications welcomed via email or website.

InsideOut – The Toronto Lesbian and Gay Film and Video Festival

219–401 Richmond Street West, Toronto, Ontario, M5V 3A8
⊤ 416 977 6847 Ⓕ 416 977 8025
inside@insideout.ca
www.insideout.on.ca

Held over 11 days in May, promoting film and video by or about lesbian, gay, bisexual and transsexual people. Hosts screenings, artist talks, installations and discussions, featuring more than 275 films and videos from across the world.

Montreal International Festival of New Cinema & Media (Festival du nouveau cinéma de Montréal)

3805 Boulevard Saint-Laurent, Montréal, Quebec, H2W 1X9
⊤ 514 282 0004 Ⓕ 514 282 6664
info@nouveaucinema.ca
www.nouveaucinema.ca

Founded in 1971, aims to be a forum for audiences to discover original new works, particularly in cinéma d'auteur and digital creation. Features over 300 works from more than 40 countries.

Montreal World Film Festival

1432 de Bleury Street, Montréal, Québec, H3A 2J1
⊤ 514 848 3883 Ⓕ 514 848 3886
info@ffm-montreal.org
www.ffm-montreal.org

Aims to be a forum for quality cinema from around the world. Includes jury-chosen awards and People's Choice awards (for most popular Canadian film, best film from Latin America, best documentary and best Canadian short).

Nickel Independent Film Festival

PO Box 1644, Stn. C, St John's, Newfoundland, A1C 5P3
⊤ 709 576 FEST
nickelfestival@yahoo.ca
www.nickelfestival.com
President *Baptiste Neis*
Administrator *Laura Churchill*

Festival for independent film founded in 2000. Showcases film and video of all lengths and genres once a year, usually in the last week of June for five days. Includes screenings, workshops, awards, Q&A, musical entertainment, master classes and readings.

Submissions welcomed via www.withoutabox.com.

NSI FilmExchange Canadian Film Festival

Suite 400, 141 Bannatyne Avenue, Winnipeg, Manitoba, R3B 0R3
⊤ 204 956 7800 Ⓕ 204 956 5811
info@nsi-canada.ca
www.nsi-canada.ca/filmexchange

Annual celebration of Canadian screen achievement, bringing together writers, producers, directors and executives with emerging talent. Shows features and shorts and runs numerous events including masterclasses, competitions, discussions and networking events. Also operates SnowScreen, an outdoor 10 x 8 foot movie screen made of snow.

Ottawa International Animation Festival

2 Daly Avenue, Suite 120, Ottawa, Ontario, K1N 6E2
T 613 232 8769 F 613 232 6315
info@animationfestival.ca
www.ottawa.awn.com

Largest animation event of its kind in North America. Events include screenings, panels, workshops and parties.

Planet in Focus: International Environmental Film and Video Festival

55 Mill Street, Case Goods Warehouse Building 74 Studio 402, Toronto, Ontario, M5A 3C4
T 416 531 1769 F 416 531 8985
information@planetinfocus.org
www.planetinfocus.org
Administative and Development Co-ordinator
Kim Haladay
Educational and Outreach Officer *Myan Marcen Gaudaur*
Festival Director *Candida Paltiel*

Environmental film and video festival founded in 1999. Features over 50 screenings along with workshops, round table discussions, pitch sessions, awards ceremonies and talks. Recently screened films include *Conflict Tiger* (animal rights documentary) and *Terra* (short animation). Runs annually for one week. Scheduled for 24–28th October 2008.
    Unsolicited applications welcomed. Details on website.

ReelWorld Film Festival

438 Parliament Street, Suite 300, Toronto, Ontario
T 416 598 7933 ext. 28 F 416 585 2524
www.reelworld.ca

Festival celebrating diversity in international filmmaking. Provides screenings, seminars, workshops and special events. Run in conjunction with the ReelWorld Foundation.

Toronto After Dark Film Festival

3219 Yonge Street, Suite 346, Toronto, Ontario, M4N 3S1
info@torontoafterdark.com
www.torontoafterdark.com

A week-long showcase of new international horror, sci-fi, fantasy and thrillers (features and shorts). Held in October.

Toronto International Film Festival

2 Carlton Street, Suite 1600, Toronto, Ontario, M5B 1J3
T 416 967 7371
www.tiffg.ca

Founded in the 1970s and now a major festival on the world stage.
    Films may be submitted via form available on website.

Toronto International Latin Film Festival

100 Seaton Street, Toronto, Ontario, M5A 2T3
T 416 364 3131 F 416 359 9238
info@tilff.com
www.tilff.com

Festival presenting mostly independent films from Latin America, Southern Europe and Quebec. Held annually in October, lasting eight days.
    Welcomes submissions through the website.

Toronto Jewish Film Festival

17 Madison Avenue, Toronto, Ontario, M5R 2S2
T 416 324 8226 F 416 324 8668
tjff@tjff.ca
www.tjff.ca

Aims to showcase Jewish culture, heritage and the diversity of the Jewish experience in Canada and around the world via the medium of cinema. Includes feature films, documentaries and shorts.

Toronto Reel Asian International Film Festival

401 Richmond Street West, Suite 309, Toronto, Ontario, M5V 3A8
T 416 703 9333 F 416 703 9986
info@reelasian.com
www.reelasian.com

Established in 1997, showcasing contemporary Asian cinema and work from the Asian diaspora. Annual event lasting five days, offering screenings, forums, workshops and galas. Festival also spotlights one Canadian artist from the Asian filmmaking community each year.

Vancouver Asian Film Festival

110 Keefer Street, Vancouver, British Columbia, V6A 1X4
F 604 251 6828
www.vaff.org

Aims to promote independent North American Asian filmmakers and to act as a springboard to bigger film festivals. Committed to supporting both emerging and established artists. Held annually in November.

Vancouver International Film Festival

1181 Seymour Street, Vancouver, British Columbia, V6B 3M7
T 604 685 0260 F 604 688 8221
e.viff@viff.org
www.viff.org

Founded in 1981, over 150,000 people attend almost 600 screenings of films from more than 50 countries. Also runs a film and television forum with many leading names in the field.
    Submission forms available from website.

Vancouver International Jewish Film Festival

6184 Ash Street, Vancouver, British Columbia, V5Z 3G9
T 604 266 0245 F 604 266.0244
vjff@telus.net
www.vijff.com

Screens around fifty films from across the world, including new films and classics of Jewish interest. Also features awards ceremony, panel discussions and gala events.

Vancouver Queer Film Festival

Out On Screen, 405–207 West Hastings Street, Vancouver, British Columbia, V6B 1H7
T 604 844 1615 F 604 844 1698
www.outonscreen.com

Festival promoting the production and exhibition of independent queer media art. Hosts free professional development workshops.

Vancouver Student Film Festival

info@vsff.com
www.vsff.com

Established in 2005, screening films by student filmmakers and graduates from Vancouver. Aims to raise awareness of young talent throughout the area.

The Victoria Film Festival

1225 Blanchard Street, Victoria, British Columbia, V8W 3J4
T 250 389 0444 F 250 389 0406
festival@victoriafilmfestival.com
www.victoriafilmfestival.com
Programmer *Donovan Aikman*

Founded in 1995. Runs annually for 10 days. Screens over 140 films including 50 features.

Includes an international film co-production financing conference (Trigger Points Pacific) alongside parties, gala openings and master classes. Features narrative, documentary, experimental and animation films from Canada and around the world.
    Unsolicited applications welcomed with official entry form.

World of Comedy Film Festival

T 416 487 7574
info@worldcomedyfilmfest.com
www.worldcomedyfilmfest.com

Held in Toronto during March, celebrating the comedy film. Mixture of new movies from around the world and old classics. Also serves as an industry marketplace and networking event.

Worldwide Short Film Festival

shortfilmfest@cfccreates.com
www.worldwideshortfilmfest.com

Major festival of short films from Canada and around the world.

Ireland

Cork Film Festival

Emmet House, Emmet Place, Cork
T 021 427 1711 F 021 427 5945
info@corkfilmfest.org
www.corkfilmfest.org
Festival Manager *Eimear O'Herlihy*

Founded in 1956. Includes screenings, features, documentaries, shorts, workshops and seminars, an awards ceremony and 'meet the film-maker' events. Shows films of all genres, including fiction, documentary, experimental and student. Special focus on gay and lesbian films.
    Held annually and lasts for 8 days.
    Welcomes unsolicited applications. Entry forms are available on-line each year from January to June.

Cork Youth International Film Arts Festival

16 Fr. Matthew Street, Cork City
T 021 430 6019
hprout@europe.com

Established in 1980, a festival of film for the young, especially those involved in youth clubs, schools and community groups.

Dark Light Digital Festival

69 Dame Street, Dublin 2
T 01 670 9017
linda@darklight-filmfestival.com
www.darklight-filmfestival.com

Established in 1999, screening short films, feature films, animation, mobile films and documentaries. Ranges from student films to household-name directors.

Dublin International Film Festival

Temple Lane, Cecilia Street, Temple Bar, Dublin 2
ⓣ 01 635 0290
info@dubliniff.com
www.dubliniff.com

Founded in 2002, screening films from around the world.

Fresh Film Festival

c/o Belltable Arts, 69 O'Connell Street, Limerick
ⓣ 061 319 555 ⓕ 061 319 555
info@freshfilmfestival.net
www.freshfilmfestival.net

Founded in 1997, aimed at young people up to the age of 18. Includes a competition for films made by children aged 7 to 18, and screens feature films for schools with study guides.

Held for a week during spring.

Welcomes unsolicited applications and prefers to be contacted via email.

Galway Film Fleadh

Cluain Mhuire, Monivea Road, Galway
ⓣ 091 751 655 ⓕ 091 735 831
gafleadh@iol.ie
www.galwayfilmfleadh.com
Managing Director *Miriam Allen*
Programme Director *Felim Mac Dermott*
Administrator *Cathy o' Connor*

Founded in 1988, The Fleadh is a filmmakers' festival, attracting directors and actors from all cultural backgrounds. Guests have include Robert Towne, Kathy Bates and Wood Harrelson. Activities including actors, directors and screenwriters masterclasses, public interviews, debates and workshops. Runs alongside Ireland's only Film Fair Market which coordinates prescheduled meeting between filmmakers and industry professionals.

A 6-day event held annually.

Does not accept unsolicited applications.

Kerry Film Festival

Samhlaiocht, The Old Presbytery, Lower Castle Street, Tralee, County Kerry
ⓣ 066 712 9934 ⓕ 066 712 0934
info@samhlaiocht.com
www.kerryfilmfestival.com
Artistic Director *Jason O'Mahony*

Held annually in the winter. Screens features and shorts. The main feature of the festival is a short film competition, which focuses on young filmmakers.

A synopsis of 100 words must be submitted with the application form.

Movies on the Square

12 East Essex Street, Temple Bar, Dublin 2
ⓣ 01 677 2255 ⓕ 01 677 2525
info@templebar.ie
www.templebar.ie
Head of Cultural Development *Grainne Millar*
Communications Manager *Roisin McCarthy*
Venue and Events Manager *Declan Greaney*
Cultural Programme Administrator *Lorraine Maye*

Founded in 1999, a 12-week programme of free outdoor movies screened at the Temple Bar on Saturday nights. Also runs a short film award competition. Screens an eclectic mix of films ranging from musicals to black and white classics.

Welcomes unsolicited applications. Contact the Cultural Department on 01 677 2255 or via email.

Australia

Adelaide Film Festival

12 King William Road, Unley, SA 5061
ⓣ 08 8271 1029 ⓕ 08 8271 9905
info@adelaidefilmfestival.org
www.adelaidefilmfestival.org

Founded in 2003, screening films from over 40 countries. Shows feature films, documentaries, shorts, music videos and animations. Introduced a juried Best Feature Film award in 2007 (cash prize of A$25,000).

The Alliance Française French Film Festival

301 George Street, Sydney, NSW 2000
ⓣ 02 9292 5700 ⓕ 02 9279 1640
enquiries@afsydney.com.au
www.afsydney.com.au
Cultural Events Assistant *Pascale Reuter*

Francophone festival with over 250 screenings including a wide range of new French films never seen before in Australia. Fosters interaction between Australian and French cultures.

Lasts one week in March/April.

Brisbane International Film Festival

Level 3, The Regent, 167 Queen Street, Brisbane, QLD 4000
ⓣ 07 3007 3003
biff@biff.com.au
www.biff.com.au

Executive Director *Anne Demy-Geroe*
Festival Manager *Andrew Rose*

Established in 1991, events include screenings, seminars and awards. Covers international films, films on filmmaking, experimental work and shorts. In previous years the festival has featured films such as *A Prairie Home Companion*, *The Wind that Shakes the Barley* and *U-Carmen eKhayelitsha*.

Held annually, the festival lasts three weeks.

Welcomes unsolicited applications. Tickets are available from website or box office.

FlickerFest

PO Box 7416, Bondi Beach, Sydney, NSW 2026
☏ 02 9365 6888
info@flickerfest.com.au
www.flickerfest.com.au

Founded in 1992, Australia's premiere international short film festival. There is a main competitive programme, a short documentary competition, Australian competition and a number of additional programmes and forums out of competition.

Films must be no more than 30 minutes in length. All productions must have been completed within two years of the closing date and must be in English or have English subtitles.

Melbourne International Animation Festival

PO Box 1024, Collingwood, VIC 3066
☏ 03 9375 1490 ☒ 03 9376 9995
info@miaf.net
www.miaf.net

Founded in 2005, screening about 200 animated films from over 30 countries. Highlights of the festival include many guest artists and visiting animators, both local- and foreign-based. Held annually. See website for submission information.

Melbourne International Film Festival

PO Box 4982, Melbourne, VIC 3001
☏ 03 8660 4888 ☒ 03 9654 2561
miff@melbournefilmfestival.com.au
www.melbournefilmfestival.com.au

Founded in 1951, showcasing new Australian cinema alongside films from over 50 countries. Held over 19 days each winter.

Submission of material is subject to an entry fee. Films made in languages other than English must have English subtitles. No training or advertising films. Films 30 minutes or less are eligible for entry in the Best MIFF Shorts Competition with the exception of the documentary category, which accepts films of

60 minutes or less. Feature films must be Victorian premieres in order to be eligible for screening.

Melbourne Queer Film Festival

MQFF Festival Office, 6 Claremont Street, South Yarra, VIC 3141
☏ 03 9827 2022 ☒ 03 9827 1622
info@melbournequeerfilm.com.au
www.melbournequeerfilm.com.au

Founded in 1991, the festival includes programmes such as the 'Queeries: Bent On Film Youth Program'. Shows international shorts and feature films, documentaries, foreign language films and experimental works. Presents the City of Melbourne Emerging Filmmaker Award and hosts a series of forums and lectures. Held annually.

The Realm of the Senses

simon@realmofsenses.com
www.myspace.com/realmofsenses07

Founded in 2001, an outdoor festival with a primary focus on screening short films from Australia and New Zealand. Offers prizes of A$35,000 to the best three films in competition (films have to be made in the past two years). Held annually.

Email for submission information.

Revelation: Perth International Film Festival

PO Box 30, Fremantle, WA 6959
☏ 08 9335 3904
info@revelationfilmfest.org
www.revelationfilmfest.org
Director *Megan Spencer*
Administration *Mary Lusted*
Sponsorship *Rebecca Matthews*

Established in 1997, activities include workshops, masterclasses, screenings and talks. Recently featured films include *Spellbound*, *Amubulance* and *The Aura*.

Annual event lasting 10 days in July.

Welcomes unsolicited applications. The festival is ticketed, available from venues.

St Kilda Film Festival

Private Bag 3, Post Office, St Kilda, VIC 3182
☏ 03 9209 6490
filmfest@portphillip.vic.gov.au
www.stkildafilmfestival.com.au

Founded in 1983, showcasing both local and international short films. It screens 100 Australian short films and hosts international sessions from the Interfilm Berlin International Short Film Festival (Germany), Clermont Ferrand Short Film

Festival (France) and the Sao Paulo International Short Film Festival (Brazil). Held annually over 6 days.

Sydney International Film Festival

Suite 102, 59 Marlborough Street, Surry Hills, NSW 2010
T 02 9318 0999 F 02 9319 0055
info@sydneyfilmfestival.org
www.sydneyfilmfestival.org

Founded in 1954, screening a variety of films across genres from over 50 countries. Held over 17 days in June. See website for submission information.

Trasharama A-go-go

PO Box 376, Goodwood, SA 5034
jero@trasharama.com.au
www.trasharama.com.au
Festival *Jero Cocksmith*
Festival Director *Dick Dale*

Touring short film festival established in 1997. Films under 15 minutes in length, including 'schlock horror, cheesy sci-fi, bad taste comedy and other filmic disasterpieces'. Tours across Australia, visiting more than 15 cities and regional areas.

Welcomes unsolicited international applications by email. Event is ticketed and can be bought at venues.

Tropfest

62-64 Riley Street, East Sydney, NSW 2010
T 02 9368 0434 F 02 9360 1594
mail@tropfest.com.au
www.tropinc.com

Founded in 1993, a free, public, outdoor, short film festival held in locations across Australia. From more than 800 entries, 16 finalists are selected each year to premiere to an audience of more than 150,000 people via simultaneous satellite screenings in Sydney, Melbourne, Brisbane, Canberra, Perth, Hobart and eight regional locations around Australia. Held annually in February.

For submission films must be no longer than seven minutes, must never have been shown publicly before and must contain the Tropfest Signature Item (varies each year; in 2007 it was 'sneeze').

New Zealand

Allshorts Film Festival

PO Box 217, Takaka 7172, Golden Bay
allshorts@paradise.net.nz
www.allshorts.org.nz

Free festival held in the Village Theatre at Takaka each October. Restricted to New Zealand residents who are thirteen years and over.

Any visual content will be considered but should not exceed five minutes and should be submitted on a pal region 0 DVD (along with completed entry form). No prize money but judges choose 20 entries for the 'Festival Selection'.

Auckland International Film Festival

PO Box 9544, Marion Square, Wellington 6141
T 04 385 0162 F 04 801 7304
festival@nzff.co.nz
www.nzff.telecom.co.nz

Founded in 1970 as a component of the Auckland Festival. Has over time become a fund-raising event subsidizing live arts. Entry forms and regulations for submitting films are available for download on the website. Volunteer positions are available on application to assist with the festival.

Belladonna Canterbury Short Film Festival

PO Box 1351, Christchurch 8015
T 03 365 6151
admin@belladonna.org.nz
www.belladonna.org.nz

Founded in 2002, held annually in July. Showcases works across genres including documentaries, media art, experimental works, narrative and dance films. All films must be the work of a New Zealand citizen or resident.

Drama/documentary films should be no more than 15 minutes in duration and dance/experimental films no more than 10 minutes.

Big Mountain Short Film Festival

PO Box 221, Ohakune 5461
info@bigmountain.co.nz
www.bigmountain.co.nz
Director *Jeff Bollow*
Director *Bret Gibson*

Founded in 2006, events include screenings of 'budget' and 'no budget' short films from around the world, seminars, Q&As and an awards ceremony.

Held annually in October, lasting 3 days.

Does not welcome unsolicited applications. Festival is free to attend.

DOCNZ Film Festival

Level 2, 1 College Hill, Freemans Bay, Auckland 1011
T 09 309 2613 F 09 309 4084
info@docnz.org.nz
www.docnz.org.nz

Founded in 2004, an annual international documentary film festival with a competitive element.

Accepts documentary films in three different categories: short (less than 30 minutes); medium (30–70 minutes); feature (over 70 minutes). Entries must include a short and long written description of the film (including credits) plus entrant filmography.

Human Rights Film Festival

PO Box 24 423, Wellington
Ⓣ 04 496 9616
filmfestival@humanrights.net.nz
www.humanrightsfilmfest.net.nz

Founded in 2005, featuring international and local documentaries and dramatic films with a strong human rights theme. Screenings are followed by a speakers' forum.

Accepts submissions of feature films, documentaries and shorts with a strong human rights theme. A synopsis of 80 to 100 words should be included with the entry.

Latin American Film Festival

www.rialto.co.nz

Founded in 2002, screening new and classic Latin American films. The festival is programmed by the ambassadors of relevant countries. All films are screening for the first time in New Zealand.

Magma Short Film Festival

PO Box 6211, Whakarewarewa, Rotorua,
Ⓣ 07 343 9653
info@magmafilm.org.nz
www.magmafilm.org.nz
Director *Kiri Jarden*
Director *Juliet Boone*

Founded in 2006 as a forum for short film in digital media. Held over four days in late November and features contributions from leading NZ actors, filmmakers and production companies. An awards ceremony is held on the last night of the festival. Tickets available on-line, at selected outlets or on the door.

Film entries must be accompanied by an official entry form for selection consideration. Showcase films may be submitted by producers/directors not wishing their film to be part of the competition.

Show Me Shorts Film Festival Trust

PO Box 6685, Wellesley Street, Auckland 1141
Ⓣ 21 427 553

gina@showmeshorts.co.nz
www.showmeshorts.co.nz

Focus on short films with a strong New Zealand/ Australia component.

Films of any genre qualify but they should be no less than 3 minutes and no more than 30 minutes long.

The Wairoa Māori Film Festival

PO Box 85, Nuhaka 4165
Ⓣ 06 837 8854 Ⓕ 06 837 8761
maorimovies@gmail.com
www.manawairoa.com
Festival Director *Leo Koziol*
Executive Secretary *Huia Koziol*
Chairperson *Pauline Tangiora*

Indigenous festival founded in June 2005. Screens films featuring Māori cast and crew. Also workshops and talks with filmmakers, awards night presentations and scholarships for emerging talent. Annual event in June lasting four days.

Encourages submissions of feature films, short films and documentaries with an indigenous theme. Applications via email.

Wellington Film Festival

PO Box 9544, Marion Square, Wellington 6141
Ⓣ 04 385 0162 Ⓕ 04 801 7304
michael@nzff.co.nz
www.enzedff.co.nz

Annual festival founded in 1972, with a long-standing tradition of supporting New Zealand filmmakers. Presents a non-competitive, invited season of new films, encompassing a wide variety of genres, themes and styles.

Films must not have been publicly screened in New Zealand before (requirement waived for invited, archival and retrospective programmes). The organizers prefer 35mm, original language prints (with English subtitles if foreign-language).

Wellington Fringe Film Festival

PO Box 9207, Te Aro, Wellington
Ⓣ 04 938 0219
fibartlett@paradise.net.nz
www.fringefilmfest.co.nz

Founded in 1987, one of the leading forums for New Zealand filmmakers to show their work. The festival also hosts a special competition to promote originality and innovation within a given theme and time frame.

15-minute limit for all drama, animation and experimental films; 30-minute limit for documentaries.

Awards and Prizes

So what can we learn from the classic scenes? Perhaps that the best (and most memorable) movie moments are when the audience's emotions are engaged as much as their intellect.

Pete Daly in 'Learn from the classics'

It is a frequent complaint of mainstream movie making that it rarely focuses on serious issues or engages in ideas that are anything but mundane. This could change with the opportunity of harnessing the computer generation who are forever looking for something new and compelling to download.

Barry Turner in 'An ever changing screen'

Awards and Prizes

USA

Academy Awards (The Oscars)

Academy of Motion Picture Arts and Sciences
Academy Foundation, 8949 Wilshire Boulevard,
Beverly Hills, CA 90211
T 310 247 3000
www.oscars.org

The world's most prestigious film awards,
presented annually since 1929. Ceremony held
in Los Angeles in January. Voting by Academy
members is conducted by secret ballot. Includes
awards for best original and best adapted
screenplays.

Acclaim Film Screenwriting Award

300 Central Avenue, Suite 501, St Petersburg,
FL 33701
acclaimfilms@go.com
acclaimtv.netfirms.com
Co-ordinator *Frank Drouzas*

Founded in 2001, running two awards a
year (Spring/Summer and Fall/Winter) for
feature-length scripts. $1,000 cash prize for
winner; all finalists receive consideration from
established production companies. Also offers TV
scriptwriting contests.

Awards are open to all writers worldwide.
Scripts must not have been optioned or sold.

African Film Commission Storytelling Screenwriting Competition

468 North Camden Drive, Suite 200, Beverly
Hills, CA 90210
T 310 860 7527 F 310 860 7500
www.africanfilmcommission.org

Conceived to create greater international
cooperation in bringing African subject matter
and locals to the screen. Promotes exceptional
feature-length screenplays and documentary
subjects (or projects) that combine African and
world experiences into stories that use, to any
extent, African authentic location.

American Accolades Screenwriting Competition

2118 Wilshire Boulevard, Suite 160B,
Santa Monica, CA 90403
F 310 576 6026
info@AmericanAccolades.com
www.americanaccolades.com

Founded in 2000, this screenplay contest aims
to find and usher emerging talent into the
Hollywood mainstream marketplace. Over $5,000
in cash and prizes plus meetings with Hollywood
executives. There are 5 category winners and
1 Grand Prize winner.

Past Grand Prize winners include Lon Harris,
David Zorn and Danny Howell.

The competition is open to most writers but
screenplays must not be produced, optioned, or
purchased at the time of submission. Submission
is subject to an entry fee.

American Gem Short Script Contest

PO Box 4678, Mission, Viejo,
CA 92690
info@filmmakers.com
www.filmmakers.com

Founded in 2001, offering cash prizes to the value
of $1,000 for the first place winner, as well as cash
prizes for 4 runners-up.

Previous winners of the competition include
Michael Ugulini (*Parched*), Ronald Bramhall
(*The Veteran*) and Marc Calderwood (*The
Christmas Card*).

Entry is open to anyone (except employees).
Submissions must be short film (narrative), up
to 45 pages in length in any genre, and must be
possible to produce on a low budget. Submission
is subject to an entry fee.

American Screenwriters Association/ Gotham Writer's Workshop

Annual ASA International Screenplay
Competition c/o Gotham Writer's Workshop,
555 Eighth Avenue, Suite 1402, NY 10018-4358
T 866 265 9091

asa@goasa.com
www.goasa.com

Founded in 1999, this screenplay competition offers the grand prize winner a financial reward of $5,000, as well as industry recognition and writer development. The grand prize winner is awarded the Mick Caswell Award.

Previous grand prize winners include Anita Skibski (*Mine*) and Mike Miller (*Ten and a Half*).

Submission of material is subject to an entry fee. Scripts must be between 80 and 130 pages in length. It must be original work that is not sold or optioned at the time of entry. A 1-page synopsis must be attached to the script.

American Screenwriting Competition

311 N. Robertson Boulevard, Suite 172, Beverly Hills, CA 90211
Ⓣ 888 299 0234
www.flatshoe.com

Founded in 2003, offering prizes to the value of $50,000 and has a grand prize of $12,000 cash.

Past grand prize winners include Bill Balas (*The Pros and Cons of Breathing*) and Kieran Shea and David Osorio (*The Outcast*).

All feature length English language scripts are eligible for the competition. Each screenplay submission must include the appropriate entry fee, an official entry form, a short synopsis and the complete screenplay, which should be 70 to 150 pages in length.

American Zoetrope Screenplay Contest

916 Kearny Street, San Francisco, CA 94133
Ⓣ 415 788 7500 Ⓕ 415 989 7910
contests@zoetrope.com
www.zoetrope.com

Founded in 2003, this annual contest provides the grand prize winner with $5,000. The winner and finalists will also be considered for representation by leading agencies.

Past winner Michael Cahill wrote and directed *King of California*, which premiered at Sundance 2007 and starred Michael Douglas and Evan Rachel Wood.

Submission of material is subject to an entry fee. Entrants must be 18 years and older and not have earned over $5,000 as a screenwriter. Screenplay submissions should be in the US Motion Picture industry standard screenplay format and approximately 87 to 130 pages in length. Submissions should be written in English.

Asian American International Film Festival Screenplay Competition

133 West 19th Street, Suite 300, New York, NY 10011

Ⓣ 212 989 1422 Ⓕ 212 727 3584
info@asiancinevision.org
www.asiancinevision.org

Annual competition open to all writers of Asian descent of any nationality, held in July as part of the film festival. Only unproduced full-length screenplays are eligible. Staged reading of winning script, along with other prizes.

Augusta Pictures Screenplay Contest

PO Box 5085, Glendale, CA 91221
Ⓣ 818 445 9013
diana@augustapictures.com
www.augustapictures.com
Producer/Script Doctor *Diana Osberg*

Founded in 2005, competition offering winner one-on-one advice and guidance on developing the winning screenplay from screenwriter/producer Diana Osberg.

Screenplays must not be optioned, purchased or produced at time of submission. Subject to an entry fee.

BIFF Screenplay to Production Contest

PO Box 240023, Boston, MA 02124
Ⓣ 617 482 3900
info@bifilmfestival.com
www.bifilmfestival.com/biffscreenplaycontest.html

Part of the Boston International Film Festival. Offers an opportunity for screenwriters and filmmakers to produce a film entirely from pre-production to completed project on High Definition Format only. Selected projects may be produced anywhere in the United States.

Big Bear Lake International Screenwriting Competition

PO Box 1981, Big Bear Lake, CA 92315
Ⓣ 909 866 3433
karsten33@earthlink.net
www.bigbearlakefilmfestival.com

The festival has been hosting the award since 2000.

Past winner Iris Yamashita has worked with Paul Haggis and Clint Eastwood to write the screenplay for the Academy Award-nominated film *Letters From Iwo Jima*.

Submission of material is subject to an entry fee. Screenplays should be 90 to 130 pages in length and should be written in English. Screenplays should not have been produced or optioned.

Big Break: A Final Draft Contest

26707 W. Agoura Road, Suite 205, Calabasas, CA 91302

Ⓣ 888 245 2228 ext. 202 Ⓕ 410 592 8062
green@finaldraft.com
www.finaldraft.com

Founded in 2000, the competition offers prizes of $30,000 and consideration for representation from AEI.

Submission of material is subject to an entry fee. Screenplays must be feature length (90 to 120 pages) and written in English.

BlueCat Screenplay Competition

PO Box 2630, Hollywood, CA 90028
info@bluecatscreenplay.com
www.bluecatscreenplay.com

Founded in 1998, the competition offers a $10,000 grand prize and $1,500 for each of the finalists. Every writer who enters the competition will receive written script analysis.

Submission of material is subject to an entry fee. All entries must be in English and between 80 and 145 pages in length. Screenplays must not have been purchased or optioned.

Brass Brad Screenwriting Mentorship Award

PO Box 892410, Temecula, CA 92589
Ⓣ 951 587 3890 Ⓕ 951 587 6350
brassbradmentor@aol.com
www.brassbrad.com
Founder *Kimberley Seilhamer*
Co-ordinator *Brenda McKoy*

Founded in 2005, mentorship scheme offering advice and guidance to help the winner move forward in the industry. Winner receives one year of bi-weekly phone conferences and script analysis, as well as prizes such as screenwriting software and online subscriptions.

Past winners include Donald Cager and Elizabeth Pfeiffer.

Open to residents of Canada and the USA. No collaborative work is accepted. Submission is subject to an entry fee.

Century City Cell Phone Festival Competition

PO Box 67132, Century City, CA 90067
Ⓣ 310 652 0271
www.centurycityfilmfestival.com

Annual contest for 15 second-3 minute films made specifically for mobile phones.

Critics' Choice Awards

info@bfca.org
www.bfca.org

Presented by the Broadcast Film Critics Association, the largest film critics organization in the United States and Canada. Includes best writer category.

Cynosure Screenwriting Awards

3699 Wilshire Boulevard, Suite 850, Los Angeles, CA 90010
Ⓣ 310 855 8730
cynosure@broadmindent.com
www.broadmindent.com

Founded in 1999, the competition offers two $2,500 prizes to quality character- and concept-driven screenplays showcasing female and minority protagonists.

The competition is open to writers from any country. Submission of material is subject to an entry fee. Screenplays may not have been previously optioned, purchased or produced. Screenplays must be in English only, and may be of any genre. Screenplays must be in standard US screenplay format, between 90 and 130 numbered pages.

Duke City Shootout

PO Box 37080, Albuquerque, NM
info@shootdfi.net
www.dukecityshootout.org

Annual competition for scripts of 12 minutes or shorter. Winners are taken to Albuquerque, New Mexico for a week with a cast, crew, cameras, catering, equipment and an industry mentor to turn the script into a movie ($50,000 value).

Emmy Awards

Academy of Television Arts & Sciences, 5220 Lankershim Boulevard, North Hollywood, CA 91601 3109
Ⓣ 818 754 2800 Ⓕ 818 761 2827
webmaster@emmys.org
www.emmys.org

The leading awards for American television, voted for by over 12,000 members of the Academy. The Primetime Awards are for excellence in national prime time programming. There are also Daytime Emmys, awarded by the National Television Academy in New York, and regional and international Emmys. Categories in the Primetime division include outstanding writing for a comedy series, for a drama series, for a variety, music or comedy programme and for a mini-series, movie or dramatic special.

Fade In Award

287 South Robertson Boulevard, Suite 467, Beverly Hills, CA 90211
Ⓣ 310 275 0287
inquiries@fadeinonline.com
www.fadeinonline.com
Administration *Heather Millerton*
Editorial Assistant *Gabrielle Palmatier*

Introduced in 1996 with over $15,000 prize money each year. Recent winner Jon Bokenkamp since received sole credit on *Taking Lives*, starring Angelina Jolie.

A Feeding Frenzy – Screenplay Competition with Feedback

AMLC Productions, 703 Pier Avenue, Suite B #687, Hermosa Beach, CA 90254
amlcprods@afeedingfrenzy.com
www.afeedingfrenzy.com

Offers feedback to all entries. Top three scripts are distributed to production companies and industry contacts. Feedback is 'encouragingly blunt'.
    Entries should be submitted by mail or email.

Film Columbia's First Screenwriters' Contest

The Chatham Film Club, PO Box 181, East Chatham, NY 12060
www.filmcolumbia.com/ScreenwritersContest.htm
Coordinator *Mark Dickerman*

The winning script is performed at the FilmColumbia script reading event. Grand prize of $750. Second prize $500. Third prize $250.

Film Genesis

PO Box 6038, Long Beach, CA 90806
contest@filmgenesis.com
www.filmgenesis.com

Competition for scripts in any genre of 5–15 pages. Must have beginning, middle and end. Grand prize includes roundtrip airfare to Los Angeles plus hotel accommodations, and opportunity to direct your film. Also up to $10,000 of production services to produce your film and a variety of smaller prizes.

Filmaka, LLC

7955 W. 3rd Street, Los Angeles, CA 90048
www.filmaka.com

An on-line network created by a group of independent producers, industry professionals, and film financiers including Deepak Nayar, Thomas Augsberger, Christopher Sharp, Robert Herman, Tim Levy, Mahesh Mathai, Kurt Woolner and Colin Firth. Aims to give undiscovered filmmakers an opportunity to show custom-made short films to industry professionals, award prizes and find 'one truly talented director' who will win a feature film deal with Filmaka at the end of the year.
    Contests are divided into three levels, Entry, Jury and Final. Entry level contests are held periodically and up to 20 winners get $500.

At the Jury level contest, all winners from the Entry Level are given $1,000 to make a second clip on the same subject. Two winners are selected. The first winner is given a $3,000 cash prize and a runner-up is given a $2,000 cash prize. The winner and runner-up move into the final level. A special Jury prize will also be announced at the Jury level which wins no cash but qualifies to compete in the yearly contest in addition to the Jury Level winners and runner ups. Final level contest happens yearly. The winners, runners-up and special jury prize winners from each jury level contest compete for the grand feature film contract and the runner-up wins $5,000 cash.

Find the Funny

531-A N. Hollywood Way, Suite 260, Burbank, CA 91505
T 818 679 3845
screenplaycontest@findthefunny.com
www.findthefunny.com

Founded in 2005, this competition is dedicated to comedy and offers a first place prize of $2,000.
    Competition is open to all screenwriters 18 years of age and older. Submission of material is subject to an entry fee. Only feature-length comedy screenplays will be accepted (all genres and sub genres are acceptable – romantic comedy, action comedy, dark comedy etc). All scripts must be in English and must be between 90 and 115 pages in length. Screenplays must not have been previously optioned or sold.

Gimme Credit Screenplay Competition

4470 W. Sunset Boulevard, Suite 278, Los Angeles, CA 90027
T 310 734 8516
submissions@gimmecreditcompetition.com
www.gimmecreditcompetition.com
Director/Founder *Erica Engelhardt*
Co-founder *Monica Winter Vigil*

Founded in 2006, triennial contest created with the aim of giving exposure to and development advice on talented writers' short scripts. The Grand Prize-winning screenplay is produced. Awards categories for short scripts (up to 30 pages) and 'Super Shorts' of 5 pages or less.
    Previous winners include Alan R. Baxtor, Menotti and Benjamin Harrison.
    All screenplays must be in English and registered with the Writers Guild or the US Copyright Office. Open to all writers over 18.

Golden Globes

The Hollywood Foreign Press Association, 646 N. Robertson Boulevard, West Hollywood, CA 90069

info@hfpa.org
www.hfpa.org

Prestigious awards for the motion picture and television industries, administered since 1944 by the Hollywood Foreign Press Association. Includes an award for best screenplay.

Great Lakes Film Association Scriptwriting Competition

PO Box 346, Erie, PA 16512
screenplays@greatlakesfilmfest.com
www.greatlakesfilm.org

Competition for domestic and international short- and feature-length screenplays, stage plays, and teleplays of all genres and niches. Top 10 finalists get exposure to literary agencies, film studios and Hollywood producers and have names and contact information posted on GLFA website.

Film/TV scripts should be: feature length (70+ pages); shorts (69 pages or less); teleplay feature (no more than 60 pages); teleplay short (no more than 30 pages).

Hollywood Nexus Screenwriting Contest

2554 Lincoln Boulevard, Suite 401, Marina Del Rey, CA 90291
☎ 310 663 0656
info@hollywoodnexus.com
www.hollywoodnexus.com
Founder *Cynthia Cree*

Founded in 2002, the contest offers a grand prize of $5,000. Every entry receives feedback from the judges. Winner, finalists and semi-finalists announced each March.

Submission of material is subject to an entry fee. Entrants must be at least 18 years of age. Entries are accepted from all locations, including internationally, except Colorado and Vermont. The screenplay must not have been produced or sold.

International Horror & Sci-Fi Film Festival Screenplay Competition

1700 N 7th Avenue, Suite 250, Phoenix, AZ 85007
☎ 602 955 6444
info@HorrorSciFi.com
www.horrorscifi.com

Winners receive cash prizes and the opportunity to have their screenplays reviewed by major entertainment companies.

Ivy Film Festival Screenplay Competition

screenplay@ivyfilmfestival.com
www.ivyfilmfestival.com

Caters for both short- and long-form student-written screenplays in separate competitions. A panel of qualified student judges read each submission and select twenty semi-finalists, which are sent to industry professionals. Winner announced during the festival in April and receive cash prizes and a reading of their work during the festival. Additional prizes for outstanding work in comedy, drama and genre films.

Lighthouse Screenplay Competition

lighthhousescreenplay@yahoo.com
www.lighthousescreenplay.com

Established to allow screenwriters another avenue to get scripts read. Currently has forty-seven award winning producers affiliated with the competition. First prize of $5,000, runner-ups get $1,500 and honorable mentionees, $50.

Any screenplay may be entered in any genre. Screenplays will be judged on creative concept, character development, script arch, and overall writing.

Los Angeles Film Critics Association Awards

LaelL@aol.com
www.lafca.net

Presented annually since 1975. Includes award for best screenplay.

The Manhatten Short Film Festival's Screenplay Competition

22 Prince Street, PO Box #110, New York, NY 10012
☎ 212 529 8640
info@msfilmfest.com
www.msfilmfest.com
Executive Director *Nicholas Mason*
Assistant Director *Tom See*

Short film competition running since 1997. Festival takes place in 200 locations across Europe, North, South and Central America during the last week of September. Finalists' films are judged by the audience at each venue, with votes tallied and emailed to Union Square Park in New York, where the winner is announced.

Entry is via the website. Tickets to screenings available from participating cinemas.

McKnight Screenwriting Fellowship

The McKnight Foundation, 710 South Second Street, Suite 400, Minneapolis, MN 55401
☎ 651 644 1912 ext. 106

llipold@ifpmn.org
www.ifpmsp.org/mcknight.html

Fellowship worth $25,000 awarded to two
screenwriters annually, judged by two panels
of industry professionals. Open to mid-career
Minnesota-based writers, the fellowship
is intended to provide financial assistance,
professional encouragement and industry
recognition.

Applicants should send one completed full-
length screenplay.

Monterey Screenplay Competition

Monterey County Film Commission,
PO Box 111, Monterey, CA 93942
T 831 646 0910 F 831 655 9250
info@filmmonterey.org
www.filmmonterey.org

Founded in 1996, the competition awards a grand
prize of $2,000 for the winning feature-length
screenplay and the winning screenwriters
have the opportunity to have their screenplays
reviewed by a Monterey County-based film
investor.

Submission of material is subject to an
entry fee. Submissions must be completed,
feature-length scripts between 90 and 120 pages.
A submitted screenplay must not have been
optioned or sold at the time of submission.

National Society of Film Critics

Presented annually since 1966. The panel
includes 48 critics from leading US publications
and media bodies. Includes award for best
screenplay.

New Jersey Short Screenplay Festival

PO Box 595, Cape Bay, NJ 08204
T 609 823 9159 F 609 884 6722
bsaks47@comcast.net
www.njstatefilmfestival.com/screenplay.htm
Chair *William Sokolic*

Founded in 2002, offering software of varying
cost as a prize for screenplays of 45 pages or less.
The writer must be from New Jersey or the state
should be the primary setting. Recent winners
include Susan Pelligrino.

Scripts must be received by 1 November and
should include an application form. Scripts
should be in industry-standard format.

New York Film Critics Circle Awards

www.nyfcc.com

Presented annually since 1935. Includes award for
best screenplay.

Nicholl Fellowships in Screenwriting

1313 Vine Street, Hollywood,
CA 90028
T 310 247 3010
nicholl@oscars.org
www.oscars.org/nicholl
Program Co-ordinator *Greg Beal*

Founded in 1985, up to 5 fellowships worth $30,000
are awarded each year to emerging writers. During
the fellowship year winners complete an original
screenplay 100–130 pages long.

2007 fellows were Amy Garcia and Cecilia
Contreras, Michael L. Hare, Sidney King, David
Mango and Andrew Shearer, and Nicholas J.
Sherman.

Screenplays or teleplays which have been
optioned are not eligible. Applicants may not
have earned more than $5,000 as a screenwriter.
Applications are only accepted by mail and are
subject to an entry fee.

Nickelodeon Productions Fellowship Program

Nickelodeon Animation Studios, Attn. Nick
Writing Fellowship, 231 W. Olive Avenue,
Burbank, CA 91502
T 818 736 3663
info.writing@nick.com
www.nickwriting.com

Founded in 2000, the writing fellowship provides
a salaried position for up to one year and offers
hands-on experience writing spec scripts and
pitching story ideas in both live action and
animation television.

The fellowship is only available to US applicants
who are 18 years or older. Previous writing
experience is not necessary. Applicants must
submit an application form, two copies of a spec
script, a one-page resume, a half-page biography,
plus a submission release form and schedule A
(allowing material to be read).

One in Ten Screenplay Competition

c/o Cherub Productions, PO Box 540,
Boulder, CO 80306
T 303 629 3072
Cherubfilm@aol.com
www.OneInTenScreenplayContest.com
Contact *Mike Dean*

Founded in 1999, screenplay contest dedicated
to the positive portrayal of gay, lesbian, bisexual
and transgender individuals in film. Winners
receive cash prizes and submission to a major
film studio.

Past winners include Denise P. Meyer and Jon
Zelazny.

Screenplays must not have been optioned. Entries are subject to a fee.

The PAGE International Screenwriting Awards

7510 Sunset Boulevard, Suite 610, Los Angeles, CA 90046

Ⓕ 323 969 0993
info@internationalscreenwritingawards.com
www.internationalscreenwritingawards.com
Administrative Director *Jennifer Berg*
Contest Coordinator *Zoe Simmons*

Founded in 2003, the mission of the PAGE Awards is to discover the most exciting new screenplays by up-and-coming writers from around the world. Over $30,000 in cash and prizes are awarded each year, including a $10,000 Grand Prize and Gold, Silver and Bronze prizes in each of ten categories.

Previous Grand Prize winners include Laurie Weltz and Jay C. Key, Larry Postel, Scott LaCagnin and John Arends.

Open to all writers over the age of 18 who have not previously earned more than $25,000 writing for film and/or television. Entries accepted by mail or online.

Rhode Island Film Festival Screenplay Competition

Rhode Island International Film Festival, PO Box 162, Newport, RI 02840
Ⓣ 401 861 4445 Ⓕ 401 490 6735
adams@film-festival.org
www.film-festival.org

Award open to screenplays in all genres. Winner receives prizes valued at over $15,000. Screenplays judged on 'creativity, innovation, vision, originality and the use of language'. Also contests for lesbian, gay, bisexual and transsexual writers and for short scripts.

Entry forms available online.

The Santa Barbara Script Competition

PO Box 4068, Santa Barbara, CA 93140
Geoff@santabarbarascript.com
www.santabarbarascript.com

Offers a $750 first prize for the best screenplay and has a grand prize of a $4,000 option and a staged script-reading by professional actors.

Submission of material is subject to an entry fee. Screenplays should be between 85–120 pages in length.

Satellite Awards

IPA Production Office, 5525 Halbrent Avenue, Sherman Oaks, CA 91411

Ⓣ 818 989 1589 Ⓕ 818 787 3627
info@pressacademy.com
www.pressacademy.com

Awarded by the International Press Academy. Includes categories for best screenplay (adapted) and best screenplay (original).

Screenplay Festival

11693 San Vicente Boulevard, Suite 806, Los Angeles, CA 90049
Ⓣ 310 801 7896 Ⓕ 310 820 2303
info@screenplayfestival.com
www.screenplayfestival.com

Founded in 2002, the feature length screenplay competition has 5 categories of competition: action/adventure, comedy, drama, family and thriller/horror. In each of the five categories, a $1,000 grand prize will be awarded.

Submission of material is subject to an entry fee. Screenplays must be submitted in English and must be longer than 60 pages in length. Screenplays that have been previously optioned, sold or produced are not eligible.

Screenplay Shootout

c/o Hangman Productions Inc., 1338 South Foothill Drive, Suite 200, Salt Lake City, UT 84108
information@hangmanproductions.com
www.screenplayshootout.com

Designed to help young and undiscovered writers break into the filmmaking industry. Grand prize of $2,500; 2nd Place $500; 3rd Place $250.

Screenwriter Showcase Screenwriting Contest

897 Oak Park Boulevard, PMB 263, Pismo Beach, CA 93449
contact@screenwritershowcase.com
www.screenwritershowcase.com

Established in 2002, competition for feature length screenplays. Prizes include screenwriting software, exposure to industry professionals and promotion on dedicated website.

Previous winners include *Cinders* by Kate Pellettiere and *Bloodletting* by William D. Prystauk.

Submission is subject to an entry fee.

Screenwriter's Challenge

NYC Midnight, 130 7th Avenue, Suite 358, New York, NY 10011
Ⓣ 212 647 9653
www.nycmidnight.com

Founded in 2002, an international online screenwriters' competition. The competition has

2 rounds. In the first round writers are given 1 week to create an original short screenplay (20 pages maximum) based on an assigned genre and subject. In the second round, finalists are chosen to compete for thousands of prizes by writing a short screenplay in 24 hours.

Submission of material is subject to an entry fee. Entry is open to everyone.

Script Nurse Mini-Movie Screenwriting Contest

writingcontest@scriptnurse.com
www.scriptnurse.com

Prizes are a selection of 'productive tools to assist you to write better'.

All scripts must be original (not adaptations from another work). Can be in any genre but no anime, cartoon silents or narrative voice-over only scripts. Scripts should be 3 pages minimum and 5 pages maximum plus title page; live action only (no anime, cartoon silents or narrative voice-over only scripts). No entry fee. Entrants must be a registered user in the Script Nurse Forum (registration free).

Script Pimp Screenwriting Competition

8033 W. Sunset Boulevard, Suite 3000, Hollywood, CA 90046
Ⓣ 310 401 1155 Ⓕ 310 564 2021
comp@scriptpimp.com
www.scriptpimp.com
Contest Director *Chadwick Clough*

Founded in 2003, the competition offers a grand prize of $2,500 and guaranteed circulation to over 20 production companies.

Submission of material is subject to an entry fee. Entrants must be at least 18 years of age. The script must not have been produced, sold or optioned.

Scriptapalooza Screenwriting Competition

7775 Sunset Boulevard, Suite 200, Hollywood, CA 90046
Ⓣ 323 654 5809
info@scriptapalooza.com
www.scriptapalooza.com

Running since 1998. Grand prize of $10,000. Winning entries read by extensive range of production companies and literary representatives. Also runs an Annual Television Writing Competition.

The Rod Serling Conference Short Feature Scriptwriting Competition

c/o Ithaca College Los Angeles Program, 3800 Barham Boulevard, Suite 305, Los Angeles, CA 90068
www.ithaca.edu/rhp/serling/script.html

Open to non-produced, or non-optioned writers only. First Place: $250; Second Place: $150; Third Place: $100. Scripts must be between 10 to 20 pages and should display traits of either a horror or a science fiction genre, while exhibiting strong social themes.

Set in Philadelphia Screenwriting Competition

Greater Philadelphia Film Office, 100 S. Broad Street, Suite 600, Philadelphia, PA 19110
Ⓣ 215 686 2668
joanb@film.org
www.film.org
Director *Joan Bressler*

Contest for screenplays set in the Greater Philadelphia region. Prizes include the Parisi Award for writers under 25 worth $1,000, a regional award worth $2,500 and a Grand Prize worth $10,000. Awarded during the Philadelphia Film Festival.

Previous prize winners include Bruce Graham, Brian Loshiavo and David Hoag.

Submission is subject to an entry fee.

Slamdance Screenwriting Competition

5634 Melrose Avenue, Los Angeles, CA 90038
Ⓣ 323 466 1786 Ⓕ 323 466 1784
screenplay@slamdance.com
www.slamdance.com

Founded in 1995, dedicated to new writers. Has a grand prize of $7,000. Accepts screenplays in every genre, on any topic and from every country.

Submission of material is subject to an entry fee. Screenplays must not have been previously optioned, purchased or produced. Material must be in English and not have received awards from other competitions over a value of $500.

TalentScout TV Writing Contest

1484 1/2 Robertson Boulevard, Los Angeles, CA 90035
contact@aTalentScout.com
www.atalentscout.com

Founded in 1991, this TV writing competition offers a $3,000 grand prize and representation. Only 1,000 entries will be accepted.

Past winner Marcus Folmar had his script *I'm Perfect* produced, starring Wayne Brady, Brad Dourif, Illeana Douglas and Malcolm-Jamal Warner.

Submission of material is subject to an entry fee. Material submitted must not have been bought or optioned.

The John Templeton Foundation Kairos Prize

5620 Paseo de Norte, #127c-308, Carlsbad, CA 92008
℡ 760 434 3070
contact@kairosprize.com
www.kairosprize.com

Founded in 2006, offering a $25,000 grand prize for scripts that teach lessons in ethics and morality, and which are primarily spiritual. The scripts submitted must be suitable for a G or PG rating and must be free of foul language (and must refer specifically to the Bible).

Submission of material is subject to an entry fee. Entrants are required to submit a 500 word synopsis or treatment in addition to the script. The script must be in English and must be 87 to 130 pages in length. Entrants must be 18 years of age or older, and must not have earned money as a screenwriter.

TVWriter.com People's Pilot

Cloud Creek Ranch, 3767 Marion County 5026, St Joe, AZ 72675
℡ 870 449 2488
larrybrody@tvwriter.com
www.peoplespilot.com

Founded in 2000, the winner of this competition receives $500 and a 1-week writing retreat at the Cloud Creek Institute for the Arts. This is not a screenplay competition and requires only the title of the series idea, the main thrust of the storyline, the characters and the setting. The competition is held bi-annually.

Submission of material is subject to an entry fee. The competition is open to writers at all levels and from all countries. Pilot submissions must contain a summary of the basic idea behind the show, including the setting, theme and profession. It must also have a summary of the main continuing characters, their background, appearance, personalities, interaction and a statement of the kinds of situation in which the characters will be placed.

TVWriter.com Spec Scriptacular

Cloud Creek Ranch, 3767 Marion County 5026, St Joe, AZ 72675
℡ 870 449 2488
larrybrody@tvwriter.com
www.tvwriter.com

This competition has four categories—sitcom, action/drama, pilot/MOW/special and new media. The grand prize is awarded to the best entry regardless of category. The winner receives $500 and a 1-week writing retreat at the Cloud Creek Institute For The Arts.

Submission of material is subject to an entry fee. The competition is open to all writers in all countries.

Sir Peter Ustinov Television Scriptwriting Award

The International Academy of Television Arts and Sciences Foundation, 888 7th Avenue, 5th Floor, New York, NY 10019
℡ 212 489 6969 ℻ 212 489 6557
awardsdept@iemmys.tv
www.iemmys.tv
Manager, Foundation and Special Projects *Tracy Oliver*

Administered by the International Academy of Television Arts and Sciences Foundation since 1998. A competition for young writers of English-language television drama for a family audience. Winner receives $2,500 and a trip to New York City. A reading of the winning script is staged with professional actors.

Entrants must submit a complete script lasting between 30 minutes and an hour. Open to non-US citizens under the age of 30. Email for entry form.

Visionfest (Domani Vision Film Society)

PMB 155, 6402 18th Avenue, Brooklyn, NY 11204
℡ 718 837 5736 ℻ 718 554 0030
contact@domanivision.org
www.domanivision.org

Founded in 2004, this feature-length screenplay competition awards winners with either cash or special services prizes.

Submission of material is subject to an entry fee. Screenplays should be 80 to 130 pages in length and must be in English.

Walt Disney Studios/ABC Writers Fellowship

Talent Development Programs, 500 South Buena Vista Street, Burbank, CA 91521-4016
℡ 818 560 6894
abc.fellowships@abc.com
www.abctalentdevelopment.com

Started in 1991, offering fellowships in the feature film and television sectors with The Walt Disney Studios and ABC Entertainment, respectively. No previous experience is necessary but writing samples are required. Fellows will each be provided a salary of $50,000 for a one-year period.

The programme is open to all writers.

Recent fellowship graduates include Sean Diviny, who was hired as a staff writer on *Alias*, and Sonia Steele, who was staffed on *ER*.

Warner Bros. TV Writers Workshop

4000 Warner Boulevard, Buillding 36, Room 155, Burbank, CA 91522
www2.warnerbros.com/writersworkshop

Running for over 30 years, a scheme to foster new talent and provide opportunity for aspiring television writers. Two programmes, in comedy and drama.

To apply for the comedy, applicants must send a half-hour writing sample (a spec script) based on a primetime network or cable comedy series that aired new episodes during the previous year. For drama, applicants supply a one-hour writing sample (a spec script) based on a primetime network or cable drama series that aired new episodes during the previous year.

Waterfront Film Festival Screenwriting Competition

PO Box 387, Saugatuck, MI 49453
screenplay@waterfrontfilm.org
www.waterfrontfilm.org

$250 cash prize for first place, along with an all access pass to the festival, lodging, and sponsors' prizes. Also an industry party held at the festival in honour of the winner.

Entries accepted only through Withoutabox.

WinFemme Monthlies

2118 Wilshire Boulevard, Suite 144, Santa Monica, CA 90403
Ⓣ 310 229 5365
info@winfemme.com
www.thewinawards.com

This monthly competition runs 3 distinct categories: films or videos created by men and women which feature a female protagonist; films with a lesbian story; and films which are directed or produced by a woman. Scripts may be TV specs, TV movies, feature films, theatre plays and/or novellas.

Submission of material is subject to an entry fee.

Worldfest Houston Screenplay Competition

PO Box 56566, Houston, TX 77256
Ⓣ 713 965 9955 Ⓕ 713 965 9960
mail@worldfest.org
www.worldfest.org

Founded in 1968, a screenplay competition attached to the festival with a $2,500 first prize. First place also includes a 1-year production option.

Submission of material is subject to an entry fee. Screenplays should be between 70 to 150 pages in length. Screenplays should not have been produced.

WriteMovies.com International Writing Contest

11444 Washington Boulevard, Suite C-227, Los Angeles, CA 90066
Ⓕ 206 203 1256
author@writemovies.com
www.writemovies.com
Contact *Chris Elley*

Founded in 1999, award for scriptwriters from around the world. Grand prize includes $3,000 cash and guaranteed representation by TalentScout Management. Winning scripts that have been produced include *Valley of Angeles* and *The List*.

Entry is open to any script which has not been bought or optioned. Scripts in English, French and German are welcome. Subject to entry fees.

Writer's Arc Screenwriting Fellowship

8332 Melrose Avenue, Second Floor, Los Angeles, CA 90069
info@writersarc.org
www.writersarc.org

This is a 10-week fellowship programme with a stipend of $3,000. Fellows are required to spend a minimum of 10 hours a week in the Arc Hub to work on their script and improve on their writing skills, as well as 4 hours a week at community events to broaden their industry awareness and experience. Offer about 5–10 fellowships a year depending on the applicant pool.

No applicant may have earned money or other consideration as a screenwriter for theatrical films or television, or for the sale of, or sale of an option to, any original story, treatment, screenplay, or teleplay for more than $5,000. Applicants may not have received a screenwriting fellowship or prize that includes a 'first look' clause, an option or any other quid pro quo involving the writer's work.

Writers Guild of America Awards

www.wga.org/awards or www.wgaeast.org/awards

Prestigious annual awards for outstanding achievements in writing for film, TV and radio. Ceremonies held simultaneously by the East Guild in New York and West Guild in Los Angeles. Categories include best original movie screenplay, adapted movie screenplay, documentary screenplay and numerous television awards.

The Writers Network Annual Screenplay & Fiction Competition

287 South Robertson Boulevard, Suite 467,
Beverly Hills, CA 90211
writersnet@aol.com
www.fadeinonline.com/Contests/
TWN_Competition/
Program Director *Sarah Kelly*

Co-sponsored by WGA signatory literary
agencies in Los Angeles and New York. Each
year, up to ten winners in each category are
chosen to participate. Two winning original
feature-length screenplays, half-hour teleplays or
plays are distributed to top producers, directors
and production companies. Prizes worth over
$10,000.

Writers On The Storm Screenplay Contest

6404 Wilshire Boulevard, Suite 105, Los Angeles,
CA 90048
☎ 323 207 4110
writerstorm@gmail.com
www.writerstorm.com

Founded in 2006, this competition encourages
submissions for any and all genres. The winning
scriptwriter will receive $10,000 in cash.

Submission of material is subject to an entry
fee. The competition is open to all writers in all
countries. All submissions should be in English.
Applicants must not have made $10,000 or more
on feature options or sales. Scripts should be
between 90 to 120 pages in length.

The Writers Place Screenplay Contest

525 East 72nd Street, Suite 18A, New York,
NY 10021
☎ 212 249 0255
contact2@thewritersplace.org
www.thewritersplace.org

This competition accepts full-length features
and MOWs, short film scripts and TV scripts
(1/2 an hour). The first prize consists of money
or screenwriting software.

Submission of material is subject to an entry
fee. The competition is open to everyone.
Screenplays must be in English and must be
registered with the WGA or copyrighted with the
Library of Congress.

WriteSafe Present-A-Thon

422 Carlisle Road, Westlake Village,
CA 91361
☎ 805 495 3659 ☏ 805 495 0099
admin@writesafe.com
www.writesafe.com

Founded in 1999, this quarterly competition is for
all material registered with WriteSafe for public
view. The competition has no categories. The first
prize is consideration for publication, production,
or representation by a panel of entertainment
industry experts, and two free WriteSafe
registrations.

Competition is based on registered material.

Yale Summer Session and The Yale Summer Film Institute Short Film Screenwriting Competition

PO Box 208355, New Haven, CT 06520-8355
☎ 203 432 2430
ojedafilms@aol.com
yale.edu/summer/film/contest

Competition for short film scripts, with the
winner(s) produced and exhibited in the Intensive
Filmmaking and Acting in Film Workshops.

Each submission should be in written in
industry standard screenplay form and be
between 12–15 pages in length.

UK

BAFTAs

195 Piccadilly, London, W1J 9LN
☎ 020 7734 0022 ☏ 020 7734 1792
info@bafta.org
www.bafta.org

Among the most prestigious film and TV
awards available in the UK. Administers several
ceremonies including the British Academy
Film Awards, British Academy Video Games
Awards (which includes a scriptwriting award),
British Academy Television Awards and the
British Academy Children's Film &Television
Awards.

BBC College of Comedy

collegeofcomedy@bbc.co.uk
www.bbc.co.uk/writersroom/opportunity/
comedy_college.shtml

A scheme taking six writers and training them
over a year by attaching them to sitcoms and
sketch shows, giving each a mentor for original
work, and running a series of masterclasses in
all aspects of comedy writing. Financial support
for the scheme from BBC Worldwide. Designed
for people who have already begun their careers,
and can demonstrate some achievement, such
as broadcast material, a script commission or
performance of their work.

BBC Drama Writers Academy

www.bbc.co.uk/writersroom/opportunity/
writers_academy.shtml

Guarantees writers the opportunity to work
on prime time television on programmes
such as *EastEnders, Casualty, Holby City* and
Doctors. Alongside training on all aspects of
drama production from editing to scheduling,
students will receive direct writing experience on
continuing dramas, with the aim of transmitting
their finished scripts on BBC One.

Applicants must have had at least one
professional commission in either television,
theatre, radio, or film. Applications via BBC jobs
website.

British Independent Film Awards

81 Berwick Street, London, W1F 8TW
☎ 020 7287 3833 ⊕ 020 7439 2243
info@bifa.org.uk
www.bifa.org.uk

Created in 1998 to reward achievement in
independently-funded British filmmaking.
Ceremony held annually in central London.
Among the prizes on offer are Best British 15
Second Short, British Documentary, British
Independent Film, British Short Film, Foreign
Independent Film and Screenplay.

British Short Screenplay Competition

c/o Kaos Films, Pinewood Studios, Pinewood
Road, Iver Heath, Bucks, SL0 0NH
info@kaosfilms.co.uk
www.kaosfilms.co.uk

Annual competition managed by Kaos Films in
association with the National Film and Television
School. Three winning scripts are produced by
Kaos Films and directed by new NFTS graduates,
then premiered at BAFTA and entered into film
festivals around the world before being screened
in selected cinemas in the UK. A variety of
runners-up prizes also available. All finalists are
guaranteed an interview with the panel of the
NFTS for an MA in screenwriting, by-passing
preliminary rounds.

All genres accepted. Scripts must be 5–15
minutes screen-time. The entered screenplay
must not have been previously optioned, sold or
produced. Other rules and guidelines outlined on
website.

Coming Up

www.iwcmedia.co.uk/television/drama_prog_2.
html

Offers emerging filmmakers the opportunity
to make an authored drama with a guaranteed
network broadcast. Jointly run by IWC and
Channel 4. Makes up to seven 30-minute films.
Seeks 'films that push boundaries in a way that
wouldn't / couldn't be done in mainstream drama'
and that can be shot in 4 days on a limited budget.
Designed for writers and directors building a track
record and a career in film and television drama.

Open to all writers who have not had an
original single, series or serial broadcast on UK
television. Writers who have contributed episodes
to series and serials (eg a long-running soap) are
now eligible to apply.

Euroscript Screenwriting Competition

64 Hemingford Road, London, N1 1DB
☎ 07958 244656
ask@euroscript.co.uk
www.euroscript.co.uk

Offers six-months script development worth
£1,000 and other consultancy prizes. Seeking
writers with powerful story ideas and original
voices.

Submit a two page prose outline of the story
you want to develop, plus ten pages of sample
script, which can either be a complete short film
or an extract from a full length screenplay (either
the one you wish to develop through Euroscript
or another screenplay you have completed); the
sample must include a mixture of action and
dialogue. Price per entry: £35 (2007). All rights
in the winning screenplay and related materials,
such as the outline, remain with the writer.

Evening Standard British Film Awards

Prestigious cinema awards including categories
for best screenplay.

Every 1's a Critic

info@every1sacritic.com
www.every1sacritic.com

Run by producers Declan Hill and Simon Wright
in conjunction with the Arts Theatre in London.
Aims to provide a platform for new comedy
writing.

Seeks sitcoms that can be staged and last no
longer than 15 minutes.

The Grierson Awards

77 Wells Street, London, W1T 3QJ
☎ 020 7580 7526
awards@griersontrust.org
www.griersontrust.org

The Grierson Awards commemorate the
pioneering Scottish documentary maker John
Grierson (1898–1972), famous for *Drifters* and
Night Mail and the man widely regarded as the

father of the documentary. Established in 1972, the awards recognize documentaries that have significantly contributed to the genre from Britain and abroad. Categories include: Best documentary on a contemporary issue; Best documentary on the arts; Best historical documentary; Best science documentary; the Jonathan Gili Award for most entertaining documentary.

Downloadable online entry forms available on the website.

Make Your Mark in Film

info@makeyourmarkinfilm.org
www.makeyourmarkinfilm.org

Coordinated by the Make Your Mark campaign, a business-led, government-backed campaign to encourage people in their teens and twenties to develop creative ideas. Works in partnership with CobraVision, which runs a monthly competition to give aspiring film makers the opportunity to show their work on national TV (see www.cobravision.co.uk).

PAWS Drama Fund

OMNI Communications Limited, First Floor, 155 Regent's Park Road, London, NW1 8BB
Ⓣ 020 7483 4545 Ⓕ 020 7483 0934
pawsomni@btconnect.com
www.pawsdrama.co.uk
Director *Andrew Millington*
Director *Andree Molyneux*

A fund administered by PAWS (founded in 1993) to raise the profile of science and technology in television drama. Offers research and development grants to writers to develop an idea based on science or engineering into a viable treatment for television drama. Experienced writers may receive up to £5,000 while less experienced writers (who can demonstrate production company-backing) get up to £2,000. For 2007–8 most grants went to ideas with a women scientist or engineer in a prominent role. Other activities administered by PAWS include visits for screenwriters to important scientific establishments (such as the Rutherford space laboratory) and evening events where producers, writers and scientists can get together. The European arm of PAWS, EuroPAWS awards the Midas Prizes for the best science in TV and New Media broadcasts in Europe. In 2007 they were for the best TV drama, the best presentation in general TV programming, and the best in a New Media production.

A Pitch in Time

pitch@screenwritersfestival.com
www.screenwritersfestival.com/a-pitch-in-time.php

Organized by the Screenwriters' Festival in conjunction with Channel 4's 4Talent. Offers short-listed entrants the opportunity to pitch a movie or TV idea to an industry panel and a live festival audience. An expert will offer pitching guidance.

All genres accepted. Ten writers will be short-listed. Entries should include a 25 word pitch on one page, expanded to 150 words on a second page. Further details on the website.

Red Planet Prize

13 Doolittle Hill, Ampthill, Froghall Road, Bedford, MK45 2ND
Ⓣ 01525 408970 Ⓕ 01525 408971
info@redplanetpictures.co.uk
www.redplanetpictures.co.uk

Annual prize for new screenwriting talent administered by the production company of writer Tony Jordan. Winner receives £5000, professional representation and a script commission for Red Planet Pictures.

Entries should consist of no more than 10 pages of screenplay and a covering page about the writer (including a half-page synopsis of writer's work). Sending complete screenplays will result in disqualification. All entries must be in English and in screenwriting format. They must also be original and the sole property of the applicant. Entries should be sent to redplanetprize@redplanetpictures.co.uk. See website for entry dates.

Satyajit Ray Foundation Short Film Competition

The British Council, 10 Spring Street, London, SW1A 2BN
info@satyajitray.org.uk
www.satyajitray.org.uk
Contact *Julian Pye*

Open to all film-makers, of any age, either resident or studying in the UK. Film should be not more than 30 minutes running time and should be informed by the experiences of South Asians (Afghanistan, Bangladesh, Bhutan, India, Nepal, Pakistan, Sri Lanka) either within their own countries or the diaspora. All submissions must have been completed within two years prior to the closing date. First prize of £1,500. See website for specific entry requirements.

Script_1 Workshop and Competition

International Screenwriters' Festival, Unit 24C, Bath Road, Stroud, Gloucester, GL5 3TL
cathi@screenwritersfestival.com
www.screenwritersfestival.com/script1.php
Contact *Cathi Beloe*

A competition offering aspiring writers based in Gloucestershire the opportunity to develop their stories into scripts for the big or small screen.

Submit an outline of your idea (one-page maximum) for a ten-minute fiction film spelling out the essential story elementno scripts at this stage. The story can be on any subject but must have a point of view and show a significant moment in life where you live.

The Dylan Thomas Prize

The Dylan Thomas Centre, Somerset Place, Swansea, SA1 1RR

ⓣ 01792 474051 ⓕ 01792 463993

tim@dylanthomasprize.com
www.dylanthomasprize.com
Contact *Tim J. Prosser*

Founded 2004. An award of £60,000 will be given to the winner of this prize, which was established to encourage, promote and reward exciting new writing in the English-speaking world and to celebrate the poetry and prose of Dylan Thomas. Entrants should be the author of a published book (in English), under the age of 30 (when the work was published), writing within one of the following categories: poetry, novel, collection of short stories by one author, play that has been professionally performed, a broadcast radio play, a professionally produced screenplay that has resulted in a feature-length film.

Authors need to be nominated by their publishers, or producers in the case of performance art. Final entry date: 30 April. The award is presented at the Dylan Thomas Literary Festival in Swansea in November. Further entry details can be found on the website.

UK Film Council's Development Fund

10 Little Portland Street, London, W1W 7JG

ⓣ 020 7861 7861 ⓕ 020 7861 7862

development@ukfilmcouncil.org.uk
www.ukfilmcouncil.org.uk

Aims to broaden the quality, range and ambition of UK film projects and talent being developed. The annual budget for the Development Fund is £4 million. Funding decisions are taken by the Head of the Development Fund.

The First Feature Film Development Programme awards up to £25,000. to writers/directors to write and develop a feature film. See website for further details and submission guidelines.

Winchester Writers' Conference Prizes

Faculty of Arts, University of Winchester, Winchester, SO22 4NR

ⓣ 01962 827238

christian.francis@winchester.ac.uk
www.writersconference.co.uk

A series of fifteen writing competitions sponsored by major publishers. Includes a category for Scriptwriting.

Open for all writers in January; closing date: 30 May. Phone or write for booklet giving details of each competition or access them on the website.

Canada

Canadian Independent Film & Video Fund (CIFVF)

Suite 203, 666 Kirkwood Avenue, Ottawa, Ontario, K1Z 5X9

ⓣ 613 729 1900

info@cifvf.ca
www.cifvf.ca

A private sector funding body that supports non-theatrical film, videos and new media projects created by Canadian independent producers to enable lifelong learning. Various genres considered including documentary, docu-drama, drama and animation. Has provided CA$17.9 million since 1991.

Applications for funding are normally accepted at two specified deadlines each year.

Canadian Screenwriting Awards

c/o Writer's Guild of Canada, 366 Adelaide Street West, Suite 401, Toronto, Ontario, M5V 1R9

ⓣ 416 979 7907 ⓕ 416 979 9273

awards@wgc.ca
www.wgc.ca
Director of Communications *Barb Farwell*

Announced annually in spring in Toronto. 'Best Script' awards in several categories including feature film, one-hour TV drama, documentary and radio drama. Established in 1997. All Writer's Guild of Canada members eligible to apply. More details available on the website.

COGECO Program Development Fund

2 Carlton Street, Suite 1709, Toronto, Ontario, M5B 1J3

ⓣ 416 977 8966 ⓕ 416 977 0694

info@ipf.ca
www.ipf.ca

Encourages the development of new Canadian drama by Canadian writers to be produced by independent Canadian producers, in English or French, or (preferably) in both languages. Runs three funding programs: Development Programme (loans for the development of dramatic television programming, including television series, MOWs, mini-series, and animated series); Production Programme (equity investments for the production of MOWs, mini-series and pilots); Theatrical Feature Film Development Programme (corporate loans for companies with a slate of theatrical feature films in development).

Application deadlines are March 1st, July 1st and October 1st.

CTV Writer Only Drama Development Benefit

kmeek@ctv.ca
www.ctv.ca
Manager, BCE Drama Development & Production
Contact *Kathleen Meek*

Administered by CTV Inc., one of Canada's leading broadcast communications companies. Runs a CA$5 million programme for professional Canadian screenwriters, which allows for the development of dramatic screenplays for TV movies, mini-series or dramatic series. In the initial phase of development, CTV acts as producer of the project to facilitate the business aspects of the project. Working with the writer, CTV will approve the development budget, contract with story consultants and researchers, and negotiate and option underlying rights. At the appropriate stage, the project will be assigned to an independent third party producer.

Normally screenwriters will have a minimum credit of 60 minutes of produced drama to be eligible for consideration.

Geminis

172 King Street E., Toronto, Ontario, M5A 1J3
☏ 416 366 2227 ⒻAX 416 366 8454
info@academy.ca
www.geminiawards.ca

First broadcast in 1986, the awards celebrate excellence in Canadian English-language television in 87 categories, 6 of which are for writing.

Any production which qualifies as a television production under the Canadian Radio-Television and Telecommunications Commission or the Canadian Audio-Visual Certification Office, and has had its first Canadian commercial release in English within the qualifying period on an English-language or ethnic telecaster licensed by the CRTC may enter. The programme may not be a rerun and must not have been previously entered in either the Gemini or Genie Awards. Programmes previously entered for the Prix Gémeaux are eligible, subject to meeting all other Gemini eligibility criteria.

Genies

172 King Street East, Toronto, Ontario, M5A 1J3
☏ 416 366 2227 ⒻAX 416 366 8454

info@academy.ca
www.genieawards.ca

Created in 1980 to replace the Canadian Film Awards, recognizing the best of Canadian cinema. Dramatic feature-length films (including animated features), theatrical shorts and documentaries are eligible for awards. Has an award for best original screenplay and for best adapted screenplay.

Films may be submitted in either English or French. Films must be Canadian productions or co-productions, and must have had its first commercial release in Canada in a theatrical venue. A detailed synopsis of the film in both official languages must accompany the official entry form. For the screenplay awards, a complete production script must be submitted with the entry form.

Grants to Film and Video Artists Program

Media Arts Section, Canada Council for the Arts, PO Box 1047, 350 Albert Street, Ottawa, Ontario, K1P 5V8
☏ 1 800 263 5588
www.canadacouncil.ca/grants/mediaarts

Supports Canadian professional independent artists who use cinema and video as a mode of artistic expression. Scriptwriting Grants cover the direct costs of scriptwriting (including research for artists' documentaries) for independent film and video artworks. Grants range from $3,000 to $20,000. Eligible expenses include subsistence costs for the time spent working on the project (up to $2,000 per month per applicant); rental costs for equipment, studios, and other facilities; costs for training and/or development; fees for consultants, technicians and other participants; travel costs; contingency funds.

Application forms available from website.

The Harold Greenberg Fund

BCE Place, 181 Bay Street, PO Box 787, Toronto, Ontario, M5J 2T3
☏ 416 956 5431 ⒻAX 416 956 2018
hgfund@tv.astral.com
www.astral.com/en/theharoldgreenbergfund

Founded in 1986 by Astral Media, the first Canadian broadcaster to set up a national, non-profit private fund to support the development of film and made-for-pay-TV productions. An additional French-language programme was added in 1996. Collectively invested over CA$50 million to date. The English-language programme (sponsored by The Movie Network, Family and Viewers Choice) offers feature film script development loans as well as equity investments for both feature film and television productions. The French-language programme (sponsored by Astral Media, Super Écran, Family and Canal Indigo) offers the same for feature films, documentaries, special events, youth dramas and music programmes.

Leo Awards

700–1155 West Pender Street, Vancouver, British Columbia, V6E 2P4
℡ 604 688 4875 🅕 604 669 2288
walter@leoawards.com
www.leoawards.com
President *Walter Daroshin*

Presented each May since 1999 by the Motion Picture Arts and Sciences Foundation of British Columbia to honour the best in British Columbian television and film production.

Praxis Film Script Competitions

Suite 3120, 515 W. Hastings Street, Vancouver, British Columbia, V6B 5K3
℡ 778 782 7880 🅕 778 782 7882
praxis@sfu.ca
www.praxisfilm.com

Two feature film script competitions held each year (November and June). 4–6 winners have their scripts workshopped with a veteran story editor or screenwriter.
　Scripts can be of any genre.

Prix Gemeaux

172 King Street East, Toronto, Ontario, M5A 1J3
℡ 416 366 2227 🅕 416 366 8454
info@academy.ca

Founded in 1986, celebrating excellence in Canadian French-language television.

Toronto Film Critics Association Awards

www.torontofilmcritics.com

Established in 1997, the association is comprised of Toronto-based journalists and broadcasters who specialize in film criticism and commentary. Every year the TFCA gives out awards to Best Canadian Film, Best Picture, Best Director, Best Actor, Best Actress, Best Supporting Actor and Actress, Best Screenplay, Best First Feature and Best Documentary.

Writer's First Program

Telefilm Canada, 360 St. Jacques Street, Suite 500, Montréal, Quebec, H2Y 1P5
℡ 514 283 6363 🅕 514 283 8212
info@telefilm.gc.ca
www.telefilm.gc.ca
Contact *Carrie Earl (Ontario)*

Contact *Isabelle Picard (Quebec)*
Contact *Jennifer Porter (Western Region)*
Contact *Patricia Voogt (Atlantic Region)*

Administered by Telefilm Canada, a federal cultural agency dedicated to the development and promotion of the Canadian audiovisual industry. This scheme focuses resources on promising screenwriters for the creation of screenplays likely to lead to success at the box office.
　Guidance and application forms available via the website. Regional contacts can be emailed via a webform on the site.

Ireland

Arts Council Bursary

70 Merrion Square, Dublin 2
℡ 01 618 0200 🅕 01 676 1302
www.artscouncil.ie

The council offers a bursary award to a maximum value of €15,000 to artists. Film artists and writers are eligible for the award, but the award does not go towards the funding of a short or feature film production.

Cork Film Centre/RTE Short Script Awards

Civic Trust House, 50 Pope's Quay, Cork
℡ 021 421 5160
info@corkfilmcentre.com
www.corkfilmcentre.com

Has disbursed an annual award to emerging filmmakers since 1999.

FilmBase/RTE Short Film Awards

Curved Street Building, Temple Bar, Dublin 2
℡ 01 679 6716 🅕 01 679 6717
info@filmbase.ie
www.filmbase.ie

Awards funding for six short films a year. Cash awards of €10,000 (approximately) are made to the successful candidates, along with full production facilities.
　Awards are open to drama, animation and experimental works to be completed on film or broadcast quality video. Scripts should be 8–15 minutes in duration. Applicants for the award must be fully paid-up members of Filmbase as of the deadline. Full-time students are ineligible.

FilmBase/TG4 Lasair Awards

Curved Street Building, Temple Bar, Dublin 2
℡ 01 679 6716 🅕 01 679 6717
info@filmbase.ie
www.filmbase.ie

Funded by TG4, the awards are open to drama, animation and other fictional works in the Irish language. Cash awards averaging €10,000 are made to the successful candidates, and there is a further award of equipment and facilities.

Submissions should be between 8–26 minutes duration. Awards are made to directors. Full time students are ineligible. The applicant must be a fully paid up member of Filmbase before applying.

Galway Film Centre/RTE Short Script Awards

Cluain Mhuire, Monivea Road, Galway
Ⓣ 091 770748 Ⓕ 091 770746
info@galwayfilmcentre.ie
www.galwayfilmcentre.ie

Founded in 1997, the Short Script Awards are open to emerging filmmakers with original scripts that display a strong cinematic vision and a fresh Irish perspective. Four awards are allocated from a total production fund of €38,000. Use of the facilities of Galway Film Centre is a key constituent of the awards.

Films should be no more than 15 minutes in duration. Applicants must be members of the Galway Film Centre. Full-time students are ineligible.

Irish Film and Television Awards

Irish Film and Television Academy,
First Floor, Palmerstown Centre, Kennelsfort Road, Palmerstown, Dublin 20
Ⓣ 01 620 0811 Ⓕ 01 620 0810
info@ifta.ie
www.ifta.ie

Founded in 2003, the awards are presented annually by the Academy to honour and celebrate outstanding Irish creativity, talent and achievement. Includes an award for best script.

Open to all Irish people working in the film and television industries in Ireland and internationally.

Oscailt (TG4 and Bord Scannán na héireann)

Bord Scannán na héireann, Queensgate, 23 Dock Road, Galway
Ⓣ 091 561 398 Ⓕ 091 561 405
info@irishfilmboard.ie
www.irishfilmboard.ie

Awards for short fiction film-making in Irish. Has been broadened to include fictional pieces in animation, flash animation or docu-drama projects. Although priority will be given to first-time directors and writers, the scheme is also open to film-makers with more experience. A maximum fund of €300,000 is available.

Films should be between 10 and 25 minutes in duration. Applicants for the fund must normally be of Irish residence. Film-makers must have had some previous experience or training.

Stella Artois Pitching Awards

www.galwayfilmfleadh.com/pitching.html

Screenwriters are invited to submit a one-page story idea for the screen. Five finalists are chosen to appear at the Galway Film Fleadh to make their pitch to a panel of industry experts and a public audience. The best pitch wins €5,000.

Australia

AFI Awards

236 Dorcas Street, South Melbourne, VIC 3205
Ⓣ 03 9696 1844 Ⓕ 03 9696 7972
awards@afi.org.au
www.afi.org.au
Awards Manager *Justine Beltrame*

Awards ceremony run by the Australian Film Institute, held in November/December each year.

Open to Australian productions or co-productions and Australian practitioners.

AWGIE Awards

8/50 Reservoir Street, Surry Hills, NSW 2010
Ⓣ 02 9281 1554 Ⓕ 02 9281 4321
admin@awg.com.au
www.awg.com.au
Executive Director *Jacqueline Woodman*
Communications and Membership Manager *Stephen Asher*
Membership Officer *Rebecca Freeman*

The AWGIE Awards are the Australian Writers' Guild annual awards recognizing excellence in performance writing including film, television, theatre, radio and new media. They are the only peer-assessed awards in Australia and are judged on the script alone - the writers' intention, not the completed production. The Short Form and Long Form Monte Miller Awards are presented to an associate member of the AWG for unproduced scripts.

2007 Major Winner - Keith Thompson for *Clubland*;2006 Major Winner - Barbara Samuels and Katherine Thomson for *Answered by Fire*;2005 Major Winner - Melissa Reeve for *The Spook*.

To be eligible to enter writers must be a financial member of the AWG and must have been produced between 1 January and 31 December the previous year.

Film Critics Circle of Australia Awards

PO Box 673, Crows Nest, NSW 1585
fcca@optusnet.com.au
www.fcca.com.au

Annual awards ceremony promoting excellence in Australian film and supporting the advancement of Australian film culture. Presented at gala dinner held in October.

IF Award for Best Script

PO Box 55, Glebe, NSW 2113
clare@if.com.au
www.ifawards.com

ZTudio and InsideFilm co-sponsor this competition for the best unproduced screenplay in Australia. Seeks screenwriters who have been proactive in seeking screenplay development but who have yet to see their work produced. The winner is flown to a country of their choice to pitch to agents or producers of their own choice.

Entry forms available from the website.

The Logie Awards

c/o Kerry O'Brien Publicity
℡ 03 9597 9792
publicity@kob.com.au

Prestigious awards for the Australian television industry awards, running since 1959. Sponsored by *TV Week* magazine.

Monte Miller Awards

8/50 Reservoir Street, Surry Hills, NSW 2010
℡ 02 9281 1554 ℻ 02 9281 4321
admin@awg.com.au
www.awg.com.au

Awarded to an associate member of the Australian Writers' Guild for an unproduced script (one prize for a script under 30 minutes and one for over 30 minutes).

New Zealand

Air New Zealand Screen Awards

C/- SDGNZ PO Box 47-294, Ponsonby, Auckland
℡ 09 360 2102

office@sdgnz.co.nz
www.sdgnz.co.nz

Awards for television and film. In the television category there are awards for best script in comedy and in drama, and in the film category there is an award for best script for a short film.

A synopsis of 100 words or less must be included with the entry. Films must be less than 20 minutes duration and shot on film or tape. To be eligible it must have been produced in New Zealand or by a producer normally resident in New Zealand and the film cannot have been entered into the awards previously.

The Richies

Richmond Road Short Film Festival,
113 Richmond Road, Ponsonby, Auckland,
℡ 09 376 1091
entries@therichies.co.nz
homepages.paradise.net.nz/urumold/rrsff/home.html

Founded in 2005, this short film competition is open to any New Zealand short film (drama, documentary, experimental or animated).

Films should run for 20 minutes or less. Either the writer, director, or producer must be a New Zealand citizen or a New Zealand permanent resident. All shooting formats are accepted. Music videos are not eligible.

Signature Film

PO Box 7015, Wellesley Street,
Auckland 1141
℡ 09 307 0835 ℻ 09 366 0503
trevor@thefilm.co.nz
www.thefilm.co.nz
Executive Producer *Trevor Haysom*

A joint initiative funded by the New Zealand Film Commission, New Zealand On Air and Television New Zealand. Managed and executive produced by Trevor Haysom. Aims to facilitate the development, production and broadcast of entertaining, original and innovative films, rooted in New Zealand stories that reflect the country's multi-racial culture. Aim is to create auteur films of between 65 to 85 minutes in length, for television broadcast. A scheme for filmmakers with significant experience and not for entry-level talent.

For full submission details see website.

Recommended Reading

Adaptations: From Other Works into Films: From Other Media into Films by Ronald Harwood (Guerilla Books Limited, 2007)

Advanced Screenwriting: Raising Your Script to the Academy Award Level by Linda Seger (Silman-James Press, 2003)

Adventures in the Screentrade by William Goldman (Warner Books, 1983)

Alternative Scriptwriting: Successfully Breaking the Rules by Ken Dancyger and Jeff Rush (Focal Press, 2001)

An Introduction to Writing for Electronic Media: Scriptwriting Essentials Across the Genres by Robert B. Musburger (Focal Press, 2007)

"Bambi" Vs. "Godzilla": On the Nature, Purpose, and Practice of the Movie Business by David Mamet (Simon & Schuster, 2007)

Blueprint for Screenwriting: A Writer's Guide to Creativity, Craft and Career by Rachel Ballon (Lawrence Erlbaum Assocs, 2004)

Breakfast with Sharks: A Screenwriter's Guide to Getting the Meeting, Nailing the Pitch, Signing the Deal and Navigating the Murky Waters of Hollywood by Michael Lent (Three Rivers Press, 2004)

Characters and Conflict: the Cornerstones of Screenwriting by Mark Axelrod (James Bennett Pty, 2004)

Classic American Films: Conversations with the Screenwriters by William Baer (Greenwood Press, 2007)

Crafty Screenwriting: Writing Movies that Get Made by Alex Epstein (Owl Books, 2002)

Creating Unforgettable Characters by Linda Seger (Owl Books 1990)

Creative Screenwriting: A Practical Guide by Duncan Kenworthy (Crowood Press, 2002)

Dealmaking in the Film and Television Industry From Negotiations Through Final Contracts by Mark Litwak (Silman-James Press, 2002)

Easy Riders, Raging Bulls: How the Sex-Drugs-and-Rock 'n' Roll Generation Saved Hollywood by Peter Biskind (Simon & Schuster, 1998)

Elements of Screenwriting by Irwin R. Blacker (Scribner, 1987)

Everything Screenwriting by Robert Pollock (Adams Media Corp, 2003)

Fade in: Screenwriting Process by Robert A. Berman (Michael Wiese Productions, 1997)

Formatting Your Screenplay by Rich Reichman (Book Smiths Inc., 1994)

Gardner's Guide to Screenwriting: the Writer's Road Map by Marilyn Webber & Bonney Ford (Garth Gardner Co, 2001)

Global Scriptwriting by Ken Dancyger (Butterworth-Heinemann, 2001)

Hot Property: Screenwriting in the New Hollywood by Christopher Keane (Berkley, 2003)

How Not to Write a Screenplay: 101 Common Mistakes Most Screenwriters Make by Denny Martin Flinn (Lone Eagle, 1999)

How to Make Money Scriptwriting by Julian Friedmann (Boxtree, 1995)

How to Write: A Screenplay by Mark Evan Schwartz (Continuum International Publishing Group, 2007)

I Liked It, Didn't Love It: Screenplay Development from the Inside Out by Rona Edwards and Monika Skerbelis (Lone Eagle, 2005)

Lew Hunter's Screenwriting 434: The Industry's Top Teacher Reveals the Secrets of the Successful Screenplay by Lew Hunter (Berkley Books, 2004)

LifeTips 101 Screenwriting Tips by Alexis Niki (LifeTips.com, Inc., 2007)

Making a Good Script Great by Linda Seger (Samuel French, 1987)

Plots & Characters: A Screenwriter on Screenwriting by Millard Kaufman (Really Great Books, 2001)

Power Screenwriting: The 12 Stages of Story Development by Michael Chase Walker (Lone Eagle Publishing, US, 2002)

Raindance Writers' Lab: Write + Sell the Hot Screenplay by Elliot Grove (Focal Press, 2001)

Real Screenwriting by Ronald Suppa (Premier Press, 2005)

Romancing the A List by Christopher Keane (Michael Wiese Productions, 2006)

Save the Cat! The Only Book on Screenwriting You'll Ever Need by Blake Snyder (Michael Wiese Productions, 2005)

Scenario: The Craft of Screenwriting by Tudor Gates (Wallflower Press, 2002)

Screen Plays: How 25 Scripts Made It to a Theater Near You—For Better or Worse by David S. Cohen (HarperEntertainment, 2008)

Screenplay Story Analysis by Asher Garfinkel (Allworth Press, 2007)

Screenplay: The Foundations of Screenwriting by Syd Field (Dell, 1979)

Screenplaying: Arming Yourself for a Shot at Screenwriting by John Scott Lewinski (Xlibris Corp, 2000)

Screenwright: The Craft of Screenwriting by Charles Deemer (Xlibris Corp, 2000)

Screenwriter's Survival Guide by Max Adams (Warner Books, 2001)

Screenwriting 101: The Essential Craft of Feature Film Writing by Neill Hicks (Michael Wiese Productions, 1999)

Screenwriting by Declan McGrath (RotoVision, 2003)

Screenwriting by Felim MacDermott & Declan McGrath (Focal Press, 2003)

Screenwriting for a Global Market – Selling Your Scripts from Hollywood to Hong Kong by A. Horton (University of California Press, 2003)

Screenwriting for Dummies by L. Schellhardt (Hungry Minds Inc., 2003)

Screenwriting for Film and Television by Roger LeRoy Miller (Prentice Hall, 1997)

Screenwriting for the 21st Century by Pat Silver-Lasky (Batsford, 2004)

Screenwriting from the Heart: The Technique of the Character-driven Screenplay by James Ryan (Billboard Books, 2000)

Screenwriting from the Soul by Richard Krevolin (Renaissance Books, 1998)

Screenwriting is Storytelling: Creating an A-List Screenplay That Sells by Kate Wright (Perigee Books, 2004)

Screenwriting Life by R. Whiteside (GP Putnam, 1998)

Screenwriting Tricks of the Trade by William Froug (Silman-James Press, US, 1992)

Screenwriting Updated: New (and Conventional) Ways of Writing for the Screen by Linda Aronson (Silman-James Press, 2001)

Screenwriting: A Manual by Jonathan Dawson & Ian Stocks (OUP Australia & NZ, 2000)

Screenwriting: Step by Step by Wendy J. Henson (Allyn & Bacon, 2004)

Screenwriting: Techniques for Success by Jimmy Sangster (Reynolds & Hearn, 2003)

Screenwriting: The Art, Craft and Business of Film and Television Films by Richard Walter (Penguin Australia, 1992)

Screenwriting: The Complete Idiot's Guide to Screenwriting by Skip Press (Alpha Books, 2000)

Screenwriting: The Sequence Approach by Paul Gulino (Continuum, 2004)

Selling Rights by Lynette Owen (Routledge, 1997)

Story and Character: Interviews with British Screenwriters by Alistair Owen (Bloomsbury, 2003)

Story: Substance, Structure, Style and the Principles of Screenwriting by Robert McKee (Methuen, 1999)

Teach Yourself Screenwriting by Ray Frensham (Teach Yourself, 2003)

The 101 Habits of Highly Successful Screenwriters: Insider's Secrets from Hollywood's Top Writers by Karl Iglesias (Adams Media Corporation, 2001)

The Anatomy of Story by John Truby (North Point Press, 2007)

The Art and Science of Screenwriting by Philip Barker (Intellect Books, 2002)

The Art of Plotting: Add Emotion, Suspense, and Depth to Your Screenplay by Linda J. Cowgill (Lone Eagle, 2007)

The Art of Screenwriting by W. Packard (Thunder's Mouth Press, 1998)

The Art of Screenwriting Simplified: The Most Comprehensive Guide for Film & Television by Willie E. Mason (Bawn Publishers Inc., 1996)

The Complete Book of Scriptwriting by Michael Straczynski (Writer's Diget, 1996)

The Definitive Guide to Screenwriting by Syd Field (Ebury Press, 2003)

The Savvy Screenwriter: How to Sell Your Screenplay (and Yourself) Without Selling Out! by Susan Kouguell (St. Martin's Griffin, 2006)

The Screenwriter's Bible: A Complete Guide to Writing, Formatting and Selling Your Script by David Trottier (Silman-James Press, US, Sept. 2005)

The Screenwriter's Guide to Agents and Managers by John Scott Lewinski (Allworth Press, 2001)

The Screenwriter's Legal Guide by Stephen F. Breimer (Allworth Press, 1999)

The Screenwriter's Problem Solver by Syd Field (Dell, 1998)

The Screenwriter's Sourcebook: A Comprehensive Marketing Guide for Screen & Television Writers by Michael Haddad (Chicago Review Press, 2005)

The Screenwriter's Story Planning Guide (Or How to Begin Working on an Idea You Don't Have) by Steven R. Gottry (Priority Multimedia Group, 1999)

The Script Selling Game: A Hollywood Insider's Look at Getting Your Script Sold and Produced by Kathie Yoneda (Michael Wiese Productions, 2002)

The Secrets of Action Screenwriting by William C. Martell (First Strike Productions, 2000)

The Tools of Screenwriting by David Howard & Edward Mabley (St Martin's Press, 1995)

The Writer's Journey: Mythic Structure for Storytellers and Screenwriters by Christopher Vogler (Michael Wiese Productions, 1992)

Top Secrets: Screenwriting by Jurgen Wolff & Kerry Cox (Lone Eagle, 1999)

Vault Career Guide to Screenwriting by David Kukoff (Vault.com, Nov. 2005)

What Happens Next?: A History of Hollywood Screenwriting by Marc Norman (Aurum Press Ltd, 2008)

What Happens Next? An Introduction to Screenwriting by Charles Deemer (Booksurge, 2003)

Which Lie Did I Tell? More Adventures in the Screentrade by William Goldman (Pantheon Books, 2000)

Writer's Guide to Hollywood Producers, Directors and Screenwriter's Agents (Skip Press, 2001)

Writing for the Hollywood $$$ by Tony Blake (Xlibris Corporation, 2007)

Writing Screenplays That Sell by Maichael Hauge (Elm Tree, 1989)

Writing Television Comedy by Jerry Rannow (Allworth Press, 2000)

Writing Television Sitcoms by Evan S Smith (G P Putnam's Sons, 1999)

Writing the TV Drama Series: How to Succeed as a Professional Writer in TV by Pamela Douglas (Michael Wiese Productions, 2005)

WEBSITES

www.dailyscript.com – A collection of movie scripts and screenplays.

www.finaldraft.com – Industry-standard screenwriter's software

www.gutenburg.org – Over 25,000 out-of-US copyright titles free online, including many scripts.

www.imdb.com – The internet movie database site.

www.imsdb.com – Internet database of movie scripts.

www.iscriptdb.com – Movie script database.

www.screenplay.com – Software resources for screenwriters.

www.screenscripts.com – Platforms for writers to pitch their stories and ideas.

www.screenwriter.com – Advice from professional screenwriters.

www.scriptfly.com – Screenplays, articles and analysis.

www.script-o-rama.com – Movie scripts and screenplays.

www.scriptpimp.com – Screenwriting and movie screenplay database.

www.scriptsecrets.net – Offers advice from script guru William C. Martell.

www.simplyscripts.com – Free movie scripts and screenplays.

www.thescreenwritersstore.com – Complete online store.

www.wordplayer.com – Writing advice from professional including Terry Rossio and Ted Elliott.

www.writingtreatments.com – Guidance on the discipline of creating a treatment.

www.zoetrope.com – Includes a Virtual Studio where artists can submit work and producers can make movies using built-in production tools.

Index

10 Sec Film Fest, 229
100 Percent Film & Television Inc., 111
1492 Pictures, 37
2 Days Later Short Film Competition, 229
2000 AD Productions, 124
20th Century Fox, 37
20th Century Fox Film Co., 76
20th Century Fox Television, 37
2929 Entertainment, 37
The 3 Americas Film Festival, 241
3 Arts Entertainment, 37
3monkeyfilms Pty Ltd, 134
44 Blue Productions, 37
4th Row Films, 38
50 Cannon Entertainment, 38

A

A & B Personal Management Ltd, 158
Aardman, 76
Above the Line Agency, 151
Abrams Artists, 151
Absolutely Productions Ltd, 76
Abstract Images, 77
Abú Media, 124
Acacia Productions Ltd, 77
Academy Awards (The Oscars), 253
Academy of Canadian Cinema and Television, 207
Academy of Motion Picture Arts and Sciences, 197
Academy of Television Arts & Sciences (ATAS), 197
Acclaim Film Screenwriting Award, 253
Accomplice Television, 124
Acme Talent and Literary Agency, 151
Acrobat Television, 77
Actaeon Films Ltd, 77
Activision, Inc., 38
Actual Reality Pictures, 38
Adare Productions, 124
Addictive TV, 77

Adelaide Film Festival, 246
AEI, 38
Affinity Productions, 111
AFI Awards, 269
AFI Festival, 215
Africa In Motion, 229
African Diaspora Film Festival, 215
African Film Commission Storytelling Screenwriting Competition, 253
After Image, 77
Afterdark Productions Inc., 112
Agamemnon Films Inc., 38
Agence Omada, 167
The Agency (London) Ltd, 158
Agency for the Performing Arts (APA), 151
Agenda Film Productions, 134
Air New Zealand Screen Awards, 270
AJ Films, 142
Alberta Filmworks, 112
Alchemy, 38
All3Media, 77
Alliance for Children and Television (ACT), 207
The Alliance Française French Film Festival, 246
Alliance of Motion Picture & Television Producers, 197
Allshorts Film Festival, 248
The Alpern Group, 151, 167
Alpine Pictures, Inc., 38
Amber Films, 78
Amberwood Entertainment Corp., 112
Ambient Light Productions Ltd, 78
Ambition, 167
Ambush Entertainment, 39
AMC Pictures Ltd, 78
Amen Ra Films, 39
American Accolades Screenwriting Competition, 253
American Black Film Festival, 215

American Cinema International, 39
American Empirical Pictures, 39
American Film Institute (AFI), 173, 197
Independent Film & Television Alliance, 197
American Gem Short Script Contest, 253
American Screenwriters Association, 197
American Screenwriters Association/Gotham Writer's Workshop, 253
American Screenwriting Competition, 254
American University, 173
American Zoetrope, 39
American Zoetrope Screenplay Contest, 254
The Ampersand Agency Ltd, 159
Anagram Pictures, 112
Anaïd Productions Inc., 112
Craig Anderson Productions, 39
Darley Anderson Literary, TV & Film Agency, 159
George Andrews Productions Ltd, 142
Angel Entertainment Corporation, 112
The Angel Film Festival, 230
Angelus Student Film Festival, 215
Anglia Factual, 78
Anglo-Fortunato Films Ltd, 78
Animate the World, 230
Animex International Festival of Animation and Computer Games, 230
Animo Television, 124
Anonymous Content, 39
Anonymous Content Talent and Literary Management, 151
Antelope Valley Independent Film Festival, 215
Antidote International Films, Inc., 39

Antimatter Underground Film Festival, 241
Apartment 11 Productions, 112
Apertura, 39
Apocalypso Pictures, 78
Appian Way, 39
Aquila Pictures, 112
Arcane Pictures, 78
Archer Street, 79
Arden Entertainment, 40
Arenafilm, 134
Arenas Entertainment, 40
Arista Development, 179
Arjay Entertainment, 40
Arlington Productions Limited, 79
Art & Training Films Ltd, 79
Artist International, 151
The Artists Agency, 151
The Artists' Colony, 40
Arts Council Bursary, 268
Arts Council Ireland, 208
Arvon Foundation, 179
Ascendant Pictures, 40
The Ashford Entertainment Corporation Ltd, 79
Asian American International Film Festival, 215
Asian American International Film Festival Screenplay Competition, 254
Asian CineVision, 216
Aspects Literature Festival, 230
Aspen FilmFest, 216
Aspen Shortsfest, 216
Associated Creative Talents, 134
The Asylum, 40
Athena Media, 125
Atlanta Film Festival, 216
Atlanta Screenwriters Group, 198
Atlantic Film Festival, Halifax, 241
Auckland International Film Festival, 248
Auckland University, 193
Augusta Pictures Screenplay Contest, 254
Aurora (Norwich International Animation Festival), 230
Aurora Artists, 167
Austin Film Festival, 216
Austin Gay and Lesbian International Film Festival, 216
Australian Children's Television Foundation, 134
Australian College QED, 191
Australian Film Commission, 210
Australian Film Institute, 210

Australian Film, Television and Radio School, 191
Australian International Pictures Pty Ltd, 135
Australian Script Centre, 210
The Australian Writers' Guild, 210
Aus-TV International, 134
Author Literary Agents, 159
The Authors Guild, 198
AV Pictures, 79
A-V-A Productions, 38
Avalon, 79
Avalon Film Corporation, 135
Avanti Pictures Corporation, 112
Avatar Productions, 40
Avignon/New York Film Festival, 216
Avrio Filmworks Inc., 113
AWGIE Awards, 269
Axial Entertainment, 40

B

Babelgum Online Film Festival, 217
Baby Cow Productions, 79
BAFTAs, 263
Baker Street Media Finance, 79
Baldwin Entertainment Group, 40
Ballistic Talent Management, 152
Banff Mountain Film Festival, 241
Banff World Television Festival, 241
Christopher Banks & Associates, 167
Banksia Productions, 135
Barcelona Films, 41
Bardel Entertainment Inc., 113
Barley Films, 125
Barna-Alper Productions Inc., 113
Barnstorm Films, 41
Bath Film Festival, 230
Bath Spa University, 179
Bauer Martinez Studios, 41
BBC College of Comedy, 263
BBC Drama Writers Academy, 264
BBC Films, 80
BBC New Talent, 80
BBC Vision, BBC Audio & Music, 80
BBC Writersroom, 80
Beacon Pictures, 41
Becker Entertainment, 135
BECTU Script Registration Service, 200
Belfast Festival at Queen's, 230
Belfast Film Festival, 230
Belladonna Canterbury Short Film Festival, 248
Belladonna Productions, 41

Benderspink, 41
Berwick Film & Media Arts Festival, 231
Betting on Shorts, 231
Beyond Pix, 41
Beyond TV International Video Festival, 231
BFM International Film Festival, 231
Bicoastal Talent, 152
BIFF Screenplay to Production Contest, 254
Big and Little Films Pty Ltd, 135
Big Apple Film Festival, 217
Big Bear Lake International Film Festival, 217
Big Bear Lake International Screenwriting Competition, 254
Big Break: A Final Draft Contest, 254
Big Heart Media, 81
Big Motion Pictures Limited, 113
Big Mountain Short Film Festival, 248
Big Soul Productions Inc., 113
BioWare Corp., 113
Birds Eye View Film Festival, 231
Bite the Mango Film Festival, 231
Bitesize Productions Ltd, 142
Black Sheep Entertainment, 41
Black Walk, 113
Black Watch Productions, Inc., 42
Blackwatch Productions Limited, 81
Bleiberg Entertainment, Inc., 42
Blink Films Pty Ltd, 135
Blizzard Entertainment, 42
Blowback Productions, 42
Blue Rider Pictures, 42
Blue Sky Studios, 42
BlueCat Screenplay Competition, 255
Blueskin Films Ltd, 142
Boca Productions LLC, 42
The Bohrman Agency, 152
Bona Broadcasting Limited, 81
Bond University, 191
Bootstrap Films Limited, 125
Irish Film Board/Bord Scannán na héireann, 209
Borderless Productions Limited, 142
Boston International Film Festival, 217
Boston University, 173
Boulder Media, 125
Bournemouth University, 179
Box, 81

Bradford Animation Festival, 231
Bradford International Film
   Festival, 232
Richard Bradley Productions Pty
   Ltd, 135
Brant Rose Agency, 152
Brass Brad Screenwriting
   Mentorship Award, 255
Braun Entertainment
   Group, Inc., 42
Braverman Productions, Inc., 42
Breakthrough Films & Television
   Inc., 113
BreakThru Films Ltd, 81
Bridgend College, 179
Brightlight Pictures Inc., 114
Brillstein-Grey
   Entertainment, 152
Brillstein-Grey Entertainment, 43
Brisbane International Film
   Festival, 246
Bristol Bay Productions, 43
Bristol Screenwriters, 200
Britdoc, 232
The British Academy of Film &
   Television Arts (BAFTA), 200
British Film Institute, 200
British Independent Film
   Awards, 264
British Short Screenplay
   Competition, 264
The Broad Humor Film
   Festival, 217
Alan Brodie Representation
   Ltd, 159
Bronx Independent Film
   Festival, 217
Brooklyn College, 173
Brooklyn Films, 43
Brooklyn International Film
   Festival, 217
Brookwell/McNamara
   Entertainment, 43
Brown Bag Films, 125
Jerry Bruckheimer Films, 43
Don Buchwald and
   Associates, Inc., 152
Bunim/Murray Productions, 43
Burberry Productions Pty Ltd, 135
The Bureau, 81
Brie Burkeman, 159
Burton Manor, 180
Butchers Run Films, 43

C

C3 Entertainment Inc., 43
CAAMA (Central Australian
   Aboriginal Media
   Association), 136

CABOOM, 125
Cactus TV, 81
Calgary International Film
   Festival, 242
Caliber Talent Group, 152
California Institute of the Arts, 173
Calon TV, 81
Cambridge Film Festival, 232
Cambridge Literary
   Associates, 152
Camelot Entertainment
   Group, Inc., 43
Cameron Creswell Agency, 169
Campbell Ryan Productions
   Ltd, 125
Can Leicester International Short
   Film Festival, 232
Canadian Accents Inc., 114
Canadian Comedy Festival and
   Awards, 242
Canadian Film and Television
   Production Association
   (CFTPA), 207
Canadian Film Centre's
   Worldwide Short Film
   Festival, 242
Canadian Film Institute, 207
Canadian Filmmaker's
   Festival, 242
Canadian Independent Film &
   Video Fund (CIFVF), 266
The Canadian Screen Training
   Centre, 189
Canadian Screenwriter Collection
   Society (CSCS), 207
Canadian Screenwriting
   Awards, 266
Canadian Women in
   Communications, 207
Canamedia Film Productions
   Inc., 114
Capcom USA, Inc., 43
Capel & Land Ltd, 159
Capital Arts Entertainment, 44
Capitol Films, 81
Capitol Productions, 136
Capri Films Inc., 114
Cardiff University Centre for
   Lifelong Learning, 180
Carnival Film & Television, 82
Carolina Film and Video
   Festival, 217
Matt Carroll Films Pty Ltd, 136
Carsey-Werner, LLC., 44
Cartoon Saloon, 125
Cartwn Cymru, 82
Maria Carvainis Agency, Inc., 152
Casarotto Ramsay and Associates
   Ltd, 160

Cascade Films, 136
Castle Rock Entertainment, 44
CatchLight Films, 44
Celador Films, 82
Celador Productions, 82
Celtic Films Entertainment
   Ltd, 82
Celtic Media Festival, 232
Centaur Enterprises Pty Ltd, 136
The Central School of Speech and
   Drama, 180
Centropolis Entertainment, 44
Century City Cell Phone Festival
   Competition, 255
CFC (Canadian Film Centre), 189
CFP Productions, 44
Chameleon Television Ltd, 82
Channel 4 British Documentary
   Film Foundation, 200
Channel 4 Television, 82
Channel X, 82
Jan Chapman Films Pty Ltd, 136
Chapman University, 173
The Characters, 167
Charnjit Films, 44
The Chasin Agency, Inc., 152
Cheap and Dirty Productions, 114
Mic Cheetham Literary
   Agency, 160
Cherry Road Films, 44
Chicago City Limits Comedy Film
   Festival, 217
Chicago International Film
   Festival, 218
Chicago Underground Film
   Festival, 218
Chichester Film Festival, 232
The Children's Film & Television
   Foundation Ltd, 83
Children's Film Unit, 83
Cine Excel Entertainment, 44
Ciné Qua Non Média, 114
Cinefest – Sudbury International
   Film Festival, 242
Cinem@tic, 232
Cinema Libre Studio, 45
Cinemagic International Film
   Festival for Young People, 233
Cinequest San Jose Film
   Festival, 218
C2 Pictures, 45
CineStory, 198
Cinetel Films, 45
Cinetel Productions Pty Ltd, 136
Cineville International, 45
Circe's World Films, LLC, 45
City Lights Pictures, 45
City Lit, 180
City University, 180

Claddagh Films, 126
Clarity Pictures, 45
Dick Clark Productions, 45
Mary Clemmey Literary
Agency, 160
Cleveland International Film
Festival, 218
Cleveland Productions, 83
Click & Copyright, 198
Jonathan Clowes Ltd, 160
Club Panico@The London Film
Academy, 200
Coastal Productions Limited, 83
Elspeth Cochrane Agency, 160
Coco Television, 126
Codemasters, 83
COGECO Program Development
Fund, 266
Cogswell Polytechnical College, 173
COI, 83
Rosica Colin Ltd, 160
Collingwood O'Hare
Entertainment Ltd, 83
Columbia Tristar Motion Picture
Group, 83
Columbia Tristar Motion Picture
Group / Sony Pictures, 46
The Comedy Unit, 84
Coming Up, 264
Community Creative Writing and
Publishing, 180
Company Pictures, 84
Conbrio Media Ltd, 143
Concept Entertainment, 152
Coney Island Short Film
Festival, 218
Connection III Entertainment
Corp., 46
Contentfilm International, 84
Coolabi, 84
Co-operative Young Film-Makers
Film Festival, 233
Cooper's Town Productions, 46
Cork Film Centre, 209
Cork Film Centre/RTE Short
Script Awards, 268
Cork Film Festival, 245
Cork Youth International Film
Arts Festival, 245
Cornwall Film Festival, 233
Cosgrove Hall Films, 84
Cougar Films, 84
Cowboy Films, 84
Crackerjack, 136
Crane Wexelblatt Entertainment
Ltd, 46
Crannog Films, 126
Crawford Productions Pty
Ltd, 137

Crawley Films Ltd, 114
Creative Artists Agency, 153
Creative Convergence, 153
Creative Media Management, 160
The Creative Partnership, 84
Crescent Entertainment
Ltd, 114
Crested Butte Reel Fest, 218
Criterion Games, 85
Critics' Choice Awards, 255
The Crosley Company, 46
Cross Day Productions Ltd, 85
Crossroads Films, 85
Crystal Sky Pictures, 46
CS Films, 85
CTV Writer Only Drama
Development Benefit, 267
CTVglobemedia Inc., 114
Cucalorus Film Festival, 218
Curtin University of
Technology, 191
Curtis Brown Group, 161
Curtis Brown Ltd, 153
Cutting Edge Productions Ltd, 85
Cynosure Screenwriting
Awards, 255

D

Judy Daish Associates Ltd, 161
Dakota Films, 46
Dakota Films Ltd, 85
Dalaklis McKeown
Entertainment, Inc., 46
Damah Film Festival, 218
Dan Films, 85
Lee Daniels Entertainment, 46
Darius Films Inc., 115
Dark Horse Entertainment, 47
Dark Light Digital Festival, 245
Darkwoods Productions, 47
Dartington College of Arts, 180
Dawson City International Short
Film Festival, 242
Daybreak Pacific, 143
Dazed Film & TV, 85
De Facto Films, 85
Dino De Laurentiis Company, 47
De Montfort University, 180
Felix de Wolfe, 161
Dead by Dawn, 233
Deaffest, the Deaf Film and
Television Festival, 233
Decode Entertainment Inc., 115
DeeGee Entertainment, 47
The Dench Arnold Agency, 161
Devine Entertainment, 115
Digital Animation Media
Limited, 126
Dimension Films, 47

Dingo Production LLC, 47
Dirty Hands Productions, 85
Discovering Latin American Film
Festival, 233
Disney Interactive
Studios, Inc., 47
Distant Horizon, 47
Diva Productions, 143
Diverse Production Limited, 86
Diverse Talent Group, 153
DMS Films Ltd, 86
DNA Films Ltd, 86
DOCNZ Film Festival, 248
Document International Human
Rights Documentary Film
Festival, 233
Documentary Filmmakers
Group, 181
Documentary Organization of
Canada, 208
Done and Dusted, 86
DoubleBand Films, 86
Drake AV Video Ltd, 86
Dramatists Guild of America, 198
Dream Street Pictures Inc., 115
Dreamworks SKG, 47
Bryan Drew Ltd, 161
Dreyfuss/James Productions, 48
Drum Productions, 143
Dublin International Film
Festival, 246
Duke City Shootout, 255
Charles Dunstan
Communications Ltd, 86
Durango Independent Film
Festival, 218

E

Earth Horizon Productions, 126
East End Film Festival, 233
East Lansing Film Festival, 219
Eaton Enterprises Pty Ltd, 137
Echo Lake Productions, LLC, 48
Ecosse Films, 86
Edge Hill University, 181
Edinburgh International Film
Festival, 233
Edinburgh Mountain Film
Festival, 234
Edmonds Entertainment, 48
Edmonton International Film
Festival, 242
Eidos Interactive Ltd, 86
El Dorado Pictures, 48
Jim Eldridge Writing
Workshops, 181
Electric Entertainment, 48
Electric Pictures, 137
Electro-Fish Media LLC, 48

Electronic Arts (EA), 48
Element Films, 48
Element Pictures Ltd, 126
Elixir Films, 48
Elstree (Production) Company
    Ltd, 87
Embryo Films, 137
Emergeandsee, 234
Emerging Pictures
    Corporation, 49
Emerson College, 174
Emmanuel Productions, 143
Emmy Awards, 255
Empire Action Pty Ltd, 137
Enchanter Pty Ltd, 137
Encounters International Short
    Film Festival, 234
End of the Pier International Film
    Festival, 234
The Endeavor Talent Agency, 153
Endemol UK Productions, 87
Energy Entertainment, 49, 153
EnterAktion Studios, Inc., 49
EO Teilifís, 126
Eon Productions, 87
Epiphany Pictures, Inc., 49
Epitome Pictures Inc., 115
Esperanza Productions, 126
Essential Viewing, 137
Euroscript, 181, 200
Euroscript Screenwriting
    Competition, 264
Evamere Entertainment LLC, 49
Evatopia, 153
Evening Standard British Film
    Awards, 264
Evergreen Films LLC, 49
Every 1's a Critic, 264
Evoke Pictures, 143
Evolution Media, 49
Execam Television and Video
    Production, 143
Exodus Film Group, 49
Exposure, 143
Eye Film and Television, 87
Eyeworks Touchdown, 143

F

Fabrication Films, 50
Fade In Award, 255
Fantastic Films, 126
Fantastic Films Weekend, 234
Fantasy Worldwide Film
    Festival, 242
Farnham Film Company Ltd, 87
FÁS Screen Training
    Ireland, 190, 209
Fast Films, 87
Fast Productions Ltd, 115

Fastnet Films, 127
FatKid Films, 50
Featherstone Productions, 137
Federation of Screenwriters in
    Europe, 201
A Feeding Frenzy – Screenplay
    Competition with
    Feedback, 256
Feelgood Fiction, 87
Feenish Productions, 127
Ferndale Films, 127
Festival Film + TV Ltd, 87
Fiery Film Fest, 219
Filbert Steps Productions, 50
Janet Fillingham Associates, 161
Film Agency for Wales, 201
Film and General Productions
    Ltd, 88
Film and Television Institute, 192
Film Australia, 210
Film Base, 209
Film Columbia, 219
Film Columbia's First
    Screenwriters' Contest, 256
Film Critics Circle of Australia
    Awards, 270
Film Education, 202
Film Finance Corporation, 210
Film Garden Entertainment, 50
Film Genesis, 256
Film Independent, 198
Irish Film Institute, 209
Film London, 202
Film New Zealand, 210
Film Police!, 50
Film Rights Ltd/Laurence Fitch
    Ltd, 161
Film4 Productions, 88
Filmaka, LLC, 256
FilmBase/RTE Short Film
    Awards, 268
FilmBase/TG4 Lasair Awards, 268
Filmmaking Summer School at
    Melbourne University, 192
Filmstock International Film
    Festival, 234
Find the Funny, 256
Firebrand Productions, 50
Firecracker Showcase: London's
    East Asian Film Festival, 234
Firefly Films Limited, 88
Firelight Media, 50
The First Film Company Ltd, 88
First Light Movies, 202
First Look Studios, Inc., 50
First Street Films, 50
First Take, 181
Five Sisters Productions, 51
Flamingo Neck Pictures Inc., 115

Flannel, 88
Flashback Television Ltd, 88
Flexitoon, 51
FlickerFest, 247
A Flickering Image Festival, 219
FLIP International Animation
    Festival, 234
Florentine Films, 51
Florida Film Festival, 219
Florida State University Film
    School, 174
Flower Films, 51
Flux Animation Studio Ltd, 144
Fobia Films, LLC, 51
Focus Features, 51
Focus Films Ltd, 88
Force Four Productions, 115
Mark Forstater Productions
    Ltd, 88
Fort Lauderdale International
    Film Festival, 219
Fortis Films, 51
Fortissimo Films (UK), 89
Fortress Entertainment, 51
Forty Acres & A Mule
    Filmworks, 52
Forward Films, 89
Jill Foster Ltd, 161
FourBoys Films, 52
Fox Atomic, 52
Fox Searchlight Pictures, 52
FRA, 162
Fragile Films, 89
Robert A. Freedman Dramatic
    Agency, Inc., 153
Freeze Frame International
    Festival of Films for Kids of All
    Ages, 242
Fremantle Corporation, 116
FremantleMedia Ltd, 89
French's, 162
Fresh Film Festival, 246
Blake Friedmann Literary
    Agency, 162
Fries Film Group, 52
Frightfest, 235
Front Page Limited, 144
Front Street Pictures, 116
Frontier Films, 127
Frontier Films Pty Ltd, 138
Fulcrum TV, 89
Full Frame Documentary
    Festival, 219
Furst Films, 52
Furthur Films, 52
Future Films Limited, 89
Future Films USA, LLC, 52
Future Shorts, 235
FXF Productions Inc., 52

G

Galafilm Inc., 116
Gallagher Literary, 154
Gallowglass Pictures, 127
Galway Film Centre, 209
Galway Film Centre/RTE Short
    Script Awards, 269
Galway Film Fleadh, 246
Rachel Gardner, 144
Noel Gay, 162
Noel Gay Television, 89
Gecko Films Pty Ltd, 138
The Geddes Agency, 154
Geminis, 267
Generation Films, 138
Genies, 267
The Gersh Agency (TGA), 154
Ghost House Pictures, 52
The Gibson Group Limited, 144
Gigantic Pictures, 53
Giggleshorts, 243
Gimme Credit Screenplay
    Competition, 256
Ginty Films, 116
Glasgow Film Festival, 235
Eric Glass Ltd, 162
Glasshouse Entertainment Ltd, 89
Glowworm Media, 127
GMTV, 90
Go Girl Media, 53
Godzone Pictures Ltd, 144
Goff-Kellam Productions, 53
Gold Circle Films, 53, 90
Goldcrest Films, 53
Goldcrest Films International
    Ltd, 90
Golden Globes, 256
Goldsmiths, University of
    London, 181
Samuel Goldwyn Films, 53
The Good Film Company, 90
Gracie Films, 53
Grammnet Productions, 53
Granada Media, 90
Granada Productions, 138
Grand Pictures Ltd, 127
Granite Film & Television
    Productions, 90
Grants to Film and Video Artists
    Program, 267
Graphite Film and Television, 90
GRB Entertainment, 53
Great Lakes Film
    Association Scriptwriting
    Competition, 257
Great Lakes Independent Film
    Festival, 219
Great North Artists Management
    Inc., 167

Great Western Entertainment Pty
    Ltd, 138
Great Western Films, 127
Green Dog Films, 53
Green Light Artist Management
    Inc., 167
Green Moon Productions, 54
Green Stone Pictures, 144
Green Umbrella Ltd, 90
The Harold Greenberg Fund, 267
Greenestreet Films Inc, 54
Greenpoint Films, 90
Robert Greenwald
    Productions, 54
Greenwich Village
    Productions, 91
The Greif Company, 54
Grey Line Entertainment,
    Inc., 54
The Grierson Awards, 264
Mark Grindle Associates Ltd, 181
Gruber Films Ltd, 91
Grundy, 138
Gryphon Films, 54
The Charlotte Gusay Literary
    Agency, 154
Guy Walks into a Bar, 54
Guy Walks into a Bar
    Management, 154

H

Haddock Entertainment Inc., 116
The Rod Hall Agency
    Limited, 162
Hallmark Entertainment, 54
Hallmark Hall of Fame, 55
Hamilton-Mehta
    Productions, 116
Hammer Productions, 91
Hammerwood Film
    Productions, 91
The Hamptons International Film
    Festival, 219
Hamzeh Mystique Films, 55
Roger Hancock Ltd, 163
Handel Productions Inc., 116
Hannibal Pictures, 55
Happy Madison Productions, 55
Harbor Lights Entertainment, 55
Harbour Pictures, 91
Harcourt Films, 91
Hartswood Films Ltd, 91
Hat Trick Productions Ltd, 92
Hawaii International Film
    Festival, 220
Hawke Films, 144
Hawkeye Films, 128
John Hawkins &
    Associates, Inc., 154

Hazel Wolf Environmental Film
    Festival, 220
HBO, 55
HDNet, 55
HDNet Films, 55
Headgear Films Limited, 92
Healthcare Productions
    Limited, 92
Heartland Film Festival, 220
Hell's Kitchen International
    Ltd, 128
Jim Henson Co., 55
Heritage Theatre Ltd, 92
Heyday, 92
David Higham Associates Ltd, 163
Hit Entertainment, 55
HLA Management Pty Ltd, 169
Hollins University, 174
Hollywood Black Film
    Festival, 220
Hollywood International Film
    Festival, 220
Hollywood Nexus Screenwriting
    Contest, 257
Hollywood Pictures, 56
Hollywood Studios
    International, 56
Holmes Associates, 92
Holy Cow Films, 92
Hoodlum Active Pty Ltd, 138
Hope and Dreams Film
    Festival, 220
Johns Hopkins University, 174
Horizon Motion Pictures Inc., 116
Valerie Hoskins Associates
    Limited, 163
Hot Docs Canadian International
    Documentary Festival, 243
Hot Springs Documentary Film
    Institute, 220
Hourglass Productions, 92
Hull International Short Film
    Festival, 235
Human Rights Film Festival, 249
Humboldt International Film
    Festival, 220
Huntaway Films, 144
Hurricane Films, 93
Huston School of Film & Digital
    Media, 190

I

Iambic Productions, 93
i-blink Film Festival, 235
Icebox Films, 128
Iceman Productions, 56
ICM (International Creative
    Management), 154
ICM London, 163

Icon Films, 93
Icon Productions, 56
IF Award for Best Script, 270
The Illuminated Film
 Company, 93
Illuminati Entertainment, 56
Illusion Animated
 Productions, 128
Image and Nation: Montréal
 International LGBT Film
 Festival, 243
Images of Black Women Film
 Festival, 235
ImagiNation Film & Television
 Productions Inc., 116
Imagine Entertainment, 56
Imari Entertainment Ltd, 93
Impact Pictures, 93
Impact Pictures LLC, 56
imX Communications Inc., 117
Independent Talent Group, 163
Indigo Television Ltd, 93
Infinitum Nihil, 56
Infinity Filmed Entertainment
 Group Ltd, 117
Initialize Films, 182
Inner City Films Inc., 117
InsideOut – The Toronto Lesbian
 and Gay Film and Video
 Festival, 243
Insight Production Company
 Ltd., 117
Instinct Entertainment, 138
Institute of
 Screenwriting, 174, 182
Intermedia Films, 56
International Documentary
 Television Corporation, 117
International Festival of Fantastic
 Films, 235
The International Film School
 Sydney, 192
International Film School
 Wales, 182
International Game Developers
 Association (IGDA), 198
International Horror & Sci-Fi
 Film Festival Screenplay
 Competition, 257
International Keystone
 Entertainment, 117
International Manga and Anime
 Festival, 235
International Media, 93
International Screenwriter's
 Festival, 235
International Scripts, 163
Intrepido, 93
Ipso Facto Films, 93

Irish Film and Television
 Awards, 269
Irish Film and Television
 Academy, 209
Irish Playwrights and
 Screenwriters Guild, 209
Irish Writers Centre, 209
Isis Productions, 94
Isle of Man Film, 202
Isola Productions Ltd, 144
Isolde Films, 94
Italian Film Festival, 236
ITV Productions, 94
Ivy Film Festival, 221
Ivy Film Festival Screenplay
 Competition, 257

J
JAM Media, 128
JAM Pictures and Jane Walmsley
 Productions, 94
Carolyn Jenks Agency, 154
Jersey Films, 57
Jigsaw Entertainment, 139
Johnson & Laird and Rachel
 Gardner Managements, 169
Joyride Films Ltd, 145
Juniper Communications Ltd, 94
Justice Entertainment, 94
Kansas City Filmmakers
 Jubilee, 221
Kaos Films, 94
Kaplan Stahler Gumer Braun, 154
Kaplan/Perrone
 Entertainment, 57
Kapow Pictures, 139
Michelle Kass Associates, 163
Kavaleer Productions Ltd, 128
Bill Keating Centre, 190
Keatley Entertainment Ltd, 117
David E. Kelley Productions, 57
Kelpie Films, 95
Kendal Mountain Festival, 236
Keo Films.com Ltd, 95
Kerry Film Festival, 246
Kettledrum Films, 57
Killer Films Inc., 57
Kingfisher Television
 Productions, 95
Kinofilm - Manchester
 International Short Film
 Festival, 236
Kismet Entertainment Group, 57
Kismet Film Company, 95
Kite Entertainment Ltd, 128
Kiwa Media, 145
The Jon Klane Agency, 155
Kneerim & Williams, 155
Knight Enterprises Inc., 117

Kojo Pictures, 139
Kopelson Entertainment, 57
Kudos Film & Television, 95
Kultur International
 Films, Inc., 57

L
La Luna Studios Ltd, 145
La Trobe University, 192
Lagan Pictures Ltd, 95
LAIKA, Inc., 57
Lakeshore Entertainment
 Corp, 58
Lancaster University, 182
Landcrab Film Festival, 236
Landseer Productions Ltd, 95
Latin American Film
 Festival, 236 , 249
LAW, 163
Leaudouce Films, 58
Leeds International Film
 Festival, 236
Leeds Metropolitan
 University, 182
Legge Work, 145
Lenhoff & Lenhoff, 155
Leo Awards, 268
The Leo Media & Entertainment
 Group, 164
Levinson/Fontana Company, 58
Lifesize Film and Television Pty
 Ltd, 139
Lifetime Entertainment, 58
Light Renegade
 Entertainment, Inc., 58
Lighthouse Arts and Training, 182
Lighthouse Screenplay
 Competition, 257
Lightstorm Entertainment, 58
Like it Love It Productions, 128
Lilyville Screen Entertainment
 Ltd, 95
Lime Pictures Limited, 96
Lion Rock Productions, 58
Lionsgate Films, 58, 96
The Christopher Little Literary
 Agency, 164
Little Bird, 128
Little Ed Films Ltd, 145
LivePlanet, 58
Liverpool Film Festival, 236
Livingstone Productions Ltd, 145
LKF Productions, 129
Llanberis Mountain Film
 Festival, 236
LMNO, 58
The Logie Awards, 270
London Australian Film
 Festival, 237

London Children's Film
Festival, 237
London College of
Communication, 182
London Film Academy, 183
The London Film School, 183
London International Animation
Festival, 237
London Lesbian and Gay Film
Festival, 237
London Scientific Films Ltd, 96
London Short Film Festival, 237
Lonely Planet Television, 139
Long Island International Film
Expo, 221
Loophead Studio, 129
Loopline Film, 129
Los Angeles Film Critics
Association Awards, 257
Los Angeles Film Festival, 221
Los Angeles Film School, 174
Los Angeles International Short
Film Festival, 221
Louisiana State University, 174
Lovebytes Digital Arts
Festival, 237
Lowenbe Holdings, 168
Loyola Marymount
University, 174
Lucas Talent, 168
LucasArts, 59
Lucasfilm Ltd, 59
Lucida Productions, 96
Lumanity Productions Inc., 118
Lumina, 96
Lunah Productions, 129

M

M F Films Ltd, 145
Mace Neufeld Productions, 59
MacGillivray Freeman
Films, Inc., 59
Mad Chance, 59
Magma Films, 129
Magma Short Film Festival, 249
Make Believe Media Inc., 118
Make Your Mark in Film, 265
MakeMagic Productions, 59
Malibu Film Festival, 221
Malone Gill Productions Ltd, 96
Malpaso Productions, 59
Management 819, 155
Mandalay Pictures, 59
Mandate Pictures, 59
The Manhatten Short Film
Festival's Screenplay
Competition, 257
Andrew Mann Ltd, 164
Marchmont Films, 96

Marjacq Scripts Ltd, 164
Massey University, 193
Matador Pictures LLC, 60
Material Entertainment, 96
Matthau Company, 60
Maverick Television, 97
Maya Vision International Ltd, 97
MBA Literary Agents Ltd, 164
MBP TV, 97
McCamley Entertainment, 129
J McElroy Holdings Pty Ltd, 139
McKernan Agency, 164
McKnight Screenwriting
Fellowship, 257
Bill McLean Personal
Management, 164
Margaret Mead Film and Video
Festival, 221
M8 Entertainment, 60
Media Headquarters Film &
Television Inc., 118
Media Platform, 129
Media That Matters Film
Festival, 221
Media World Pictures, 139
MediaGuardian Edinburgh
International Television
Festival (MGEITF), 237
Meem Productions, 129
Melbourne International
Animation Festival, 247
Melbourne International Film
Festival, 247
Melbourne Queer Film
Festival, 247
Melendez Films, 97
Melodrama Pictures Pty Ltd, 139
Mendoza Film Productions, 97
Meniscus Film Festival, 237
Mensour Agency, 168
Mentorn, 97
Merchant Ivory, 97
Merchant Ivory Productions, 60
Meridian Artists Inc., 168
Met Film Productions, 98
Metro Screen, 192
Metropolitan Film School, 183
Metropolitan Talent
Agency, 155
Miami International Film
Festival, 222
Microsoft Game Studios
(MGS), 60
Mid Ulster Film Festival, 237
Midas Films (David Douglas
Productions Pty Ltd), 140
Midas Productions, 130
Middlebury College, 174
Middlesex University, 183

Midlands BFM International Film
Festival, 238
Midsummer Films
Productions, 98
Mikemon International, LLC, 60
Milagro Films Inc., 118
Monte Miller Awards, 270
Mind The Gap Films, 130
Mindfire Entertainment, 60
Minds Eye Entertainment
Ltd., 118
Minds Eye Films, 98
Minneapolis-St Paul International
Film Festival, 222
Minnesota State University, 175
Mint, 130
Mirage, 98
Mirage Films, 98
Mirage Productions Inc, 60
Miramax Films, 60
Missing In Action Films Ltd, 98
Mission Pictures, 98
Mob Film Company, 98
Monteiro Rose Agency, 155
Monterey Screenplay
Competition, 258
Montivagus Productions, 61
Montreal International Festival
of New Cinema & Media
(Festival du nouveau cinéma
de Montréal), 243
Montreal World Film Festival, 243
Moondance International Film
Festival, 222
Moondance Productions, 130
Moonstone Entertainment, 61
Moonstone Films, 98
Moonstone International Screen
Labs, 183
Morag Productions/Passage
Films, 118
Morgan Creek Productions, 61
Morley College, 183
William Morris Agency (UK)
Ltd, 165
William Morris Agency
(WMA), 155
Henry Morrison Inc., 155
Mosaic Films, 98
Mosaic Media Group, 61
Motion Picture Corporation of
America, 61
Mount Holyoke College, 175
Movies on the Square, 246
Moving Targets, 140
MPH Entertainment, Inc., 61
Mr Mudd, 61
MTV Films, 61
Mulholland Pictures, 99

Muse Entertainment Enterprises, 118
Mushroom Pictures, 140
Music Arts Dance Films Pty Ltd, 140
MVP Entertainment Inc., 119
MWG Productions, 61
Myriad Pictures, 62

N

Namco Bandai Games America Inc., 62
Namesake Entertainment, 62
Napa Sonoma Wine Country Film Festival, 222
The Narrow Road Company, 165
National Association of Writers' Groups (NAWG) Open Festival of Writing, 238
National Film and Television School, 184
National Film Board of Canada, 208
National Film School, 190
National Geographic Feature Films, 62
National Society of Film Critics, 258
National Youth Film School, 191
Natural History New Zealand, 145
NBC Entertainment, 62
Neal Street Productions, 99
B.K. Nelson Literary Agency, 155
Nelvana Limited, 119
Nemeton Television Productions, 130
Neo Art and Logic, 62
Neon, 99
Neophyte Productions Inc., 119
Network Entertainment Inc., 119
Neverland Films, Inc., 62
New Amsterdam Entertainment, Inc., 62
New Crime Productions, 63
New Decade Film & Television, 130
The New Fest, 222
New Hampshire Film Festival, 222
New Jersey Short Screenplay Festival, 258
New Line Productions Inc., 63
New Music Theatre, 140
New Orleans Film Festival, 222
New Producers Alliance, 202
New Regency Productions, 63
New Root Films, 63
New Wave Entertainment, 63
New Wave Talent Management, 156

New Writing North, 202
New Writing South, 202
The New York City Horror Film Festival, 222
New York Film Academy, 175, 184
New York Film Critics Circle Awards, 258
New York Film Festival, 223
The New York International Documentary Festival, 223
New York International Independent Film and Video Festival, 223
New York International Latino Film Festival, 223
New York University, 175
New Zealand Film Academy, 193
The New Zealand Film Archive, 211
New Zealand Film Commission, 211
New Zealand Greenroom Productions Limited, 146
New Zealand Writers Guild, 211
Newgrange Pictures, 130
Newmarket Films, 63
Newport International Film Festival, 223
Nga Aho Whakaari, Māori in Film, Video and Television Inc., 211
Nicholl Fellowships in Screenwriting, 258
Nickel Independent Film Festival, 243
Nickelodeon Movies, 63
Nickelodeon Productions Fellowship Program, 258
Nightingale Company, 119
Nihilist Film Festival, 223
Ninox Television Ltd, 146
Nintendo of America, 63
Nite Owl Productions, Inc., 63
Nomad Films International Pty Ltd, 140
Norstar Filmed Entertainment Inc., 119
North Carolina School of the Arts, 175
Northern Ireland Screen, 203
Northern Lights Film Festival, 238
Northern Screenwriters, 203
Northern Visions Media Centre, 184
Northwest Screenwriters' Guild, 199
Nova Pictures, 64
NSI FilmExchange Canadian Film Festival, 243

Nu Image, 64
Number 9 Films, 99
Lynda Obst Productions, 64

O

October Films Ltd, 99
Odd Lot Entertainment, 64
Oddworld Inhabitants, 64
Ohio University School of Film, 175
Omaha Film Festival, 223
Omni Film Productions Ltd, 119
Omnivision Ltd, 99
One in Ten Screenplay Competition, 258
One Productions, 130
Onedotzero, 238
OPEN Channel, 192
Optronica, 238
The Orange Grove Group, Inc., 156
ORCA Productions, 119
Orchard Films, 64
Organization of Black Screenwriters, 199
Origin One Films, 146
Original Pictures Inc., 119
Orlando TV Productions, 99
Orpheus Productions, 99
Oscailt (TG4 and Bord Scannán na héireann), 269
Oska Bright Film Festival, 238
Ottowa International Animation Festival, 244
Outcast Production, 100
Outlaw Productions, 64
Outsiders: Liverpool Lesbian and Gay Film Festival, 238
Overbrook Entertainment, 64
Oxdox International Documentary Film Festival, 238
Oxford Film and Television, 100
Oxford Film and Video Makers, 184

P

PACT (Producers Alliance for Cinema and Television), 203
The PAGE International Screenwriting Awards, 259
Paladin Invision, 100
Palm Beach Jewish Film Festival, 224
Palm Pictures, 64
Palm Springs International Film Festival, 224
Pan African Film Festival, 224
Panacea Entertainment, 120

Pangaryk Productions, Inc., 168
Pangea Management, 156
Paper Moon Productions, 100
Paradigm, 156
Paradox Pictures, 131
Parallax East Ltd, 100
Paramount Motion Pictures
Group, 64
Participant Productions, 65
Passion Pictures, 100
Pathé Pictures, 100
Pathfinder Pictures LLC, 65
PAWS Drama Fund, 265
Peace Arch Motion Pictures
Inc., 120
Peace Point Entertainment
Group, 120
Pelicula Films, 101
Pennsylvania State University, 175
Perdido Productions, 65
Performing Arts Labs, 184
Perpetual Motion Pictures, 140
Persistent Entertainment, 65
Stephen Pevner Inc, 156
PFD, 165
Phoenix Pictures, 65
Photoplay Productions Ltd, 101
Picture Palace Films, 101
A Pitch in Time, 265
Pixar Animation Studios, 65
Plan B, 65
Planet in Focus: International
Environmental Film and Video
Festival, 244
Planet24 Pictures Ltd, 101
Plantagenet Films Limited, 101
Player–Playwrights, 203
Plum Pictures, 65
PMA Literary & Film
Management, Inc., 156
Point Grey Pictures Inc., 120
Point Made, 65
Point of View Productions, 146
Pollinger Limited, 165
PorchLight Entertainment, 65
Porchlight Films, 140
Portland International Film
Festival, 224
Portobello Pictures Ltd, 101
Powerstone Entertainment
Ltd, 101
Pozzitive Television, 101
Praxis Centre for
Screenwriters, 189
Praxis Film Script
Competitions, 268
Prescience Film Finance Ltd, 102
Edward R. Pressman Film
Corporation, 66

Gaylene Preston Productions, 146
Princess Productions, 102
Principia Productions Ltd, 120
Priority Pictures, 102
Prix Gemeaux, 268
Production Partners, Inc., 66
Promedia TV, 131
Prospero Pictures, 120
ProtectRite, 199
Protocol Entertainment Inc., 120
Protozoa Pictures, 66
Punch Productions, 66
Purbeck Film Festival, 239
Purchase College, 175

Q

Queen Street Entertainment, 120
Queensland School of Film &
Television, 192
QuickFlick World London, 239
Qwerty Films, 102

R

Raconteur, 146
Radar Pictures, 66
Sarah Radclyffe Productions, 102
Radhart Pictures Pty Ltd, 141
Raecom Productions, 66
Ragdoll Ltd, 102
RAI International Festival of
Ethnographic Film, 239
Rainbow Media Holdings
LLC, 66
Raindance Film Festival, 239
Raindance Ltd, 184
Rainstorm Entertainment Inc., 66
Raven West Films Ltd, 120
Raynbird Productions, 146
RDF Management, 165
RDF Media Group, 102
Real Creatives Worldwide, 165
Real to Reel Film and Video
Festival, 224
The Realm of the Senses, 247
Recorded Picture Company, 102
Red Bank International Film
Festival, 224
Red Carpet Productions Pty
Ltd, 141
Red Hour Films, 66
Red Om Films, 66
Red Pepper Productions, 131
Red Planet, 103
Red Planet Prize, 265
Red Production Company, 103
Red Storm, 67
Redhammer Management
Ltd, 165
Redweather, 103

Reel Affirmations Film
Festival, 224
Reel Girls Media Inc., 121
ReelWorld Film Festival, 244
Regent Entertainment, 67
Regent University, 176
Regional Screen Agencies, 203
Remstar Productions, 121
Renaissance Films, 103
Renegade Animation, 67
Result Talent Group, 156
Revelation: Perth International
Film Festival, 247
Revolution Films, 103
Revolution Studios, 67
Rhode Island Film Festival
Screenplay Competition, 259
Rhode Island International Film
Festival, 224
Rhombus Media Inc., 121
The Lisa Richards Agency, 168
Richards Literary Agency, 169
The Richies, 270
Richmond Films &
Television, 103
Rink Rat Productions, Inc., 121
River Road Entertainment, 67
RMIT University, 192
Roberts/David Films, Inc., 67
Rochester International Film
Festival, 225
Rocket Pictures, 103
Rockstar Games, 67
Rocliffe, 103
Rocliffe Forum, 203
Rogue Pictures, 67
Alex Rose Productions, 67
Rosenstone/Wender, 156
ROSG, 131
Royal Holloway University of
London, 184
Royal Television Society, 204
RS Productions, 103
RTÉ Independent
Productions, 131
Ruby Films, 104
Scott Rudin Productions, 68
Rushes Soho Shorts Festival, 239

S

S & S Productions Inc., 121
Sacramento Film and Music
Festival, 225
Salty Features, 68
Samson Films, 131
San Diego Film Festival, 225
San Diego State University, 176
San Francisco International Film
Festival, 225

San Francisco International Gay and Lesbian Film Festival, 225
San Francisco State University, 176
Sands Films, 104
Sand-Swansea Animation Day, 239
Santa Barbara International Film Festival, 225
The Santa Barbara Script Competition, 259
Santa Fe Film Festival, 225
Satellite Awards, 259
Saturn Films, 68
Satyajit Ray Foundation Short Film Competition, 265
Savannah Film Festival, 225
Savi Media Inc., 121
Sayle Screen Ltd, 166
Jack Scagnetti Talent & Literary Agency, 156
Scala Productions, 104
Scannain Lugh, 131
Columbia University School of the Arts, 176
Joel Schumacher Productions, 68
Scope Productions Ltd, 104
Scott Free Productions, 68
Scottish Screen, 204
Scottsdale Community College, 176
Screamfest Horror Film Festival, 226
Screen Academy Scotland, 184
Screen Directors Guild of New Zealand, 211
Screen Door, 121
Screen Door Entertainment, 68
Screen First Ltd, 104
Screen Producers Ireland, 210
Screen Production and Development Association of New Zealand (SPADA), 211
Screen Ventures Ltd, 104
Screenhouse Productions Ltd, 104
Screenlab, 185
Screenplay Festival, 259
Screenplay Shootout, 259
Screenplay.com.au, 193
Screenprojex.com – a division of Screen Production Associates Ltd, 104
Screentime ShinAwiL, 131
Screenwriter Showcase Screenwriting Contest, 259
Screenwriter's Challenge, 259
Screenwriters Federation of America, 199
The Script Factory, 185, 204

Script Nurse Mini-Movie Screenwriting Contest, 260
Script Pimp Screenwriting Competition, 260
Script to Screen, 211
The Script Vault, 204
Script_1 Workshop and Competition, 265
Scriptapalooza Screenwriting Competition, 260
Scriptease, 169
Seattle International Film Festival, 226
Section Eight, 68
Sedona International Film Festival & Workshop, 226
SEGA of America, 68
September Films Ltd, 105
Serendipity Picture Company, 105
Serendipity Point Films, 121
The Rod Serling Conference Short Feature Scriptwriting Competition, 260
Set in Philadelphia Screenwriting Competition, 260
Seven Arts Pictures, 68
Seven Dimensions, 141
Severe Features, 146
John Sexton Productions, 141
Shadowhawk Films (Eire) International, 131
Shaftesbury Films, 122
The Sharland Organisation Ltd, 166
Shavick Entertainment, 122
Sheffield DocFest, 239
Sheil Land Associates, 166
Shine Entertainment, 105
ShineBox Motion Pictures, 68
ShockerFest International Film Festival, 226
Shoes Full of Feet, 122
Shooting Star Picture Company Pty Ltd, 141
Shooting Stars Productions, 147
Shoreline Entertainment, Inc., 69
Shortt Comedy Theatre/ Warehouse TV, 132
SHOTZ Film and Video Production Ltd, 147
Show Me Shorts Film Festival Trust, 249
Showcomotion Young People's Film Festival, 239
Showdog Productions, 122
Shriekfest Film Festival & Screenplay Competition, 226
Sianco Cyf, 105
Sienna Films, 122

Sierra Entertainment, 69
Signals International Short Film Festival, 240
Signals Media Arts, 185
Signature Film, 270
Signature Pictures, 69
Silent Sound Films, 105
Silver Lake Film Festival, 226
Silver Lion Films, 69
Silverdocs, 226
Silvers/Koster Productions, 69
Sirk Productions, LLC, 69
Sixteen Films, 105
SKILLSET, 204
Skyline Productions, 105
Slamdance Film Festival, 226
Slamdance Screenwriting Competition, 260
Slate Films, 106
Sly Fox Films, 106
SMA, LLC, 156
SMG Productions & Ginger Productions, 106
Sneak Preview Entertainment, 69
Snowfall Films, 69
So Television Ltd, 106
Soapbox Productions Inc., 122
Sobini Films, 69
Société des Auteurs de Radio, Télévision et Cinéma, 208
The Society of Authors, 204
Soilsiú Films, 132
SOL Productions Ltd, 132
Somethin' Else, 106
Sommers Company, 70
The Sonoma Valley Film Society Festival, 227
Sony Pictures Entertainment Inc., 70
Sony Pictures Imageworks, 70
South Pacific Pictures, 147
South Seas Film and Television School, 193
South Thames College, 185
Southampton Solent University, 185
Southern Skies/Great Norther Film and Music, 70
Southern Star Entertainment, 141
Southwest Scriptwriters, 204
SP Films, 132
Specific Films, 106
Speers Film, 132
Spellbound Productions Ltd, 106
Spice Factory, 106
Spire Entertainment Inc., 122
Spirit Dance Entertainment, 70
Spitfire Pictures, 70

Philip G. Spitzer Literary
  Agency, Inc., 157
Spoken' Image Ltd, 106
Sprout Productions, 147
Spyglass Entertainment, 70
Square Enix Co., Ltd., 70
St Kilda Film Festival, 247
Stagescreen Productions, 107
Star Entertainment
  Group, Inc., 71
Starz Denver Film Festival, 227
Tony Staveacre Productions, 107
Elaine Steel, 166
The Stein Agency, 157
Micheline Steinberg
  Associates, 166
Stella Artois Pitching
  Awards, 269
Sterling Lord Literistic,
  Inc., 157
Gloria Stern Agency, 157
Rochelle Stevens & Co., 166
Andrew Stevens
  Entertainment, 71
Sticky Pictures, 147
Stirling Film & TV Productions
  Limited, 107
Stone Manners Agency, 157
Stonebridge Communications, 71
Stoney Road Films, 132
Storm Entertainment, 71
Storyline Entertainment, 71
Storyline Entertainment
  Inc., 122
Strada Films Inc., 122
Straight Forward Film &
  Television, 107
Strand Releasing, 71
Streiffschuss Films, 71
Sudden Storm Productions
  Inc., 123
Suite A Management Talent &
  Literary Agency, 157
Sullivan Entertainment Inc., 123
Summer Pictures Inc., 123
Summit Entertainment, 71
Summit Talent & Literary
  Agency, 157
Sundance Film Festival, 227
Sundance Institute, 199
Sunset + Vine Productions
  Ltd, 107
Super Shorts, 240
Sutton Film Festival, 240
Sweet 180, 157
Sydney Film School, 193
Sydney International Film
  Festival, 248
Syracuse University, 176

T

Table Top Productions, 107
Talent Circle, 205
TalentScout Management, 157
TalentScout TV Writing
  Contest, 260
talkbackTHAMES, 107
Talking Heads Productions, 108
Tampa International Gay and
  Lesbian Film Festival, 227
Tandem TV & Film Ltd, 108
TAPS (Training and Performance
  Showcase), 185
Taxi Film Production, 141
Taylor Made Broadcasts
  Ltd, 108
Te Aratai Film & TV
  Production, 147
Telegael Media Group, 132
Telemagination, 108
Telling Pictures, 72
Telluride Film Festival, 227
Temecula Valley International
  Film & Music Festival, 227
Temple Street Productions, 123
Temple University, 176
The John Templeton Foundation
  Kairos Prize, 261
The Tennyson Agency, 166
Tern Television Productions
  Ltd, 108
Terror Film Festival, 227
Testimony Films, 108
Stacey Testro International
  Management, 169
Stacey Testro International
  Production, 141
TG4, 132
Thames Valley University, 185
Thin Man Films, 108
Thinkfilm, 72
The Dylan Thomas Prize, 266
Threshold Entertainment, 72
Tiburon International Film
  Festival, 227
TIG Productions, Inc., 72
Tiger Aspect Productions, 108
Tigerlily Films Limited, 108
Tile Films, 132
Time River Productions, 72
Timeline Films, 72
Timeline Productions, 147
The Times BFI London Film
  Festival, 240
Timesnap Limited, 133
Tollin/Robbins Productions, 72
Top Shelf Productions Ltd, 147
Toronto After Dark Film
  Festival, 244

Toronto Film Critics Association
  Awards, 268
Toronto International Film
  Festival, 244
Toronto International Latin Film
  Festival, 244
Toronto Jewish Film Festival, 244
Toronto Reel Asian International
  Film Festival, 244
Totality Pictures Ltd, 133
Touch Productions Ltd, 109
Touchstone Pictures, 74
Trancas International, 72
Trancas International
  Management, 157
Transatlantic Films Production
  and Distribution
  Company, 109
Trasharama A-go-go, 248
Travel Channel, 109
Treasure Entertainment,
  Inc., 72, 158
Tribeca Film Festival, 228
Tribeca Productions, 72
Tricon Films & Television, 123
Trigger Street Productions, 73
Trilogy Entertainment
  Group, 73
TriMax Films, 141
Triptych Media Inc., 123
Tropfest, 248
True West Films, 123
TSG Productions, 73
The Tudor Group, 158
TV Choice Ltd, 109
The TV Set Ltd, 148
TVWriter.com People's Pilot, 261
TVWriter.com Spec
  Scriptacular, 261
Twenty Twenty Television, 109
Twilight Pictures & Production
  Services Inc., 73
Twofour Broadcast, 109
Ty Newydd Writers' Centre, 186
Tyburn Film Productions
  Limited, 110
Type A Films, 73
Tyrone Productions, 133

U

Ubisoft Entertainment, 73
UCD School of Film, 191
UCLA Extension Writers'
  Program, 176
Ugly Betty Productions, Inc., 73
UK Brasilian Film Festival, 240
The UK Film and TV Production
  Company PLC, 110
UK Film Council, 205

UK Film Council's Development Fund, 266
UK Jewish Film Festival, 240
UK MEDIA Desk, 205
Underground Films, 73
Underground Management, 158
Union Square Entertainment, 73
Unitec New Zealand, 193
United Agents Ltd, 166
United Artists Entertainment LLC, 74
United Nations Association Film Festival, 228
United Talent Agency, 158
Universal Pictures, 74
Universal Studios Canada Inc., 123
University College Falmouth, 186
University of Birmingham, 186
University of Bolton, 186
University of California (Berkeley), 177
University of California (Los Angeles)–UCLA, 177
University of California (Santa Barbara), 177
University of Central Florida, 177
University of Central Lancashire, 186
University of Derby, 186
University of East Anglia, 186
University of Edinburgh, 186
University of Exeter, 187
University of Glamorgan, 187
University of Hull, 187
University of Iowa, 177
University of Kansas, 177
University of Leeds, Bretton Hall Campus, 187
University of Liverpool, 187
University of Melbourne, 193
University of Miami, 177
University of Michigan, 177
University of Nevada (Las Vegas)–UNLV, 178
University of New Orleans, 178
University of North Carolina at Greensboro, 178
University of Oklahoma, 178
University of Portsmouth, 187
University of Regina, 190
University of Salford, 188
University of Sheffield, 188
University of Southern California (USC), 178
University of Surrey, Roehampton, 188
University of Technology, Sydney, 193
University of Texas, 178

University of Toronto, 190
University of Utah, 178
University of Waikato, 193
University of Wales, Bangor, 188
University of West of England, 188
University of Westminster, 188
University of Winchester, 188
US Copyright Office, 199
US International Film and Video Festival, 228
Sir Peter Ustinov Television Scriptwriting Award, 261

V
Vancouver Asian Film Festival, 244
Vancouver Film School, 190
Vancouver International Film Festival, 245
Vancouver International Jewish Film Festival, 245
Vancouver Queer Film Festival, 245
Vancouver Student Film Festival, 245
Vanguard Animation, 74
Vanguard Documentaries, 74
Vanguarde Artists Management, 168
Venom, 133
Vera Productions, 110
Vermont International Film Festival, 228
Vertigo Films, 110
Vico Films, 133
The Victoria Film Festival, 245
Victoria University, 194
View Askew Productions, Inc., 74
Village Roadshow Limited, 142
Village Roadshow Pictures, 74
Visionfest (Domani Vision Film Society), 261
Vistas Film Festival, 228
Vitascope Filmed Entertainment, 142
Viva! Spanish Film Fest, 240
Vulcan Productions, 74

W
Waddell Media, 110
The Wairoa Māori Film Festival, 249
Walden Media, 74
Wall to Wall, 110
Walsh Bros Ltd, 110
Walt Disney Animation Studios, 75
Walt Disney Pictures, 75

Walt Disney Studios/ABC Writers Fellowship, 261
Warden, White & Associates, 158
Cecily Ware Literary Agents, 167
Warner Bros Entertainment, Inc., 75
Warner Bros TV Writers Workshop, 262
WarpFilms, 110
Warp X, 110
Washington Square Films, 75
Washington, D.C. Independent Film Festival, 228
Waterfront Film Festival, 228
Waterfront Film Festival Screenwriting Competition, 262
Watershed, 240
Watershed Films, 75
AP Watt Ltd, 167
Irene Webb Literary Film Representation & Publishing, 158
Jim Wedaa Productions, 75
The Weinstein Company, 75
Jerry Weintraub Productions, 75
Weintraub/Kuhn Productions, 75
Weller Grossman, 75
Wellington Film Festival, 249
Wellington Fringe Film Festival, 249
Westchester Film Festival, 228
Western Michigan University, 178
Westminster Adult Education Service, 189
Westwood Creative Artists, 168
White Iron Pictures Inc., 123
WhiteLight Entertainment, 76
Whitireia Community Polytechnic, 194
Wide Eye Films, 133
Wider Vision, 133
Wildfire Films & Television Productions Limited, 133
Wildscreen Festival, 240
Williamstown Film Festival, 229
Winchester Writers' Conference, 241
Winchester Writers' Conference Prizes, 266
WinFemme Monthlies, 262
Winkler Films, 76
Withoutabox, 229
Wizz Films Inc., 124
Women in Film & Television, 212
Women in Film and Television Vancouver, 208

Women in Film and Television Toronto, 208
Women in Film and Television UK (WFTV), 206
Wood Green International Short Film Festival, 241
The Woods Hole Film Festival, 229
Woodstock Film Festival, 229
Working Artists, 158
Working Title Films Ltd, 111
World 2000 Productions, 133
World of Comedy Film Festival, 245
World of Wonder, 76
World Productions, 111
Worldfest Houston International Film Festival, 229
Worldfest Houston Screenplay Competition, 262
Worldwide Pants Inc., 76
Worldwide Short Film Festival, 245
Wortman UK / Polestar Pictures, 111
Wow! Wales One World Festival, 241
Ann Wright Representatives, 158

WriteMovies.com International Writing Contest, 262
Writer's First Program, 268
Writernet, 206
Writer's Arc Screenwriting Fellowship, 262
The Writers Bureau, 189
Writers Copyright Association, 206
Writers Guild of America, 199
Writers Guild of America Awards, 262
Writers Guild of America East Foundation/Columbia School of the Art Screenwriting Workshop, 179
Writers Guild of Canada, 208
Writers' Guild of Great Britain, 206
Writers' Holiday at Caerleon, 189
The Writers Network Annual Screenplay & Fiction Competition, 263
Writers On The Storm Screenplay Contest, 263
The Writers Place Screenplay Contest, 263

WriteSafe Present-A-Thon, 263
Writing Movies, 191
Wye Valley Arts Centre, 189

X

XGenStudios Inc., 124
Xstream Pictures Limited, 134

Y

Yale Summer Session and The Yale Summer Film Institute Short Film Screenwriting Competition, 263
Yari Film Group, 76
York University (Graduate Program in Film), 190
Yorkshire Art Circus, 189
Young Irish Film Makers, 134

Z

The Saul Zaentz Company, 76
Zanzibar Films, 134
Zephyr Films Limited, 111
Zero Pictures, 76
ZIJI Film & Television Productions Ltd, 124
Zoot Film Tasmania, 142